Springer Undergraduate Mathematics Series

Springer
London
Berlin
Heidelberg
New York
Barcelona
Hong Kong
Milan
Paris
Singapore
Tokyo

Advisory Board

Other books in this series

Analytic Methods for Partial Differential Equations *G. Evans, J. Blackledge, P. Yardley*
Applied Geometry for Computer Graphics and CAD *D. Marsh*
Basic Linear Algebra *T.S. Blyth and E.F. Robertson*
Basic Stochastic Processes *Z. Brzeźniak and T. Zastawniak*
Elementary Number Theory *G.A. Jones and J.M. Jones*
Elements of Logic via Numbers and Sets *D.L. Johnson*
Groups, Rings and Fields *D.A.R. Wallace*
Hyperbolic Geometry *J.W. Anderson*
Information and Coding Theory *G.A. Jones and J.M. Jones*
Introduction to Laplace Transforms and Fourier Series *P.P.G. Dyke*
Introduction to Ring Theory *P.M. Cohn*
Introductory Mathematics: Algebra and Analysis *G. Smith*
Introductory Mathematics: Applications and Methods *G.S. Marshall*
Linear Functional Analysis *B.P. Rynne and M.A. Youngson*
Measure, Integral and Probability *M. Capińksi and E. Kopp*
Multivariate Calculus and Geometry *S. Dineen*
Numerical Methods for Partial Differential Equations *G. Evans, J. Blackledge, P. Yardley*
Sets, Logic and Categories *P. Cameron*
Topics in Group Theory *G. Smith and O. Tabachnikova*
Topologies and Uniformities *I.M. James*
Vector Calculus *P.C. Matthews*

Andrew Pressley

Elementary Differential Geometry

Springer

Andrew Pressley
Department of Mathematics, King's College, The Strand, London WC2R 2LS, UK

Cover illustration elements reproduced by kind permission of:
Aptech Systems, Inc., Publishers of the GAUSS Mathematical and Statistical System, 23804 S.E. Kent-Kangley Road, Maple Valley, WA 98038, USA. Tel: (206) 432 - 7855 Fax (206) 432 - 7832 email: info@aptech.com URL: www.aptech.com
American Statistical Association: Chance Vol 8 No 1, 1995 article by KS and KW Heiner 'Tree Rings of the Northern Shawangunks' page 32 fig 2
Springer-Verlag: Mathematica in Education and Research Vol 4 Issue 3 1995 article by Roman E Maeder, Beatrice Amrhein and Oliver Gloor 'Illustrated Mathematics: Visualization of Mathematical Objects' page 9 fig 11, originally published as a CD ROM 'Illustrated Mathematics' by TELOS: ISBN 0-387-14222-3, german edition by Birkhauser: ISBN 3-7643-5100-4.
Mathematica in Education and Research Vol 4 Issue 3 1995 article by Richard J Gaylord and Kazume Nishidate 'Traffic Engineering with Cellular Automata' page 35 fig 2. Mathematica in Education and Research Vol 5 Issue 2 1996 article by Michael Trott 'The Implicitization of a Trefoil Knot' page 14.
Mathematica in Education and Research Vol 5 Issue 2 1996 article by Lee de Cola 'Coins, Trees, Bars and Bells: Simulation of the Binomial Process page 19 fig 3. Mathematica in Education and Research Vol 5 Issue 2 1996 article by Richard Gaylord and Kazume Nishidate 'Contagious Spreading' page 33 fig 1. Mathematica in Education and Research Vol 5 Issue 2 1996 article by Joe Buhler and Stan Wagon 'Secrets of the Madelung Constant' page 50 fig 1.

British Library Cataloguing in Publication Data
Pressley, Andrew
Elementary differential geometry. - (Springer undergraduate mathematics series)
1. Geometry, Differential
I. Title
516.3'6
ISBN 1852331526

Library of Congress Cataloging-in-Publication Data
Pressley, Andrew
Elementary differential geometry / Andrew Pressley.
p. cm. – (Springer undergraduate mathematics series, ISSN 1615-2085)
Includes index.
ISBN 1-85233-152-6 (alk. paper)
1. Geometry, Differential. I. Title. II. Series.
QA641 .P68 2000
516.3'6—dc21 00-058345

Springer Undergraduate Mathematics Series ISSN 1615-2085
ISBN 1-85233-152-6 Springer-Verlag London Berlin Heidelberg
a member of BertelsmannSpringer Science+Business Media GmbH
http://www.springer.co.uk

Printed in Great Britain
2nd printing 2002

Typesetting: Camera ready by author
Printed and bound at the Athenæum Press Ltd., Gateshead, Tyne & Wear
12/3830-54321 Printed on acid-free paper SPIN 10853918

Preface

The Differential Geometry in the title of this book is the study of the geometry of curves and surfaces in three-dimensional space using calculus techniques. This topic contains some of the most beautiful results in Mathematics, and yet most of them can be understood without extensive background knowledge. Thus, for virtually all of this book, the only pre-requisites are a good working knowledge of Calculus (including partial differentiation), Vectors and Linear Algebra (including matrices and determinants).

Many of the results about curves and surfaces that we shall discuss are prototypes of more general results that apply in higher-dimensional situations. For example, the Gauss–Bonnet theorem, treated in Chapter 11, is the prototype of a large number of results that relate 'local' and 'global' properties of geometric objects. The study of such relationships has formed one of the major themes of 20th century Mathematics.

We want to emphasise, however, that the *methods* used in this book are *not* necessarily those which generalise to higher-dimensional situations. (For readers in the know, there is, for example, no mention of 'connections' in the remainder of this book.) Rather, we have tried at all times to use the simplest approach that will yield the desired results. Not only does this keep the pre-requisites to an absolute minimum, it also enables us to avoid some of the conceptual difficulties often encountered in the study of Differential Geometry in higher dimensions. We hope that this approach will make this beautiful subject accessible to a wider audience.

It is a cliché, but true nevertheless, that Mathematics can be learned only by doing it, and not just by reading about it. Accordingly, the book contains over 200 exercises. Readers should attempt as many of these as their stamina permits. Full solutions to all the exercises are given at the end of the book, but these should be consulted only after the reader has obtained his or her own solution, or in case of desperation. We have tried to minimise the number of instances of the latter by including hints to many of the less routine exercises.

Contents

1
Curves in the Plane and in Space

In this chapter we discuss two mathematical formulations of the intuitive notion of a curve. The precise relation between them turns out to be quite subtle, so we shall begin by giving some examples of curves of each type and practical ways of passing between them.

1.1. What is a Curve?

If asked to give an example of a curve, you might give a straight line, say $y - 2x = 1$ (even though this is not 'curved'!), or a circle, say $x^2 + y^2 = 1$, or perhaps a parabola, say $y - x^2 = 0$.

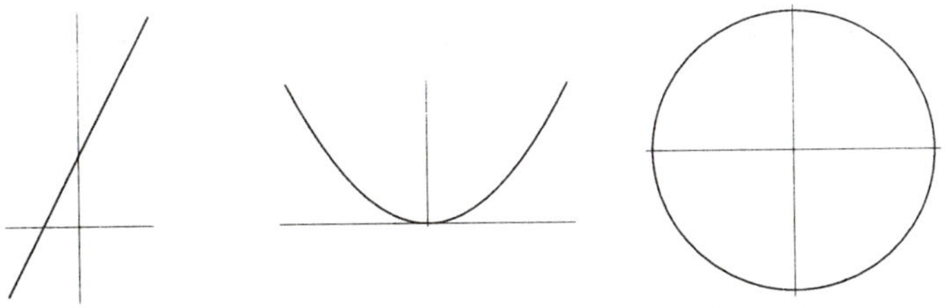

$y - 2x = 1$ $\qquad$ $y - x^2 = 0$ $\qquad$ $x^2 + y^2 = 1$

All of these curves are described by means of their cartesian equation

$$f(x, y) = c,$$

where f is a function of x and y and c is a constant. From this point of view, a curve is a set of points, namely

$$\mathcal{C} = \{(x, y) \in \mathbf{R}^2 \mid f(x, y) = c\}. \tag{1}$$

These examples are all curves in the plane $\mathbf{R}^2$, but we can also consider curves in $\mathbf{R}^3$ – for example, the x-axis in $\mathbf{R}^3$ is the straight line given by

$$\{(x, y, z) \in \mathbf{R}^3 \mid y = z = 0\},$$

and more generally a curve in $\mathbf{R}^3$ might be defined by a pair of equations

$$f_1(x, y, z) = c_1, \quad f_2(x, y, z) = c_2.$$

Curves of this kind are called *level curves*, the idea being that the curve in Eq. (1), for example, is the set of points (x, y) in the plane at which the quantity $f(x, y)$ reaches the 'level' c.

But there is another way to think about curves which turns out to be more useful in many situations. For this, a curve is viewed as the path traced out by a moving point. Thus, if $\boldsymbol{\gamma}(t)$ is the position vector of the point at time t, the curve is described by a function $\boldsymbol{\gamma}$ of a scalar parameter t with vector values (in $\mathbf{R}^2$ for a plane curve, in $\mathbf{R}^3$ for a curve in space). We use this idea to give our first formal definition of a curve in $\mathbf{R}^n$ (we shall be interested only in the cases $n = 2$ or 3, but it is convenient to treat both cases simultaneously):

Definition 1.1

A *parametrised curve* in $\mathbf{R}^n$ is a map $\boldsymbol{\gamma} : (\alpha, \beta) \to \mathbf{R}^n$, for some α, β with $-\infty \leq \alpha < \beta \leq \infty$.

The symbol (α, β) denotes the open interval

$$(\alpha, \beta) = \{t \in \mathbf{R} \mid \alpha < t < \beta\}.$$

A parametrised curve whose image is contained in a level curve $\mathcal{C}$ is called a *parametrisation* of (part of) $\mathcal{C}$. The following examples illustrate how to pass from level curves to parametrised curves and back again in practice.

Example 1.1

Let us find a parametrisation $\boldsymbol{\gamma}(t)$ of the parabola $y = x^2$. If $\boldsymbol{\gamma}(t) = (\gamma_1(t), \gamma_2(t))$, the components γ_1 and γ_2 of $\boldsymbol{\gamma}$ must satisfy

$$\gamma_2(t) = \gamma_1(t)^2 \tag{2}$$

for all values of t in the interval (α, β) where $\boldsymbol{\gamma}$ is defined (yet to be decided), and ideally every point on the parabola should be equal to $(\gamma_1(t), \gamma_2(t))$ for some value of $t \in (\alpha, \beta)$. Of course, there is an obvious solution to Eq. (2): take

$\gamma_1(t) = t, \gamma_2(t) = t^2$. To get every point on the parabola we must allow t to take every real number value (since the x-coordinate of $\boldsymbol{\gamma}(t)$ is just t, and the x-coordinate of a point on the parabola can be any real number), so we must take (α, β) to be $(-\infty, \infty)$. Thus, the desired parametrisation is

$$\boldsymbol{\gamma} : (-\infty, \infty) \to \mathbf{R}^2, \quad \boldsymbol{\gamma}(t) = (t, t^2).$$

But this is not the only parametrisation of the parabola. Another choice is $\boldsymbol{\gamma}(t) = (t^3, t^6)$ (with $(\alpha, \beta) = (-\infty, \infty)$). Yet another is $(2t, 4t^2)$, and of course there are (infinitely many) others. So the parametrisation of a given level curve is not unique.

Example 1.2

Now we try the circle $x^2 + y^2 = 1$. It is tempting to take $x = t$ as in the previous example, so that $y = \sqrt{1 - t^2}$ (we could have taken $y = -\sqrt{1 - t^2}$). So we get the parametrisation

$$\boldsymbol{\gamma}(t) = (t, \sqrt{1 - t^2}).$$

But this is only a parametrisation of the upper half of the circle, because $\sqrt{1 - t^2}$ is always ≥ 0. Similarly, if we had taken $y = -\sqrt{1 - t^2}$, we would only have covered the lower half of the circle.

If we want a parametrisation of the whole circle, we must try again. We need functions $\gamma_1(t)$ and $\gamma_2(t)$ such that

$$\gamma_1(t)^2 + \gamma_2(t)^2 = 1 \tag{3}$$

for all $t \in (\alpha, \beta)$, and such that *every* point on the circle is equal to $(\gamma_1(t), \gamma_2(t))$ for some $t \in (\alpha, \beta)$. There is an obvious solution to Eq. (3): $\gamma_1(t) = \cos t$ and $\gamma_2(t) = \sin t$ (since $\cos^2 t + \sin^2 t = 1$ for all values of t). We can take $(\alpha, \beta) = (-\infty, \infty)$, although this is overkill: any open interval (α, β) whose length is greater than 2π will suffice.

The next example shows how to pass from parametrised curves to level curves.

Example 1.3

Take the parametrised curve (called an *astroid*)

$$\boldsymbol{\gamma}(t) = (\cos^3 t, \sin^3 t).$$

Since $\cos^2 t + \sin^2 t = 1$ for all t, the coordinates $x = \cos^3 t$, $y = \sin^3 t$ of the point $\boldsymbol{\gamma}(t)$ satisfy

$$x^{2/3} + y^{2/3} = 1.$$

This level curve coincides with the image of the map $\boldsymbol{\gamma}$.

In this book, we shall be studying curves (and later, surfaces) using methods of calculus. To differentiate a vector-valued function such as $\boldsymbol{\gamma}(t)$ (as in Definition 1.1), we differentiate componentwise: if

$$\boldsymbol{\gamma}(t) = (\gamma_1(t), \gamma_2(t), \ldots, \gamma_n(t)),$$

then

$$\frac{d\boldsymbol{\gamma}}{dt} = \left(\frac{d\gamma_1}{dt}, \frac{d\gamma_2}{dt}, \ldots, \frac{d\gamma_n}{dt}\right),$$
$$\frac{d^2\boldsymbol{\gamma}}{dt^2} = \left(\frac{d^2\gamma_1}{dt^2}, \frac{d^2\gamma_2}{dt^2}, \ldots, \frac{d^2\gamma_n}{dt^2}\right), \quad \text{etc.}$$

To save space, we often denote $d\boldsymbol{\gamma}/dt$ by $\dot{\boldsymbol{\gamma}}(t)$, $d^2\boldsymbol{\gamma}/dt^2$ by $\ddot{\boldsymbol{\gamma}}(t)$, etc.

We say that $\boldsymbol{\gamma}$ is *smooth* if each of the components $\gamma_1, \gamma_2, \ldots, \gamma_n$ of $\boldsymbol{\gamma}$ is smooth, i.e. if all the derivatives $d\gamma_i/dt$, $d^2\gamma_i/dt^2$, $d^3\gamma_i/dt^3, \ldots$ exist, for $i = 1, 2, \ldots, n$. *From now on, all parametrised curves studied in this book will be assumed to be smooth.*

Definition 1.2

If $\boldsymbol{\gamma}(t)$ is a parametrised curve, its first derivative $d\boldsymbol{\gamma}/dt$ is called the *tangent vector* of $\boldsymbol{\gamma}$ at the point $\boldsymbol{\gamma}(t)$.

To see the reason for this terminology, note that the vector

$$\frac{\boldsymbol{\gamma}(t+\delta t) - \boldsymbol{\gamma}(t)}{\delta t}$$

is parallel to the chord joining the two points $\boldsymbol{\gamma}(t)$ and $\boldsymbol{\gamma}(t+\delta t)$ of the image C of $\boldsymbol{\gamma}$:

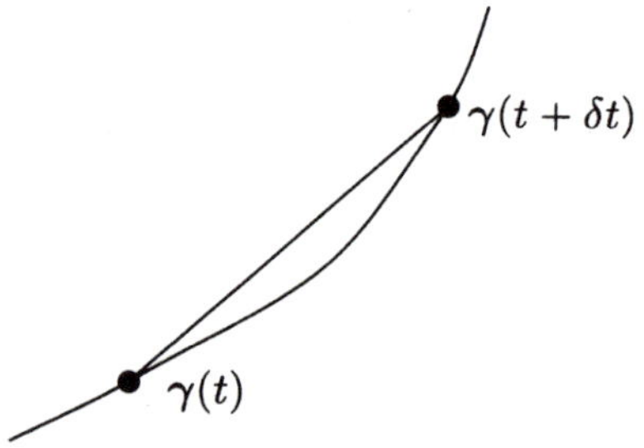

We expect that, as δt tends to zero, the chord becomes parallel to the tangent to C at $\boldsymbol{\gamma}(t)$. Hence, the tangent should be parallel to

$$\lim_{\delta t \to 0} \frac{\boldsymbol{\gamma}(t+\delta t) - \boldsymbol{\gamma}(t)}{\delta t} = \frac{d\boldsymbol{\gamma}}{dt}.$$

The following result is intuitively clear:

Proposition 1.1

If the tangent vector of a parametrised curve is constant, the image of the curve is (part of) a straight line.

Proof 1.1

If $\dot{\boldsymbol{\gamma}}(t) = \mathbf{a}$ for all t, where $\mathbf{a}$ is a constant vector, we have, integrating componentwise,

$$\boldsymbol{\gamma}(t) = \int \frac{d\boldsymbol{\gamma}}{dt} dt = \int \mathbf{a} dt = t\mathbf{a} + \mathbf{b},$$

where $\mathbf{b}$ is another constant vector. If $\mathbf{a} \neq \mathbf{0}$, this is the parametric equation of the straight line parallel to $\mathbf{a}$ and passing through the point with position vector $\mathbf{b}$:

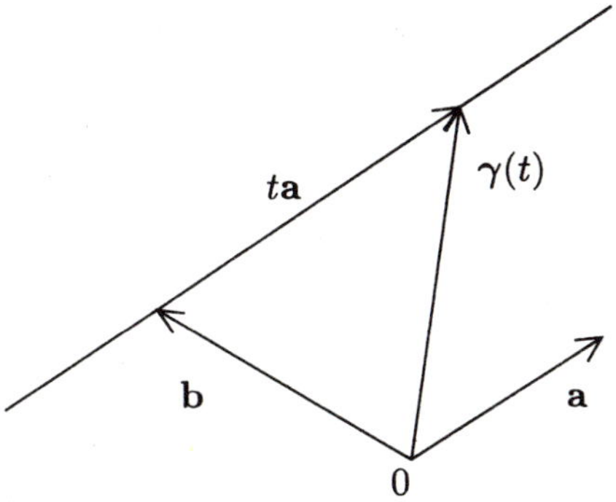

If $\mathbf{a} = \mathbf{0}$, the image of $\boldsymbol{\gamma}$ is a single point (namely, the point with position vector $\mathbf{b}$). □

EXERCISES

1.1 Is $\boldsymbol{\gamma}(t) = (t^2, t^4)$ a parametrisation of the parabola $y = x^2$?

1.2 Find parametrisations of the following level curves:
(i) $y^2 - x^2 = 1$;
(ii) $\frac{x^2}{4} + \frac{y^2}{9} = 1$.

1.3 Find the cartesian equations of the following parametrised curves:
(i) $\boldsymbol{\gamma}(t) = (\cos^2 t, \sin^2 t)$;
(ii) $\boldsymbol{\gamma}(t) = (e^t, t^2)$.

1.4 Calculate the tangent vectors of the curves in Exercise 1.3.

1.5 Sketch the astroid in Example 1.3. Calculate its tangent vector at each point. At which points is the tangent vector zero?

1.6 If P is any point on the circle $\mathcal{C}$ in the xy-plane of radius $a > 0$ and centre $(0, a)$, let the straight line through the origin and P intersect the line $y = 2a$ at Q, and let the line through P parallel to the x-axis intersect the line through Q parallel to the y-axis at R. As P moves around $\mathcal{C}$, R traces out a curve called the *witch of Agnesi*. For this curve, find

(i) a parametrisation;

(ii) its cartesian equation.

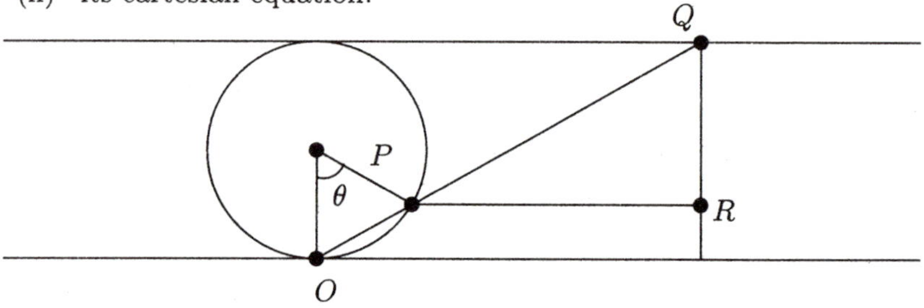

1.7 A *cycloid* is the plane curve traced out by a point on the circumference of a circle as it rolls without slipping along a straight line. Show that, if the straight line is the x-axis and the circle has radius $a > 0$, the cycloid can be parametrised as

$$\boldsymbol{\gamma}(t) = a(t - \sin t, 1 - \cos t).$$

1.8 Generalise the previous exercise by finding parametrisations of an *epicycloid* (resp. *hypocycloid*), the curve traced out by a point on the circumference of a circle as it rolls without slipping around the outside (resp. inside) of a fixed circle.

1.9 Show that $\boldsymbol{\gamma}(t) = (\cos^2 t - \frac{1}{2}, \sin t \cos t, \sin t)$ is a parametrisation of the curve of intersection of the circular cylinder of radius $\frac{1}{2}$ and axis the z-axis with the sphere of radius 1 and centre $(-\frac{1}{2}, 0, 0)$. (This is called *Viviani's Curve*).

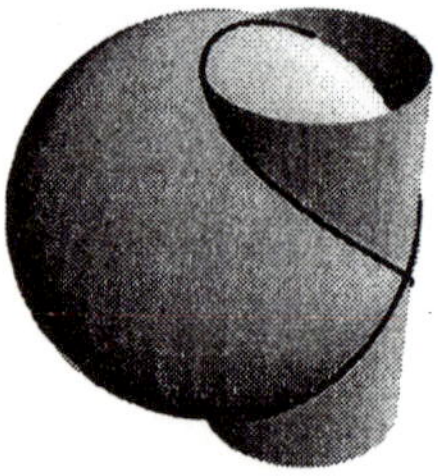

1.10 For the *logarithmic spiral* $\gamma(t) = (e^t \cos t, e^t \sin t)$, show that the angle between $\gamma(t)$ and the tangent vector at $\gamma(t)$ is independent of t. (There is a picture of the logarithmic spiral in Example 1.4.)

1.2. Arc-Length

If $\mathbf{v} = (v_1, \ldots, v_n)$ is a vector in $\mathbf{R}^n$, its *length* is

$$\| \mathbf{v} \| = \sqrt{v_1^2 + \cdots + v_n^2}.$$

If $\mathbf{u}$ is another vector in $\mathbf{R}^n$, $\| \mathbf{u} - \mathbf{v} \|$ is the length of the straight line segment joining the points in $\mathbf{R}^n$ with position vectors $\mathbf{u}$ and $\mathbf{v}$.

To find a formula for the length of any parametrised curve γ, note that, if δt is very small, the part of the image C of γ between $\gamma(t)$ and $\gamma(t + \delta t)$ is nearly a straight line, so its length is approximately

$$\| \gamma(t + \delta t) - \gamma(t) \|.$$

Again, since δt is small, $(\gamma(t + \delta t) - \gamma(t))/\delta t$ is nearly equal to $\dot{\gamma}(t)$, so the length is approximately

$$\| \dot{\gamma}(t) \| \, \delta t. \tag{4}$$

If we want to calculate the length of (a not necessarily small) part of C, we can divide it up into segments, each of which corresponds to a small increment δt in t, calculate the length of each segment using (4), and add up the results. Letting δt tend to zero should then give the exact length.

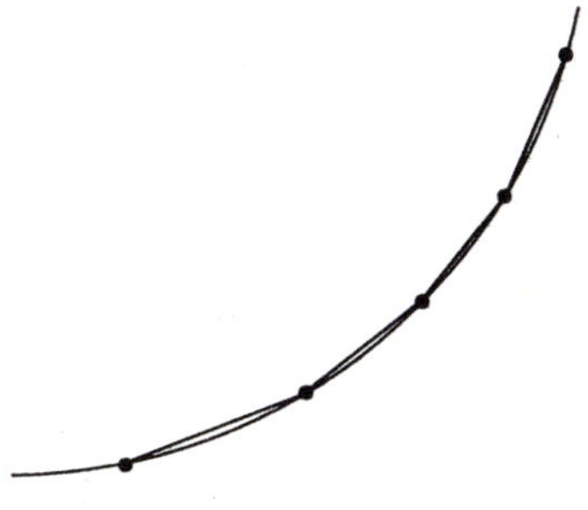

This motivates the following definition:

Definition 1.3

The *arc-length* of a curve γ starting at the point $\gamma(t_0)$ is the function $s(t)$ given by

$$s(t) = \int_{t_0}^{t} \parallel \dot{\boldsymbol{\gamma}}(u) \parallel du.$$

Thus, $s(t_0) = 0$ and $s(t)$ is positive or negative according to whether t is larger or smaller than t_0. If we choose a different starting point $\boldsymbol{\gamma}(\tilde{t}_0)$, the resulting arc-length $\tilde{s}$ differs from s by the constant $\int_{t_0}^{\tilde{t}_0} \parallel \dot{\boldsymbol{\gamma}}(u) \parallel du$.

Example 1.4

For a logarithmic spiral

$$\boldsymbol{\gamma}(t) = (e^t \cos t, e^t \sin t),$$

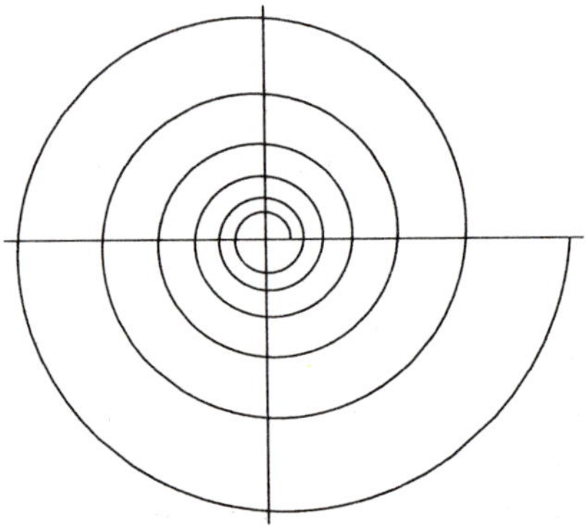

we have

$$\dot{\boldsymbol{\gamma}} = (e^t(\cos t - \sin t), e^t(\sin t + \cos t)),$$
$$\therefore \quad \parallel \dot{\boldsymbol{\gamma}} \parallel^2 = e^{2t}(\cos t - \sin t)^2 + e^{2t}(\sin t + \cos t)^2 = 2e^{2t}.$$

Hence, the arc-length of $\boldsymbol{\gamma}$ starting at $\boldsymbol{\gamma}(0) = (1, 0)$ (for example) is

$$s = \int_0^t \sqrt{2e^{2u}}du = \sqrt{2}(e^t - 1).$$

If s is the arc-length of a curve $\boldsymbol{\gamma}$ starting at $\boldsymbol{\gamma}(t_0)$, we have

$$\frac{ds}{dt} = \frac{d}{dt}\int_{t_0}^{t} \parallel \dot{\boldsymbol{\gamma}}(u) \parallel du = \parallel \dot{\boldsymbol{\gamma}}(t) \parallel. \tag{5}$$

Thinking of $\boldsymbol{\gamma}(t)$ as the position of a moving point at time t, ds/dt is the speed of the point (rate of change of distance along the curve). For this reason, we make

Definition 1.4

If $\boldsymbol{\gamma} : (\alpha, \beta) \to \mathbf{R}^n$ is a parametrised curve, its *speed* at the point $\boldsymbol{\gamma}(t)$ is $\| \dot{\boldsymbol{\gamma}}(t) \|$, and $\boldsymbol{\gamma}$ is said to be a *unit-speed curve* if $\dot{\boldsymbol{\gamma}}(t)$ is a unit vector for all $t \in (\alpha, \beta)$.

We shall see many examples of formulas and results relating to curves that take on a much simpler form when the curve is unit-speed. The reason for this simplification is given in the next proposition. Although this admittedly looks uninteresting at first sight, it will be extremely useful for what follows.

Proposition 1.2

Let $\mathbf{n}(t)$ *be a* unit vector *that is a smooth function of a parameter* t. *Then, the dot product*

$$\dot{\mathbf{n}}(t).\mathbf{n}(t) = 0$$

for all t, *i.e.* $\dot{\mathbf{n}}(t)$ *is zero or perpendicular to* $\mathbf{n}(t)$ *for all* t.

In particular, if $\boldsymbol{\gamma}$ *is a unit-speed curve, then* $\ddot{\boldsymbol{\gamma}}$ *is zero or perpendicular to* $\dot{\boldsymbol{\gamma}}$.

Proof 1.2

We use the 'product formula' for differentiating dot products of vector-valued functions $\mathbf{a}(t)$ and $\mathbf{b}(t)$:

$$\frac{d}{dt}(\mathbf{a}.\mathbf{b}) = \frac{d\mathbf{a}}{dt}.\mathbf{b} + \mathbf{a}.\frac{d\mathbf{b}}{dt}.$$

Using this to differentiate both sides of the equation $\mathbf{n}.\mathbf{n} = 1$ with respect to t gives

$$\dot{\mathbf{n}}.\mathbf{n} + \mathbf{n}.\dot{\mathbf{n}} = 0,$$

so $2\dot{\mathbf{n}}.\mathbf{n} = 0$.

The last part follows by taking $\mathbf{n} = \dot{\boldsymbol{\gamma}}$. □

EXERCISES

1.11 Calculate the arc-length of the *catenary* $\boldsymbol{\gamma}(t) = (t, \cosh t)$ starting at the point $(0, 1)$.

1.12 Show that the following curves are unit-speed:

(i) $\boldsymbol{\gamma}(t) = \left(\frac{1}{3}(1+t)^{3/2}, \frac{1}{3}(1-t)^{3/2}, \frac{t}{\sqrt{2}}\right)$;

(ii) $\boldsymbol{\gamma}(t) = \left(\frac{4}{5}\cos t, 1 - \sin t, -\frac{3}{5}\cos t\right)$.

1.13 Calculate the arc-length along the cycloid in Exercise 1.7 corresponding to one complete revolution of the circle.

1.3. Reparametrisation

We saw in Examples 1.1 and 1.2 that a given level curve can have many parametrisations, and it is important to understand the relation between them.

Definition 1.5

A parametrised curve $\tilde{\boldsymbol{\gamma}} : (\tilde{\alpha}, \tilde{\beta}) \to \mathbf{R}^n$ is a *reparametrisation* of a parametrised curve $\boldsymbol{\gamma} : (\alpha, \beta) \to \mathbf{R}^n$ if there is a smooth bijective map $\phi : (\tilde{\alpha}, \tilde{\beta}) \to (\alpha, \beta)$ (the *reparametrisation map*) such that the inverse map $\phi^{-1} : (\alpha, \beta) \to (\tilde{\alpha}, \tilde{\beta})$ is also smooth and

$$\tilde{\boldsymbol{\gamma}}(\tilde{t}) = \boldsymbol{\gamma}(\phi(\tilde{t})) \quad \text{for all} \quad \tilde{t} \in (\tilde{\alpha}, \tilde{\beta}).$$

Note that, since ϕ has a smooth inverse, $\boldsymbol{\gamma}$ is a reparametrisation of $\tilde{\boldsymbol{\gamma}}$:

$$\tilde{\boldsymbol{\gamma}}(\phi^{-1}(t)) = \boldsymbol{\gamma}(\phi(\phi^{-1}(t))) = \boldsymbol{\gamma}(t) \text{ for all } t \in (\alpha, \beta).$$

Two curves that are reparametrisations of each other have the same image, so they should have the same geometric properties.

Example 1.5

In Example 1.2, we gave the parametrisation $\boldsymbol{\gamma}(t) = (\cos t, \sin t)$ for the circle $x^2 + y^2 = 1$. Another parametrisation is

$$\tilde{\boldsymbol{\gamma}}(t) = (\sin t, \cos t)$$

(since $\sin^2 t + \cos^2 t = 1$). To see that $\tilde{\boldsymbol{\gamma}}$ is a reparametrisation of $\boldsymbol{\gamma}$, we have to find a reparametrisation map ϕ such that

$$(\cos \phi(t), \sin \phi(t)) = (\sin t, \cos t).$$

One solution is $\phi(t) = \pi/2 - t$.

As we remarked in the previous section, the analysis of a curve is simplified when it is known to be unit-speed. It is therefore important to know exactly which curves have unit-speed reparametrisations.

Definition 1.6

A point $\boldsymbol{\gamma}(t)$ of a parametrised curve $\boldsymbol{\gamma}$ is called a *regular point* if $\dot{\boldsymbol{\gamma}}(t) \neq \mathbf{0}$; otherwise $\boldsymbol{\gamma}(t)$ is a *singular point* of $\boldsymbol{\gamma}$. A curve is *regular* if all of its points are regular.

Before we show the relation between regularity and unit-speed reparametrisation, we note two simple properties of regular curves. Although these results are not particularly appealing, they will be very important for what is to follow.

Proposition 1.3

Any reparametrisation of a regular curve is regular.

Proof 1.3

Suppose that $\boldsymbol{\gamma}$ and $\tilde{\boldsymbol{\gamma}}$ are related as in Definition 1.5, let $t = \phi(\tilde{t})$, and let $\psi = \phi^{-1}$ so that $\tilde{t} = \psi(t)$. Differentiating both sides of the equation $\phi(\psi(t)) = t$ with respect to t and using the chain rule gives

$$\frac{d\phi}{d\tilde{t}}\frac{d\psi}{dt} = 1.$$

This shows that $d\phi/d\tilde{t}$ is never zero. Since $\tilde{\boldsymbol{\gamma}}(\tilde{t}) = \boldsymbol{\gamma}(\phi(\tilde{t}))$, another application of the chain rule gives

$$\frac{d\tilde{\boldsymbol{\gamma}}}{d\tilde{t}} = \frac{d\boldsymbol{\gamma}}{dt}\frac{d\phi}{d\tilde{t}},$$

which shows that $d\tilde{\boldsymbol{\gamma}}/d\tilde{t}$ is never zero if $d\boldsymbol{\gamma}/dt$ is never zero. □

Proposition 1.4

If $\boldsymbol{\gamma}(t)$ is a regular curve, its arc-length s (see Definition 1.3), starting at any point of $\boldsymbol{\gamma}$, is a smooth function of t.

Proof 1.4

We have already seen that (whether or not $\boldsymbol{\gamma}$ is regular) s is a differentiable function of t and

$$\frac{ds}{dt} = \parallel \dot{\boldsymbol{\gamma}}(t) \parallel.$$

To simplify the notation, assume from now on that $\boldsymbol{\gamma}$ is a plane curve, say

$$\boldsymbol{\gamma}(t) = (u(t), v(t)),$$

where u and v are smooth functions of t. Define $f : \mathbf{R}^2 \to \mathbf{R}$ by

$$f(u, v) = \sqrt{u^2 + v^2},$$

so that

$$\frac{ds}{dt} = f(\dot{u}, \dot{v}). \tag{6}$$

The crucial point is that f *is smooth on* $\mathbf{R}^2 \backslash \{(0,0)\}$, which means that all the partial derivatives of f of all orders exist and are continuous functions except at the origin $(0,0)$. For example,

$$\frac{\partial f}{\partial u} = \frac{u}{\sqrt{u^2 + v^2}}, \quad \frac{\partial f}{\partial v} = \frac{v}{\sqrt{u^2 + v^2}},$$

are well defined and continuous except where $u = v = 0$, and similarly for higher derivatives. Since $\boldsymbol{\gamma}$ is regular, $\dot{u}$ and $\dot{v}$ are never both zero, so the chain rule and Eq. (6) shows that ds/dt is smooth. For example,

$$\frac{d^2 s}{dt^2} = \frac{\partial f}{\partial u}\ddot{u} + \frac{\partial f}{\partial v}\ddot{v},$$

and similarly for the higher derivatives of s. □

The main result we want is

Proposition 1.5

A parametrised curve has a unit-speed reparametrisation if and only if it is regular.

Proof 1.5

Suppose first that a parametrised curve $\boldsymbol{\gamma} : (\alpha, \beta) \to \mathbf{R}^n$ has a unit-speed reparametrisation $\tilde{\boldsymbol{\gamma}}$, with reparametrisation map ϕ. Letting $t = \phi(\tilde{t})$, we have

$$\tilde{\boldsymbol{\gamma}}(\tilde{t}) = \boldsymbol{\gamma}(t),$$

$$\therefore \quad \frac{d\tilde{\boldsymbol{\gamma}}}{d\tilde{t}} = \frac{d\boldsymbol{\gamma}}{dt}\frac{dt}{d\tilde{t}},$$

$$\therefore \quad \| \frac{d\tilde{\boldsymbol{\gamma}}}{d\tilde{t}} \| = \| \frac{d\boldsymbol{\gamma}}{dt} \| \, |\frac{dt}{d\tilde{t}}|.$$

Since $\tilde{\boldsymbol{\gamma}}$ is unit-speed, $\| d\tilde{\boldsymbol{\gamma}}/d\tilde{t} \| = 1$, so clearly $d\boldsymbol{\gamma}/dt$ cannot be zero.

Conversely, suppose that the tangent vector $d\boldsymbol{\gamma}/dt$ is never zero. By Eq. (5), $ds/dt > 0$ for all t, where s is the arc-length of $\boldsymbol{\gamma}$ starting at any point of the curve, and by Proposition 1.4 s is a smooth function of t. It follows from the inverse function theorem of multivariable calculus that $s : (\alpha, \beta) \to \mathbf{R}$ is injective, that its image is an open interval $(\tilde{\alpha}, \tilde{\beta})$, and that the inverse map $s^{-1} : (\tilde{\alpha}, \tilde{\beta}) \to (\alpha, \beta)$ is smooth. (Readers unfamiliar with the inverse function theorem should accept these statements for now; the theorem will be discussed informally in Section 1.4 and formally in Chapter 4.) We take $\phi = s^{-1}$ and let $\tilde{\boldsymbol{\gamma}}$ be the corresponding reparametrisation of $\boldsymbol{\gamma}$, so that

$$\tilde{\boldsymbol{\gamma}}(s) = \boldsymbol{\gamma}(t).$$

Then,

$$\frac{d\tilde{\boldsymbol{\gamma}}}{ds}\frac{ds}{dt} = \frac{d\boldsymbol{\gamma}}{dt},$$

$$\therefore \quad \| \frac{d\tilde{\boldsymbol{\gamma}}}{ds} \| \frac{ds}{dt} = \| \frac{d\boldsymbol{\gamma}}{dt} \| = \frac{ds}{dt} \quad \text{(by Eq. (5))},$$

$$\therefore \quad \| \frac{d\tilde{\boldsymbol{\gamma}}}{ds} \| = 1.$$

□

The proof of Proposition 1.5 shows that the arc-length is essentially the only unit-speed parameter on a regular curve:

Corollary 1.1

Let $\boldsymbol{\gamma}$ be a regular curve and let $\tilde{\boldsymbol{\gamma}}$ be a unit-speed reparametrisation of $\boldsymbol{\gamma}$:

$$\tilde{\boldsymbol{\gamma}}(u(t)) = \boldsymbol{\gamma}(t) \quad \text{for all } t,$$

where u is a smooth function of t. Then, if s is the arc-length of $\boldsymbol{\gamma}$ (starting at any point), we have

$$u = \pm s + c, \tag{7}$$

where c is a constant. Conversely, if u is given by Eq. (7) for some value of c and with either sign, then $\tilde{\boldsymbol{\gamma}}$ is a unit-speed reparametrisation of $\boldsymbol{\gamma}$.

Proof 1.1

The calculation in the first part of the proof of Proposition 1.5 shows that u gives a unit-speed reparametrisation of $\boldsymbol{\gamma}$ if and only if

$$\frac{du}{dt} = \pm \| \frac{d\boldsymbol{\gamma}}{dt} \| = \pm \frac{ds}{dt} \quad \text{(by Eq. (5))}.$$

Hence, $u = \pm s + c$ for some constant c. □

Although every regular curve has a unit-speed reparametrisation, this may be very complicated, or even impossible to write down 'explicitly', as the following examples show.

Example 1.6

For the logarithmic spiral

$$\boldsymbol{\gamma}(t) = (e^t \cos t, e^t \sin t),$$

we found in Example 1.4 that

$$\| \dot{\boldsymbol{\gamma}} \|^2 = 2e^{2t}.$$

This is never zero, so $\boldsymbol{\gamma}$ is regular. The arc-length of $\boldsymbol{\gamma}$ starting at $(1, 0)$ was found to be $s = \sqrt{2}(e^t - 1)$. Hence, $t = \ln\left(\frac{s}{\sqrt{2}} + 1\right)$, so a unit-speed reparametrisation of $\boldsymbol{\gamma}$ is given by the rather unwieldy formula

$$\tilde{\boldsymbol{\gamma}}(s) = \left(\left(\frac{s}{\sqrt{2}} + 1\right) \cos\left(\ln\left(\frac{s}{\sqrt{2}} + 1\right)\right), \left(\frac{s}{\sqrt{2}} + 1\right) \sin\left(\ln\left(\frac{s}{\sqrt{2}} + 1\right)\right)\right).$$

Example 1.7

The *twisted cubic* is the space curve given by

$$\boldsymbol{\gamma}(t) = (t, t^2, t^3), \quad -\infty < t < \infty.$$

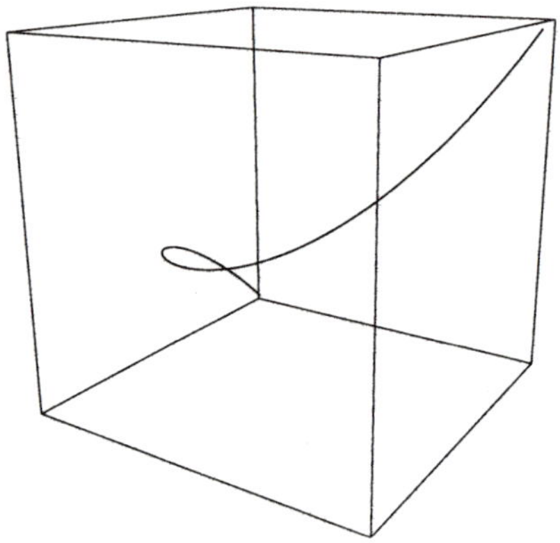

We have

$$\dot{\boldsymbol{\gamma}}(t) = (1, 2t, 3t^2),$$

$$\therefore \quad \| \dot{\boldsymbol{\gamma}}(t) \| = \sqrt{1 + 4t^2 + 9t^4}.$$

This is never zero, so $\boldsymbol{\gamma}$ is regular. The arc-length starting at $\boldsymbol{\gamma}(0) = \mathbf{0}$ is

$$s = \int_0^t \sqrt{1 + 4u^2 + 9u^4}\, du.$$

This integral cannot be evaluated in terms of familiar functions like logarithms and exponentials, trigonometric functions, etc. (It is an example of something called an *elliptic integral.*)

Our final example shows that a given level curve can have both regular and non-regular parametrisations.

Example 1.8

For the parametrisation

$$\gamma(t) = (t, t^2)$$

of the parabola $y = x^2$, $\dot{\gamma}(t) = (1, 2t)$ is obviously never zero, so γ is regular.

But

$$\tilde{\gamma}(t) = (t^3, t^6)$$

is also a parametrisation of the same parabola. This time, $\dot{\tilde{\gamma}} = (3t^2, 6t^5)$, and this is zero when $t = 0$, so $\tilde{\gamma}$ is *not* regular.

EXERCISES

1.14 Which of the following curves are regular?
 (i) $\gamma(t) = (\cos^2 t, \sin^2 t)$ for $-\infty < t < \infty$;
 (ii) the same curve as in (i), but with $0 < t < \pi/2$;
 (iii) $\gamma(t) = (t, \cosh t)$ for $-\infty < t < \infty$.
 Find unit-speed reparametrisations of the regular curve(s).

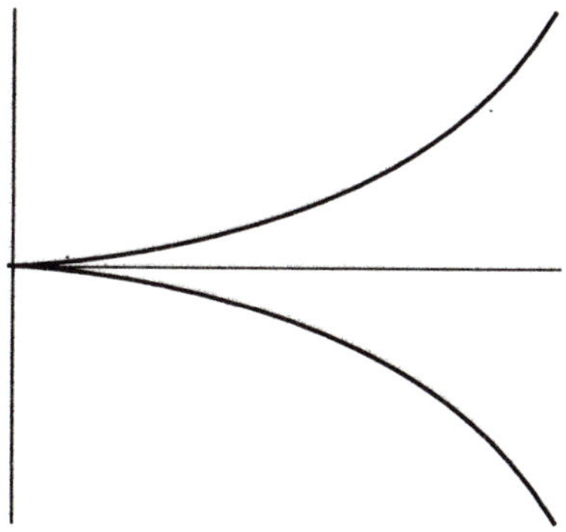

1.15 The *cissoid of Diocles* (see above) is the curve whose equation in terms of polar coordinates (r, θ) is

$$r = \sin\theta \tan\theta, \quad -\pi/2 < \theta < \pi/2.$$

Write down a parametrisation of the cissoid using θ as a parameter and show that

$$\gamma(t) = \left(t^2, \frac{t^3}{\sqrt{1-t^2}}\right), \quad -1 < t < 1,$$

is a reparametrisation of it.

1.16 Let γ be a curve in $\mathbb{R}^n$ and let $\tilde{\gamma}$ be a reparametrisation of γ with reparametrisation map ϕ (so that $\tilde{\gamma}(\tilde{t}) = \gamma(\phi(\tilde{t}))$). Let $\tilde{t}_0$ be a fixed value of $\tilde{t}$ and let $t_0 = \phi(\tilde{t}_0)$. Let s and $\tilde{s}$ be the arc-lengths of γ and $\tilde{\gamma}$ starting at the point $\gamma(t_0) = \tilde{\gamma}(\tilde{t}_0)$. Prove that $\tilde{s} = s$ if $d\phi/d\tilde{t} > 0$ for all $\tilde{t}$, and $\tilde{s} = -s$ if $d\phi/d\tilde{t} < 0$ for all $\tilde{t}$.

1.4. Level Curves vs. Parametrised Curves

We shall now try to clarify the precise relation between the two types of curve we have considered in previous sections.

Level curves in the generality we have defined them are not always the kind of objects we would want to call curves. For example, the level 'curve' $x^2 + y^2 = 0$ is a single point. The correct conditions to impose on a function $f(x, y)$ in order that $f(x, y) = c$, where c is a constant, will be an acceptable level curve in the plane are contained in the following theorem, which shows that such level curves can be parametrised. Note that we might as well assume that $c = 0$ (since we can replace f by $f - c$).

Theorem 1.1

Let $f(x, y)$ be a smooth function of two variables (which means that all the partial derivatives of f, of all orders, exist and are continuous functions). Assume that, at every point of the level curve

$$\mathcal{C} = \{(x, y) \in \mathbf{R}^2 \mid f(x, y) = 0\},$$

$\partial f/\partial x$ and $\partial f/\partial y$ are not both zero. If P is a point of $\mathcal{C}$, with coordinates (x_0, y_0), say, there is a regular parametrised curve $\gamma(t)$, defined on an open interval containing 0, such that γ passes through P when $t = 0$ and $\gamma(t)$ is contained in $\mathcal{C}$ for all t.

The proof of this theorem makes use of the inverse function theorem (one version of which has already been used in the proof of Proposition 1.5). For the moment, we shall only try to convince the reader of the truth of this theorem. The proof will be given in a later exercise (Exercise 4.31) after the inverse function theorem has been formally introduced and used in our discussion of surfaces.

To understand the significance of the conditions on f in Theorem 1.1, suppose that $(x_0+\Delta x, y_0+\Delta y)$ is a point of $\mathcal{C}$ near P, so that $f(x_0+\Delta x, y_0+\Delta y) = 0$. By the two-variable form of Taylor's theorem,

$$f(x_0 + \Delta x, y_0 + \Delta y) = f(x_0, y_0) + \Delta x\frac{\partial f}{\partial x} + \Delta y\frac{\partial f}{\partial y},$$

neglecting products of the small quantities Δx and Δy (the partial derivatives are evaluated at (x_0, y_0)). Hence,

$$\Delta x\frac{\partial f}{\partial x} + \Delta y\frac{\partial f}{\partial y} = 0. \tag{8}$$

Since Δx and Δy are small, the vector $(\Delta x, \Delta y)$ is nearly tangent to $\mathcal{C}$ at P, so Eq. (8) says that *the vector* $\mathbf{n} = \left(\frac{\partial f}{\partial x}, \frac{\partial f}{\partial y}\right)$ *is perpendicular to* $\mathcal{C}$ *at* P.

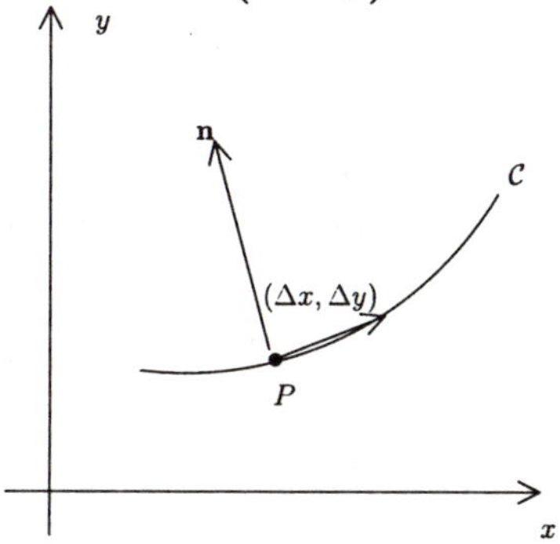

The hypothesis in Theorem 1.1 tells us that the vector $\mathbf{n}$ is non-zero at every point of $\mathcal{C}$. Suppose, for example, that $\frac{\partial f}{\partial y} \neq 0$ at P. Then, $\mathbf{n}$ is not parallel to the x-axis at P, so the tangent to $\mathcal{C}$ at P is not parallel to the y-axis.

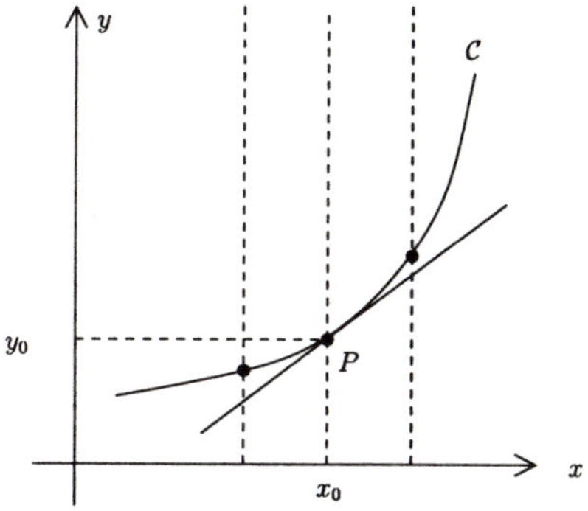

This implies that vertical lines x = constant near $x = x_0$ all intersect $\mathcal{C}$ in a unique point (x, y) near P. In other words, *the equation*

$$f(x, y) = 0 \tag{9}$$

has a unique solution y *near* y_0 *for every* x *near* x_0. Note that this may fail to be the case if the tangent to $\mathcal{C}$ at P is parallel to the y-axis:

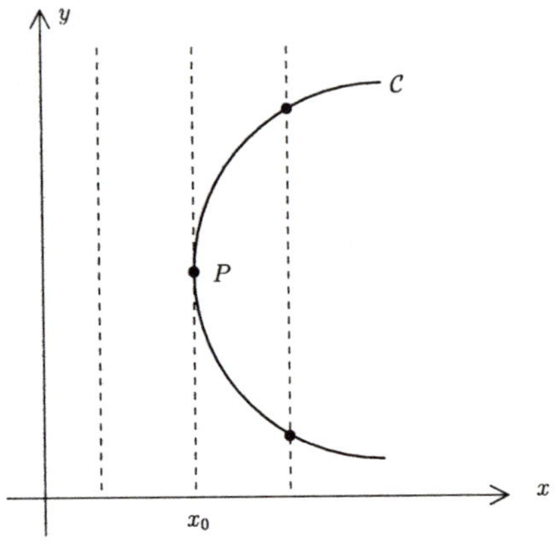

In this example, lines x = constant just to the left of $x = x_0$ do not meet $\mathcal{C}$ near P, while those just to the right of $x = x_0$ meet $\mathcal{C}$ in more than one point near P.

The italicised statement about f in the last paragraph means that there is a function $g(x)$, defined for x near x_0, such that $y = g(x)$ is the unique solution of Eq. (9) near y_0. We can now define a parametrisation $\boldsymbol{\gamma}$ of the part of $\mathcal{C}$ near P by

$$\boldsymbol{\gamma}(t) = (t, g(t)).$$

If we accept that g is smooth (which follows from the inverse function theorem), then $\boldsymbol{\gamma}$ is certainly regular since

$$\dot{\boldsymbol{\gamma}} = (1, \dot{g})$$

is obviously never zero. This 'proves' Theorem 1.1.

It is actually possible to prove slightly more than we have stated in Theorem 1.1. Suppose that $f(x, y)$ satisfies the conditions in the theorem, and assume in addition that the level curve $\mathcal{C}$ given by $f(x, y) = 0$ is *connected.* For readers unfamiliar with point set topology, this means roughly that $\mathcal{C}$ is in 'one piece'. For example, the circle $x^2 + y^2 = 1$ is connected, but the hyperbola $x^2 - y^2 = 1$ is not:

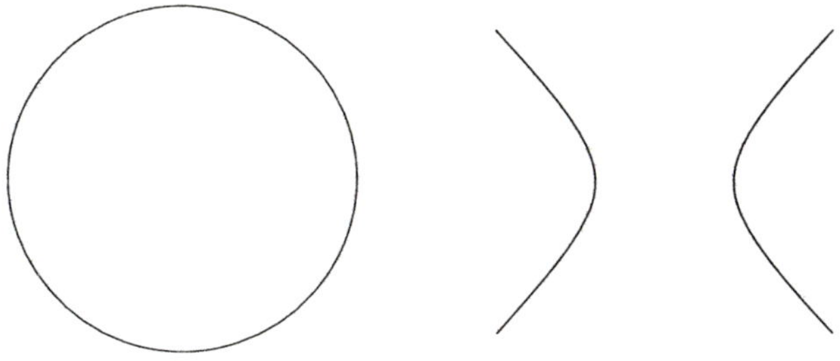

$x^2 + y^2 = 1$ $\qquad$ $x^2 - y^2 = 1$

With these assumptions on f, there is a regular parametrised curve $\boldsymbol{\gamma}$ whose image is *the whole* of $\mathcal{C}$. Moreover, if $\mathcal{C}$ does not 'close up' (like a straight line or a parabola), $\boldsymbol{\gamma}$ can be taken to be injective; if $\mathcal{C}$ does close up (like a circle or an ellipse), then $\boldsymbol{\gamma}$ maps some closed interval $[\alpha, \beta]$ *onto* $\mathcal{C}$, $\boldsymbol{\gamma}(\alpha) = \boldsymbol{\gamma}(\beta)$ and $\boldsymbol{\gamma}$ is injective on the open interval (α, β).

A similar argument can be used to pass from parametrised curves to level curves:

Theorem 1.2

Let $\boldsymbol{\gamma}$ be a regular parametrised plane curve, and let $\boldsymbol{\gamma}(t_0) = (x_0, y_0)$ be a point in the image of $\boldsymbol{\gamma}$. Then, there is a smooth real-valued function $f(x, y)$, defined for x and y in open intervals containing x_0 and y_0, respectively, and satisfying the conditions in Theorem 1.1, such that $\boldsymbol{\gamma}(t)$ is contained in the level curve $f(x, y) = 0$ for all values of t in some open interval containing t_0.

The proof of Theorem 1.2 is similar to that of Theorem 1.1. Let

$$\boldsymbol{\gamma}(t) = (u(t), v(t)),$$

where u and v are smooth functions. Since $\boldsymbol{\gamma}$ is regular, at least one of $\dot{u}(t_0)$ and $\dot{v}(t_0)$ is non-zero, say $\dot{u}(t_0)$. This means that the graph of u as a function of t is not parallel to the t-axis at t_0:

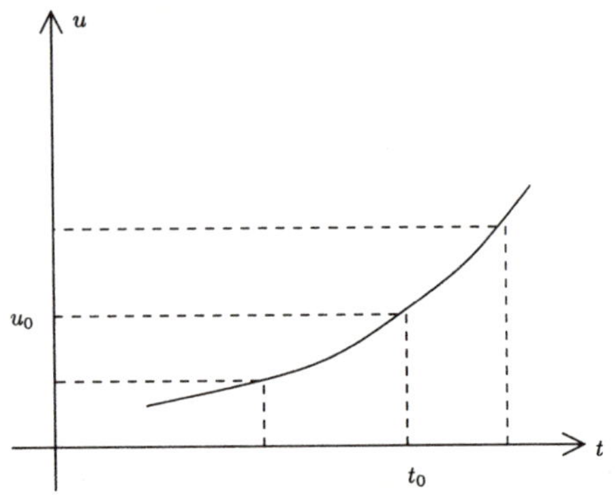

As in the proof of Theorem 1.1, this implies that any line parallel to the t-axis close to $u = x_0$ intersects the graph of u at a unique point $u(t)$ with t close to t_0. This gives a function $h(x)$, defined for x in an open interval containing x_0, such that $t = h(x)$ is the unique solution of $u(t) = x$ if x is near x_0 and t is near t_0. The inverse function theorem tells us that h is smooth. The function

$$f(x, y) = y - v(h(x))$$

has the properties we want.

It is not in general possible to find a *single* function $f(x, y)$ satisfying the conditions in Theorem 1.1 such that the image of $\boldsymbol{\gamma}$ is contained in the level curve $f(x, y) = 0$, for $\boldsymbol{\gamma}$ may have self-intersections like the *limaçon*

$$\boldsymbol{\gamma}(t) = ((1 + 2\cos t)\cos t, (1 + 2\cos t)\sin t).$$

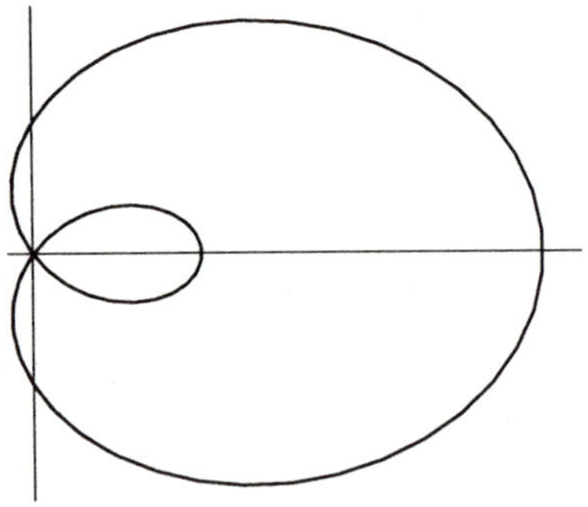

It follows from the inverse function theorem that no single function f satisfying the conditions in Theorem 1.1 can be found that describes a curve near such a self-intersection point.

EXERCISES

1.17 State a generalisation of Theorem 1.1 for level curves in $\mathbf{R}^3$ given by $f(x,y,z) = g(x,y,z) = 0$. (To guess the analogue of the condition on f in Theorem 1.1, argue that $(\frac{\partial f}{\partial x}, \frac{\partial f}{\partial y}, \frac{\partial f}{\partial z})$ is perpendicular to the surface $f(x,y,z) = 0$, and then think about the condition that two planes intersect in a line. See Exercise 4.16 for a rigorous statement.)

1.18 Generalise Theorem 1.2 for curves in $\mathbf{R}^3$ (or even $\mathbf{R}^n$). (This is easy.)

1.19 Sketch the level curve $\mathcal{C}$ given by $f(x,y) = 0$ when $f(x,y) = y - |x|$. Note that f does *not* satisfy the conditions in Theorem 1.1 because $\partial f/\partial x$ does not exist at the point $(0,0)$ on the curve. Show nevertheless that there is a smooth parametrised curve $\boldsymbol{\gamma}$ whose image is the whole of $\mathcal{C}$. (Make use of the smooth function $\theta(t)$ defined after the proof of Theorem 8.2.) Is there a *regular* parametrised curve with this property?

For the remainder of this book, we shall speak simply of 'curves', unless there is serious danger of confusion as to which type (level or parametrised) is intended.

2
How Much Does a Curve Curve?

In this chapter, we associate to any curve in $\mathbf{R}^3$ two scalar functions, called its curvature and torsion. The curvature measures the extent to which a curve is not contained in a straight line (so that straight lines have zero curvature), and the torsion measures the extent to which a curve is not contained in a plane (so that plane curves have zero torsion). It turns out that the curvature and torsion together determine the shape of a curve.

2.1. Curvature

We want to find a measure of how 'curved' a curve is. Since this 'curvature' should depend only on the 'shape' of the curve:

(i) the curvature should be unchanged when the curve is reparametrised.

Further, the measure of curvature should agree with our intuition in simple special cases, for example:

(ii) the curvature of a straight line should be zero, and large circles should have smaller curvature than small circles.

Bearing (ii) in mind, we get a clue as to what the definition of curvature should be from Proposition 1.1: this tells us that, if $\boldsymbol{\gamma}$ is a plane curve with $\ddot{\boldsymbol{\gamma}} = \mathbf{0}$ everywhere, the curve $\boldsymbol{\gamma}$ is part of a straight line, and hence should have zero curvature. So we might be tempted to define the curvature of $\boldsymbol{\gamma}$ to be $\| \ddot{\boldsymbol{\gamma}} \|$ (we take the norm because we want the curvature to be a scalar, not a vector). Unfortunately, however, this depends (in a fairly complicated way) on the parametrisation of $\boldsymbol{\gamma}$. So let us remove this freedom to reparametrise by insisting that $\boldsymbol{\gamma}$ is unit-speed, so that $\| \dot{\boldsymbol{\gamma}} \| = 1$ everywhere. (Actually, this does not quite rule out the possibility of reparametrising – see Corollary 1.1.) So we make

Definition 2.1

If $\boldsymbol{\gamma}$ is a unit-speed curve with parameter s, its *curvature* $\kappa(s)$ at the point $\boldsymbol{\gamma}(s)$ is defined to be $\| \ddot{\boldsymbol{\gamma}}(s) \|$.

The first part of condition (ii) will certainly be satisfied. As to the second part, consider the circle centred at (x_0, y_0) and of radius R. This has a unit-speed parametrisation

$$\boldsymbol{\gamma}(s) = \left(x_0 + R\cos\frac{s}{R}, y_0 + R\sin\frac{s}{R}\right).$$

We have

$$\dot{\boldsymbol{\gamma}}(s) = \left(-\sin\frac{s}{R}, \cos\frac{s}{R}\right),$$

$$\therefore \quad \| \dot{\boldsymbol{\gamma}}(s) \| = \sqrt{\left(-\sin\frac{s}{R}\right)^2 + \left(\cos\frac{s}{R}\right)^2} = 1,$$

showing that $\boldsymbol{\gamma}$ is indeed unit-speed, and hence

$$\ddot{\boldsymbol{\gamma}}(s) = \left(-\frac{1}{R}\cos\frac{s}{R}, -\frac{1}{R}\sin\frac{s}{R}\right),$$

$$\therefore \quad \| \ddot{\boldsymbol{\gamma}}(s) \| = \sqrt{\left(-\frac{1}{R}\cos\frac{s}{R}\right)^2 + \left(-\frac{1}{R}\sin\frac{s}{R}\right)^2} = \frac{1}{R},$$

so the curvature of the circle is inversely proportional to its radius.

As to condition (i), recall from Corollary 1.1 that, if $\boldsymbol{\gamma}(s)$ is a unit-speed curve, the only unit-speed reparametrisations of $\boldsymbol{\gamma}$ are of the form $\boldsymbol{\gamma}(u)$, where

$$u = \pm s + c,$$

and c is a constant. Then, by the chain rule,

$$\frac{d\boldsymbol{\gamma}}{ds} = \frac{d\boldsymbol{\gamma}}{du}\frac{du}{ds} = \pm\frac{d\boldsymbol{\gamma}}{du},$$

$$\therefore \quad \frac{d^2\boldsymbol{\gamma}}{ds^2} = \frac{d}{du}\left(\frac{d\boldsymbol{\gamma}}{ds}\right)\frac{du}{ds} = \pm\frac{d}{du}\left(\pm\frac{d\boldsymbol{\gamma}}{du}\right) = \frac{d^2\boldsymbol{\gamma}}{du^2}.$$

This shows that the curvature of the curve computed using the unit-speed parameter s is the same as that computed using the unit-speed parameter u.

But what if we are given a curve $\boldsymbol{\gamma}(t)$ that is not unit-speed? If $\boldsymbol{\gamma}$ is *regular* (see Definition 1.6), then by Proposition 1.5 $\boldsymbol{\gamma}$ has a unit-speed reparametrisation $\tilde{\boldsymbol{\gamma}}$. We define the curvature of $\boldsymbol{\gamma}$ to be that of the unit-speed curve $\tilde{\boldsymbol{\gamma}}$. But since it is not always possible to find the unit-speed reparametrisation *explicitly* (see Example 1.7), we really need a formula for the curvature in terms of $\boldsymbol{\gamma}$ and t only.

Proposition 2.1

Let $\boldsymbol{\gamma}(t)$ be a regular curve in $\mathbf{R}^3$. Then, its curvature is

$$\kappa = \frac{\parallel \ddot{\boldsymbol{\gamma}} \times \dot{\boldsymbol{\gamma}} \parallel}{\parallel \dot{\boldsymbol{\gamma}} \parallel^3}, \tag{1}$$

where the $\times$ indicates the vector (or cross) product and the dot denotes d/dt.

Of course, since a curve in $\mathbf{R}^2$ can be viewed as a curve in $\mathbf{R}^3$ whose last coordinate is zero, Eq. (1) can also be used to calculate the curvature of plane curves.

Proof 2.1

Let $\tilde{\boldsymbol{\gamma}}$ (with parameter s) be a unit-speed reparametrisation of $\boldsymbol{\gamma}$, and let us denote d/ds by a dash. Then, by the chain rule,

$$\tilde{\boldsymbol{\gamma}}' \frac{ds}{dt} = \dot{\boldsymbol{\gamma}},$$

so

$$\kappa = \parallel \tilde{\boldsymbol{\gamma}}'' \parallel = \parallel \frac{d}{ds}\left(\frac{\dot{\boldsymbol{\gamma}}}{ds/dt}\right) \parallel = \parallel \frac{\frac{d}{dt}\left(\frac{\dot{\boldsymbol{\gamma}}}{ds/dt}\right)}{ds/dt} \parallel = \parallel \frac{\ddot{\boldsymbol{\gamma}}\frac{ds}{dt} - \dot{\boldsymbol{\gamma}}\frac{d^2s}{dt^2}}{(ds/dt)^3} \parallel. \tag{2}$$

Now,

$$\left(\frac{ds}{dt}\right)^2 = \parallel \dot{\boldsymbol{\gamma}} \parallel^2 = \dot{\boldsymbol{\gamma}}.\dot{\boldsymbol{\gamma}},$$

and differentiating with respect to t gives

$$\frac{ds}{dt}\frac{d^2s}{dt^2} = \dot{\boldsymbol{\gamma}}.\ddot{\boldsymbol{\gamma}}.$$

Using this and Eq. (2), we get

$$\kappa = \parallel \frac{\ddot{\boldsymbol{\gamma}}\left(\frac{ds}{dt}\right)^2 - \dot{\boldsymbol{\gamma}}\frac{d^2s}{dt^2}\frac{ds}{dt}}{(ds/dt)^4} \parallel = \frac{\parallel \ddot{\boldsymbol{\gamma}}(\dot{\boldsymbol{\gamma}}.\dot{\boldsymbol{\gamma}}) - \dot{\boldsymbol{\gamma}}(\dot{\boldsymbol{\gamma}}.\ddot{\boldsymbol{\gamma}}) \parallel}{\parallel \dot{\boldsymbol{\gamma}} \parallel^4}.$$

Using the vector triple product identity

$$\mathbf{a} \times (\mathbf{b} \times \mathbf{c}) = (\mathbf{a}.\mathbf{c})\mathbf{b} - (\mathbf{a}.\mathbf{b})\mathbf{c}$$

(where $\mathbf{a}, \mathbf{b}, \mathbf{c} \in \mathbf{R}^3$), we get

$$\dot{\boldsymbol{\gamma}} \times (\ddot{\boldsymbol{\gamma}} \times \dot{\boldsymbol{\gamma}}) = \ddot{\boldsymbol{\gamma}}(\dot{\boldsymbol{\gamma}}.\dot{\boldsymbol{\gamma}}) - \dot{\boldsymbol{\gamma}}(\dot{\boldsymbol{\gamma}}.\ddot{\boldsymbol{\gamma}}).$$

Further, $\dot{\boldsymbol{\gamma}}$ and $\ddot{\boldsymbol{\gamma}} \times \dot{\boldsymbol{\gamma}}$ are perpendicular vectors, so

$$\| \dot{\boldsymbol{\gamma}} \times (\ddot{\boldsymbol{\gamma}} \times \dot{\boldsymbol{\gamma}}) \| = \| \dot{\boldsymbol{\gamma}} \| \| \ddot{\boldsymbol{\gamma}} \times \dot{\boldsymbol{\gamma}} \|.$$

Hence,

$$\begin{aligned} \frac{\| \ddot{\boldsymbol{\gamma}}(\dot{\boldsymbol{\gamma}}.\dot{\boldsymbol{\gamma}}) - \dot{\boldsymbol{\gamma}}(\dot{\boldsymbol{\gamma}}.\ddot{\boldsymbol{\gamma}}) \|}{\| \dot{\boldsymbol{\gamma}} \|^4} &= \frac{\| \dot{\boldsymbol{\gamma}} \times (\ddot{\boldsymbol{\gamma}} \times \dot{\boldsymbol{\gamma}}) \|}{\| \dot{\boldsymbol{\gamma}} \|^4} \\ &= \frac{\| \dot{\boldsymbol{\gamma}} \| \| \ddot{\boldsymbol{\gamma}} \times \dot{\boldsymbol{\gamma}} \|}{\| \dot{\boldsymbol{\gamma}} \|^4} \\ &= \frac{\| \ddot{\boldsymbol{\gamma}} \times \dot{\boldsymbol{\gamma}} \|}{\| \dot{\boldsymbol{\gamma}} \|^3}. \end{aligned}$$

□

If $\boldsymbol{\gamma}$ is a non-regular curve, its curvature is not defined in general. Note, however, that the formula (1) shows that the curvature is defined at all regular points of the curve (where $\dot{\boldsymbol{\gamma}}$ is non-zero).

Example 2.1

A *circular helix* with axis the z-axis is a curve of the form

$$\boldsymbol{\gamma}(\theta) = (a\cos\theta, a\sin\theta, b\theta), \quad -\infty < \theta < \infty,$$

where a and b are constants.

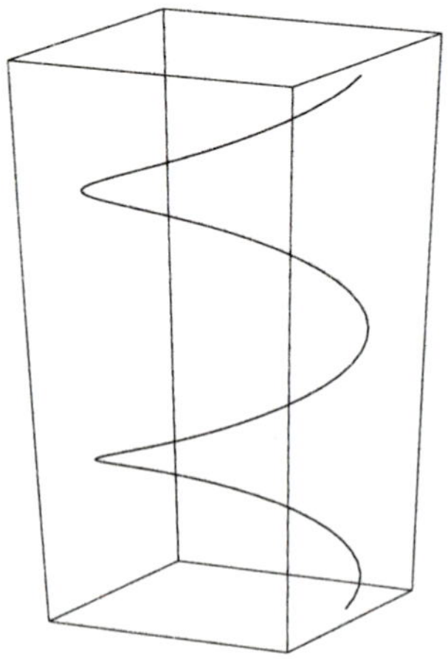

If (x, y, z) is a point on (the image of) the helix, so that

$$x = a\cos\theta, \ y = a\sin\theta, \ z = b\theta,$$

for some value of θ, then $x^2+y^2=a^2$, showing that the helix lies on the cylinder with axis the z-axis and radius $|a|$; the positive number $|a|$ is called the *radius* of the helix. As θ increases by 2π, the point $(a\cos\theta, a\sin\theta, b\theta)$ rotates once round the z-axis and moves up the z-axis by $2\pi b$; the positive number $2\pi|b|$ is called the *pitch* of the helix (we take absolute values since we did not assume that a or b is positive).

Let us compute the curvature of the helix using the formula in Proposition 2.1. Denoting $d/d\theta$ by a dot, we have

$$\dot{\boldsymbol{\gamma}}(\theta) = (-a\sin\theta, a\cos\theta, b),$$
$$\therefore \quad \| \dot{\boldsymbol{\gamma}}(\theta) \| = \sqrt{a^2+b^2}.$$

This shows that $\dot{\boldsymbol{\gamma}}(\theta)$ is never zero, so $\boldsymbol{\gamma}$ is regular (unless $a=b=0$, in which case the image of the helix is a single point). Hence, the formula in Proposition 2.1 applies, and we have

$$\ddot{\boldsymbol{\gamma}} = (-a\cos\theta, -a\sin\theta, 0),$$
$$\therefore \quad \ddot{\boldsymbol{\gamma}} \times \dot{\boldsymbol{\gamma}} = (-ab\sin\theta, ab\cos\theta, -a^2),$$
$$\therefore \quad \kappa = \frac{\| (-ab\sin\theta, ab\cos\theta, -a^2) \|}{\| (-a\sin\theta, a\cos\theta, b) \|^3} = \frac{(a^2b^2+a^4)^{1/2}}{(a^2+b^2)^{3/2}} = \frac{|a|}{a^2+b^2}. \tag{3}$$

Thus, the curvature of the helix is constant.

Let us examine some limiting cases to see if this result agrees with what we already know. First, suppose that $b=0$ (but $a \neq 0$). Then, the helix is simply a circle in the xy-plane of radius $|a|$, so by the calculation following Definition 1.1 its curvature is $1/|a|$. On the other hand, the formula (3) gives the curvature as

$$\frac{|a|}{a^2+0^2} = \frac{|a|}{a^2} = \frac{|a|}{|a|^2} = \frac{1}{|a|}.$$

Next, suppose that $a=0$ (but $b \neq 0$). Then, the image of the helix is just the z-axis, a straight line, so the curvature is zero. And (3) gives zero when $a=0$ too.

EXERCISES

2.1 Compute the curvature of the following curves:

(i) $\boldsymbol{\gamma}(t) = \left(\frac{1}{3}(1+t)^{3/2}, \frac{1}{3}(1-t)^{3/2}, \frac{t}{\sqrt{2}}\right)$;

(ii) $\boldsymbol{\gamma}(t) = \left(\frac{4}{5}\cos t, 1-\sin t, -\frac{3}{5}\cos t\right)$;

(iii) $\boldsymbol{\gamma}(t) = (t, \cosh t)$;

(iv) $\boldsymbol{\gamma}(t) = (\cos^3 t, \sin^3 t)$.

For the astroid in (iv), show that the curvature tends to ∞ as we approach one of the points $(\pm 1, 0)$, $(0, \pm 1)$. Compare with the sketch found in Exercise 1.5.

2.2 Show that, if the curvature $\kappa(t)$ of a regular curve $\boldsymbol{\gamma}(t)$ is > 0 everywhere, then $\kappa(t)$ is a smooth function of t. Give an example to show that this may not be the case without the assumption that $\kappa > 0$.

2.2. Plane Curves

For plane curves, it is possible to refine the definition of curvature slightly and give it an appealing geometric interpretation.

Suppose that $\boldsymbol{\gamma}(s)$ is a unit-speed curve in $\mathbf{R}^2$. Denoting d/ds by a dot, let

$$\mathbf{t} = \dot{\boldsymbol{\gamma}}$$

be the tangent vector of $\boldsymbol{\gamma}$; note that $\mathbf{t}$ is a unit vector. There are two unit vectors perpendicular to $\mathbf{t}$; we make a choice by defining $\mathbf{n}_s$, the *signed unit normal of* $\boldsymbol{\gamma}$, to be the unit vector obtained by rotating $\mathbf{t}$ anti-clockwise by $\pi/2$.

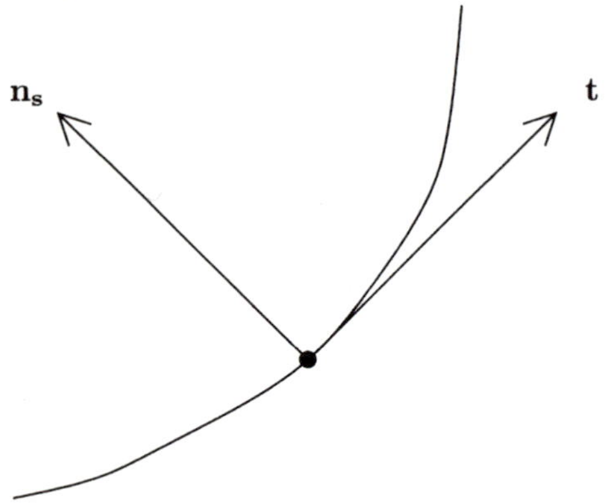

By Proposition 1.2, $\dot{\mathbf{t}} = \ddot{\boldsymbol{\gamma}}$ is perpendicular to $\mathbf{t}$, and hence parallel to $\mathbf{n}_s$. Thus, there is a number κ_s such that

$$\ddot{\boldsymbol{\gamma}} = \kappa_s \mathbf{n}_s.$$

The scalar κ_s is called the *signed curvature* of $\boldsymbol{\gamma}$ (it can be positive, negative or zero). Note that, since $\| \mathbf{n}_s \| = 1$, we have

$$\kappa = \| \ddot{\boldsymbol{\gamma}} \| = \| \kappa_s \mathbf{n}_s \| = |\kappa_s|, \tag{4}$$

so the curvature of $\boldsymbol{\gamma}$ is the absolute value of its signed curvature. The following diagrams show how the sign of the signed curvature is determined (in each case, the arrow on the curve indicates the direction of increasing s).

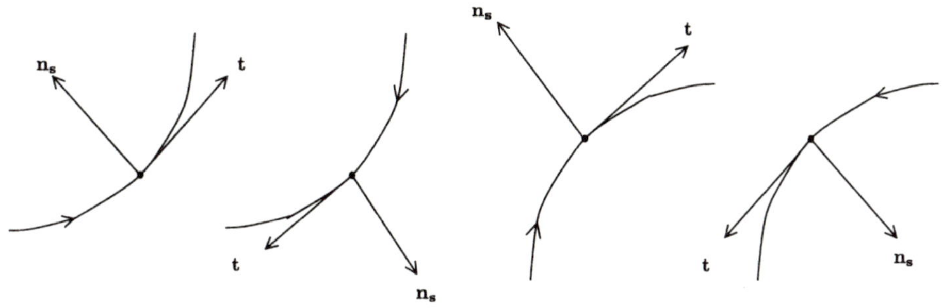

The signed curvature has a simple geometric interpretation:

Proposition 2.2

Let $\boldsymbol{\gamma}(s)$ be a unit-speed plane curve, and let $\varphi(s)$ be the angle through which a fixed unit vector must be rotated anti-clockwise to bring it into coincidence with the unit tangent vector $\mathbf{t}$ of $\boldsymbol{\gamma}$. Then,

$$\kappa_s = \frac{d\varphi}{ds}.$$

Note that, even though the angle φ is only determined up to adding an integer multiple of 2π, the derivative $d\varphi/ds$ is well defined.

Thus, the signed curvature is the rate at which the tangent vector of the curve rotates. As the above diagrams show, the signed curvature is positive or negative according as $\mathbf{t}$ rotates anti-clockwise or clockwise as one moves along the curve in the direction of increasing s.

Proof 2.2

Let $\mathbf{a}$ be the fixed unit vector and let $\mathbf{b}$ be the unit vector obtained by rotating $\mathbf{a}$ anti-clockwise by $\pi/2$. Then,

$$\mathbf{t} = \mathbf{a}\cos\varphi + \mathbf{b}\sin\varphi,$$

$$\therefore \quad \dot{\mathbf{t}} = (-\mathbf{a}\sin\varphi + \mathbf{b}\cos\varphi)\frac{d\varphi}{ds},$$

$$\therefore \quad \dot{\mathbf{t}}.\mathbf{a} = -\sin\varphi\frac{d\varphi}{ds},$$

$$\therefore \quad \kappa_s(\mathbf{n}_s.\mathbf{a}) = -\sin\varphi\frac{d\varphi}{ds} \quad (\text{since } \dot{\mathbf{t}} = \kappa_s\mathbf{n}_s). \tag{5}$$

But the angle between $\mathbf{n}_s$ and $\mathbf{a}$ is $\varphi + \pi/2$, since $\mathbf{t}$ must be rotated anti-clockwise by $\pi/2$ to bring it into coincidence with $\mathbf{n}_s$ (see the diagram below).

Hence,

$$\mathbf{n}_s.\mathbf{a} = \cos\left(\varphi + \frac{\pi}{2}\right) = -\sin\varphi.$$

Inserting this into Eq. (5) gives the required result. □

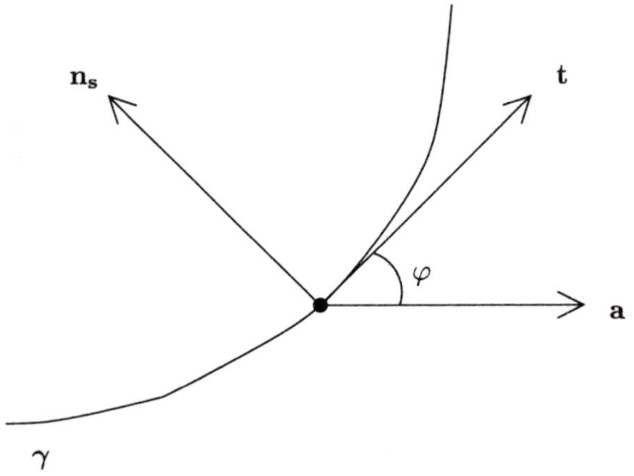

The next result shows that a unit-speed plane curve is essentially determined once we know its signed curvature at each point of the curve. The meaning of 'essentially' here is 'up to a rigid motion of $\mathbf{R}^2$'. Recall that a rigid motion of $\mathbf{R}^2$ is a map $M : \mathbf{R}^2 \to \mathbf{R}^2$ of the form

$$M = T_{\mathbf{a}} \circ R_\theta,$$

where R_θ is an anti-clockwise rotation by an angle θ about the origin,

$$R_\theta(x, y) = (x\cos\theta - y\sin\theta, x\sin\theta + y\cos\theta),$$

and $T_{\mathbf{a}}$ is the translation by the vector $\mathbf{a}$,

$$T_{\mathbf{a}}(\mathbf{v}) = \mathbf{v} + \mathbf{a},$$

for any vectors (x, y) and $\mathbf{v} \in \mathbf{R}^2$.

Theorem 2.1

Let $k : (\alpha, \beta) \to \mathbf{R}$ be any smooth function. Then, there is a unit-speed curve $\boldsymbol{\gamma} : (\alpha, \beta) \to \mathbf{R}^2$ whose signed curvature is k.

Further, if $\tilde{\boldsymbol{\gamma}} : (\alpha, \beta) \to \mathbf{R}^2$ is any other unit-speed curve whose signed curvature is k, there is a rigid motion M of $\mathbf{R}^2$ such that

$$\tilde{\boldsymbol{\gamma}}(s) = M(\boldsymbol{\gamma}(s)) \quad \text{for all } s \in (\alpha, \beta).$$

Proof 2.1

For the first part, fix $s_0 \in (\alpha, \beta)$ and define, for any $s \in (\alpha, \beta)$,

$$\varphi(s) = \int_{s_0}^{s} k(u)du, \quad \text{(cf. Proposition 2.2)},$$
$$\boldsymbol{\gamma}(s) = \left(\int_{s_0}^{s} \cos\varphi(t)dt, \int_{s_0}^{s} \sin\varphi(t)dt \right).$$

Then, the tangent vector of $\boldsymbol{\gamma}$ is

$$\dot{\boldsymbol{\gamma}}(s) = (\cos\varphi(s), \sin\varphi(s)),$$

which is a unit vector making an angle $\varphi(s)$ with the x-axis. Thus, $\boldsymbol{\gamma}$ is unit-speed and, by Proposition 2.2, its signed curvature is

$$\frac{d\varphi}{ds} = \frac{d}{ds}\int_{s_0}^{s} k(u)du = k(s).$$

For the second part, let $\tilde{\varphi}(s)$ be the angle between the x-axis and the unit tangent vector $\dot{\tilde{\boldsymbol{\gamma}}}$ of $\tilde{\boldsymbol{\gamma}}$. Thus,

$$\dot{\tilde{\boldsymbol{\gamma}}}(s) = (\cos\tilde{\varphi}(s), \sin\tilde{\varphi}(s)),$$
$$\therefore \quad \tilde{\boldsymbol{\gamma}}(s) = \left(\int_{s_0}^{s} \cos\tilde{\varphi}(t)dt, \int_{s_0}^{s} \sin\tilde{\varphi}(t)dt \right) + \tilde{\boldsymbol{\gamma}}(s_0). \tag{6}$$

By Proposition 2.2,

$$\frac{d\tilde{\varphi}}{ds} = k(s),$$
$$\therefore \quad \tilde{\varphi}(s) = \int_{s_0}^{s} k(u)du + \tilde{\varphi}(s_0).$$

Inserting this into Eq. (6), and writing $\mathbf{a}$ for the constant vector $\tilde{\boldsymbol{\gamma}}(s_0)$ and θ for the constant scalar $\tilde{\varphi}(s_0)$, we get

$$\begin{aligned}
\tilde{\boldsymbol{\gamma}}(s) &= T_{\mathbf{a}}\left(\int_{s_0}^{s} \cos(\varphi(t)+\theta)dt, \int_{s_0}^{s} \sin(\varphi(t)+\theta)dt \right) \\
&= T_{\mathbf{a}}\left(\cos\theta \int_{s_0}^{s} \cos\varphi(t)dt - \sin\theta \int_{s_0}^{s} \sin\varphi(t)dt, \right. \\
&\qquad \left. \sin\theta \int_{s_0}^{s} \cos\varphi(t)dt + \cos\theta \int_{s_0}^{s} \sin\varphi(t)dt \right) \\
&= T_{\mathbf{a}} R_\theta \left(\int_{s_0}^{s} \cos\varphi(t)dt, \int_{s_0}^{s} \sin\varphi(t)dt \right) \\
&= T_{\mathbf{a}} R_\theta(\boldsymbol{\gamma}(s)).
\end{aligned}$$

□

Example 2.2

Any regular plane curve whose curvature is a positive constant is part of a circle. To see this, let κ be the (constant) curvature of the curve $\boldsymbol{\gamma}$, and let κ_s be its signed curvature. Then, by Eq. (4),

$$\kappa_s = \pm\kappa.$$

A priori, we could have $\kappa_s = \kappa$ at some points of the curve and $\kappa_s = -\kappa$ at others, but in fact this cannot happen since κ_s is a continuous function of s (cf. Exercise 2.4), so the Intermediate Value Theorem tells us that, if κ_s takes both the value κ and the value $-\kappa$, it must take all values between. Thus, either $\kappa_s = \kappa$ at all points of the curve, or $\kappa_s = -\kappa$ at all points of the curve. In particular, κ_s is constant.

The idea now is to show that, whatever the value of κ_s, we can find a parametrised circle whose signed curvature is κ_s. The theorem then tells us that *any* curve whose signed curvature is κ_s can be obtained by applying a rigid motion to this circle. Since rotations and translations obviously take circles to circles, it follows that *any* curve whose signed curvature is constant is (part of) a circle.

A unit-speed parametrisation of the circle with centre the origin and radius R is

$$\boldsymbol{\gamma}(s) = \left(R\cos\frac{s}{R}, R\sin\frac{s}{R}\right).$$

Its tangent vector

$$\mathbf{t} = \dot{\boldsymbol{\gamma}}(s) = \left(-\sin\frac{s}{R}, \cos\frac{s}{R}\right)$$

is the unit vector making an angle $\pi/2 + s/R$ with the positive x-axis:

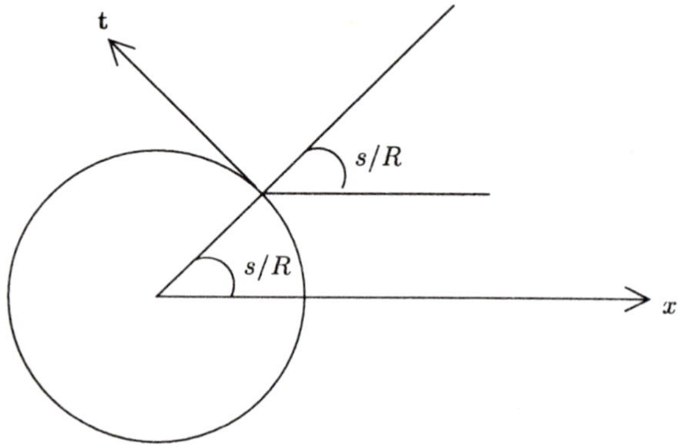

Hence, the signed curvature of $\boldsymbol{\gamma}$ is

$$\frac{d}{ds}\left(\frac{\pi}{2} + \frac{s}{R}\right) = \frac{1}{R}.$$

Thus, if $\kappa_s > 0$, the circle of radius $1/\kappa_s$ has signed curvature κ_s.

If $\kappa_s < 0$, it is easy to check that the curve

$$\tilde{\gamma}(s) = \left(R\cos\frac{s}{R}, -R\sin\frac{s}{R}\right)$$

(which is just another parametrisation of the circle with centre the origin and radius R) has signed curvature $-1/R$. Thus, if $R = -1/\kappa_s$ we again get a circle with signed curvature κ_s.

Example 2.3

Theorem 2.1 shows that we can find a plane curve with any given smooth function as its signed curvature. But simple curvatures can lead to complicated curves. For example, let the signed curvature be $\kappa_s(s) = s$. Following the proof of Theorem 2.1, and taking $s_0 = 0$, we get

$$\varphi(s) = \int_0^s u\,du = \frac{s^2}{2},$$

$$\boldsymbol{\gamma}(s) = \left(\int_0^s \cos\left(\frac{t^2}{2}\right) dt, \int_0^s \sin\left(\frac{t^2}{2}\right) dt\right).$$

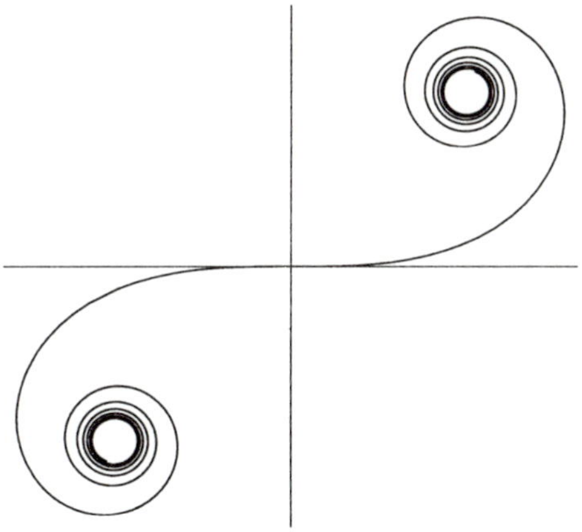

These integrals cannot be evaluated in terms of 'elementary' functions. (They arise in the theory of diffraction of light, where they are called *Fresnel's integrals*, and the curve $\boldsymbol{\gamma}$ is called *Cornu's Spiral*, although it was first considered by Euler.) The picture of $\boldsymbol{\gamma}$ above is obtained by computing the integrals numerically.

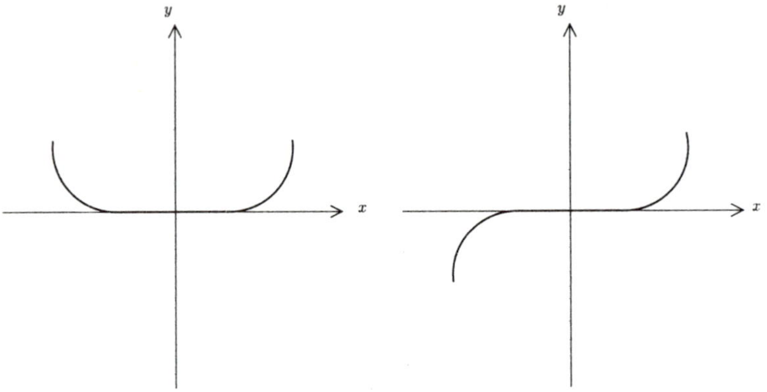

It is natural to ask whether Theorem 2.1 remains true if we replace 'signed curvature' by 'curvature'. The first part holds if (and only if) we assume that $k \geq 0$, for then $\boldsymbol{\gamma}$ can be chosen to have signed curvature k and so will have curvature k as well. The second part of Theorem 2.1, however, no longer holds. For, we can take a (smooth) curve $\boldsymbol{\gamma}$ that coincides with the x-axis for $-1 \leq x \leq 1$ (say), and is otherwise above the x-axis. (The reader who wishes to write down such a curve explicitly will find the solution of Exercise 1.19 helpful.) We now reflect the part of the curve with $x \leq 0$ in the x-axis. The new curve has the same curvature as $\boldsymbol{\gamma}$ (see Exercise 2.12), but obviously cannot be obtained by applying a rigid motion to $\boldsymbol{\gamma}$. See Exercise 2.13 for a version of Theorem 2.1 that *is* valid for curvature instead of signed curvature.

EXERCISES

2.3 Show that, if $\boldsymbol{\gamma}$ is a unit-speed plane curve,

$$\dot{\mathbf{n}}_s = -\kappa_s \mathbf{t}.$$

2.4 Show that the signed curvature of any regular plane curve $\boldsymbol{\gamma}(t)$ is a smooth function of t. (Compare with Exercise 2.2.)

2.5 Let $\boldsymbol{\gamma}(t) = (e^{kt}\cos t, e^{kt}\sin t)$, where $-\infty < t < \infty$ and k is a non-zero constant (a logarithmic spiral – see Example 1.4). Show that there is a unique unit-speed parameter s on $\boldsymbol{\gamma}$ such that $s > 0$ for all t and $s \to 0$ as $t \to \mp\infty$ if $\pm k > 0$, and express s as a function of t. Show that the signed curvature of $\boldsymbol{\gamma}$ is $1/ks$. Conversely, describe every curve whose signed curvature, as a function of arc-length s, is $1/ks$ for some non-zero constant k.

2.6 A unit-speed plane curve $\boldsymbol{\gamma}$ has the property that its tangent vector $\mathbf{t}(s)$ makes a fixed angle θ with $\boldsymbol{\gamma}(s)$ for all s. Show that

(i) if $\theta = 0$, then $\boldsymbol{\gamma}$ is part of a straight line (write $\boldsymbol{\gamma} = r\mathbf{t}$ and deduce that $\kappa_s = 0$);
(ii) if $\theta = \pi/2$, then $\boldsymbol{\gamma}$ is a circle (write $\boldsymbol{\gamma} = r\mathbf{n}_s$);
(iii) if $0 < \theta < \pi/2$, then $\boldsymbol{\gamma}$ is a logarithmic spiral (show that $\kappa_s = -1/s\cot\theta$).

2.7 Let $\boldsymbol{\gamma}(t)$ be a regular plane curve and let λ be a constant. The *parallel curve* $\boldsymbol{\gamma}^\lambda$ of $\boldsymbol{\gamma}$ is defined by

$$\boldsymbol{\gamma}^\lambda(t) = \boldsymbol{\gamma}(t) + \lambda\mathbf{n}_s(t).$$

Show that, if $|\lambda\kappa_s(t)| < 1$ for all values of t, then $\boldsymbol{\gamma}^\lambda$ is a regular curve and that its signed curvature is $\kappa_s/(1-\lambda\kappa_s)$.

2.8 Let $\boldsymbol{\gamma}$ be a unit-speed plane curve with nowhere zero curvature. Define the *centre of curvature* $\boldsymbol{\epsilon}(s)$ of $\boldsymbol{\gamma}$ at the point $\boldsymbol{\gamma}(s)$ to be

$$\boldsymbol{\epsilon}(s) = \boldsymbol{\gamma}(s) + \frac{1}{\kappa_s(s)}\mathbf{n}_s(s).$$

Prove that the circle with centre $\boldsymbol{\epsilon}(s)$ and radius $|1/\kappa_s(s)|$ is tangent to $\boldsymbol{\gamma}$ at $\boldsymbol{\gamma}(s)$ and has the same curvature as $\boldsymbol{\gamma}$ at that point. This circle is called the *osculating circle* to $\boldsymbol{\gamma}$ at the point $\boldsymbol{\gamma}(s)$. (Draw a picture.)

2.9 With the notation in Exercise 2.8, we regard $\boldsymbol{\epsilon}(s)$ as the parametrisation of a new curve, called the *evolute* of $\boldsymbol{\gamma}$ (if $\boldsymbol{\gamma}$ is any regular plane curve, its evolute is defined to be that of a unit-speed reparametrisation of $\boldsymbol{\gamma}$). Assume that $\dot{\kappa}_s(s) \neq 0$ for all values of s (a dot denoting d/ds), say $\dot{\kappa}_s > 0$ for all s (this can be achieved by replacing s by $-s$ if necessary). Show that the arc-length of $\boldsymbol{\epsilon}$ is $u_0 - \frac{1}{\kappa_s(s)}$, where u_0 is a constant, and calculate the signed curvature of $\boldsymbol{\epsilon}$.

Show that the evolute of the cycloid

$$\boldsymbol{\gamma}(t) = a(t - \sin t, 1 - \cos t), \qquad 0 < t < 2\pi,$$

where $a > 0$ is a constant, is

$$\boldsymbol{\epsilon}(t) = a(t + \sin t, -1 + \cos t)$$

(see Exercise 1.7) and that, after a suitable reparametrisation, $\boldsymbol{\epsilon}$ can be obtained from $\boldsymbol{\gamma}$ by a translation of the plane.

2.10 A string of length ℓ is attached to the point $s = 0$ of a unit-speed plane curve $\boldsymbol{\gamma}(s)$. Show that when the string is wound onto the curve while being kept taught, its endpoint traces out the curve

$$\boldsymbol{\iota}(s) = \boldsymbol{\gamma}(s) + (\ell - s)\dot{\boldsymbol{\gamma}}(s),$$

where $0 < s < \ell$ and a dot denotes d/ds. The curve $\boldsymbol{\iota}$ is called the *involute* of $\boldsymbol{\gamma}$ (if $\boldsymbol{\gamma}$ is any regular plane curve, we define its involute to be that of a unit-speed reparametrisation of $\boldsymbol{\gamma}$). Suppose that the signed curvature κ_s of $\boldsymbol{\gamma}$ is never zero, say $\kappa_s(s) > 0$ for all s. Show that the signed curvature of $\boldsymbol{\iota}$ is $1/(\ell - s)$.

2.11 Let $\boldsymbol{\gamma}$ be a regular plane curve. Show that
(i) the involute of the evolute of $\boldsymbol{\gamma}$ is a parallel curve of $\boldsymbol{\gamma}$;
(ii) the evolute of the involute of $\boldsymbol{\gamma}$ is $\boldsymbol{\gamma}$.
(These statements might be compared to the fact that the integral of the derivative of a smooth function f is equal to f plus a constant, while the derivative of the integral of f is f.)

2.12 Show that applying a reflection in a straight line to a plane curve changes the sign of its signed curvature.

2.13 Show that, if two plane curves $\boldsymbol{\gamma}(t)$ and $\tilde{\boldsymbol{\gamma}}(t)$ have the same non-zero curvature for all values of t, then $\tilde{\boldsymbol{\gamma}}$ can be obtained from $\boldsymbol{\gamma}$ by applying a rigid motion or a reflection in a straight line followed by a rigid motion.

2.3. Space Curves

Our main interest in this book will be in curves (and surfaces) in $\mathbf{R}^3$, i.e. space curves. While a plane curve is essentially determined by its curvature (see Theorem 2.1), this is no longer true for space curves. For example, a circle of radius one in the xy-plane and a circular helix with $a = b = 1/2$ (see Example 2.1) both have curvature one everywhere, but it is obviously impossible to change one curve into the other by any combination of rotations and translations. We shall define another type of curvature for space curves, called the *torsion*, and we shall prove that the curvature and torsion of a curve together determine the curve up to a rigid motion (Theorem 2.3).

Let $\boldsymbol{\gamma}(s)$ be a unit-speed curve in $\mathbf{R}^3$, and let $\mathbf{t} = \dot{\boldsymbol{\gamma}}$ be its unit tangent vector. *If the curvature $\kappa(s)$ is non-zero*, we define the *principal normal* of $\boldsymbol{\gamma}$ at the point $\boldsymbol{\gamma}(s)$ to be the vector

$$\mathbf{n}(s) = \frac{1}{\kappa(s)}\dot{\mathbf{t}}(s). \tag{7}$$

Since $\| \dot{\mathbf{t}} \| = \kappa$, $\mathbf{n}$ is a unit vector. Further, by Proposition 1.2, $\mathbf{t}.\dot{\mathbf{t}} = 0$, so $\mathbf{t}$ and $\mathbf{n}$ are actually perpendicular unit vectors. It follows that

$$\mathbf{b} = \mathbf{t} \times \mathbf{n} \tag{8}$$

is a unit vector perpendicular to both $\mathbf{t}$ and $\mathbf{n}$. The vector $\mathbf{b}(s)$ is called the *binormal* vector of $\boldsymbol{\gamma}$ at the point $\boldsymbol{\gamma}(s)$. Thus, $\{\mathbf{t}, \mathbf{n}, \mathbf{b}\}$ is an orthonormal basis of $\mathbf{R}^3$, and is right-handed, i.e.

$$\mathbf{b} = \mathbf{t} \times \mathbf{n}, \quad \mathbf{n} = \mathbf{b} \times \mathbf{t}, \quad \mathbf{t} = \mathbf{n} \times \mathbf{b}.$$

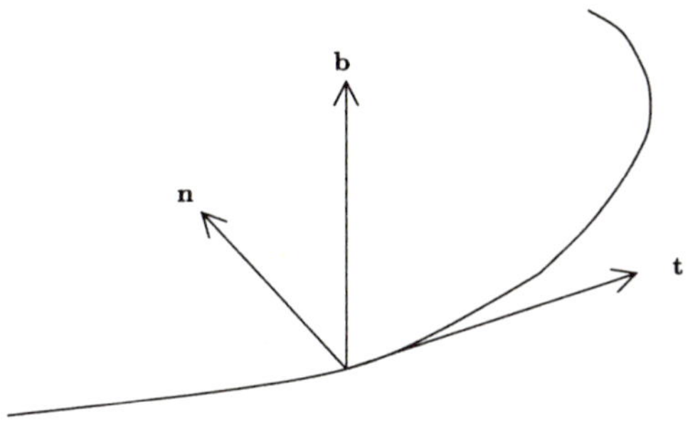

Since $\mathbf{b}(s)$ is a unit vector for all s, $\dot{\mathbf{b}}$ is perpendicular to $\mathbf{b}$. Now we use the 'product rule' for differentiating the vector product of vector-valued functions $\mathbf{u}$ and $\mathbf{v}$ of a parameter s:

$$\frac{d}{ds}(\mathbf{u} \times \mathbf{v}) = \frac{d\mathbf{u}}{ds} \times \mathbf{v} + \mathbf{u} \times \frac{d\mathbf{v}}{ds}.$$

Applying this to $\mathbf{b} = \mathbf{t} \times \mathbf{n}$ gives

$$\dot{\mathbf{b}} = \dot{\mathbf{t}} \times \mathbf{n} + \mathbf{t} \times \dot{\mathbf{n}} = \mathbf{t} \times \dot{\mathbf{n}}, \tag{9}$$

since by the definition (7) of $\mathbf{n}$,

$$\dot{\mathbf{t}} \times \mathbf{n} = \kappa \mathbf{n} \times \mathbf{n} = \mathbf{0}.$$

Equation (9) shows that $\dot{\mathbf{b}}$ is perpendicular to $\mathbf{t}$. Being perpendicular to both $\mathbf{t}$ and $\mathbf{b}$, $\dot{\mathbf{b}}$ must be parallel to $\mathbf{n}$, so

$$\dot{\mathbf{b}} = -\tau \mathbf{n}, \tag{10}$$

for some scalar τ, which is called the *torsion* of $\boldsymbol{\gamma}$ (the minus sign is purely a matter of convention). Note that the torsion is only defined if the curvature is non-zero.

Of course, we define the torsion of an arbitrary regular curve $\boldsymbol{\gamma}$ to be the torsion of a unit-speed reparametrisation of $\boldsymbol{\gamma}$. As in the case of the curvature, to see that this makes sense, we have to show that if we make a change in the unit-speed parameter of $\boldsymbol{\gamma}$ of the form

$$u = \pm s + c,$$

where c is a constant, then τ is unchanged. But this change of parameter has the following effect on the vectors introduced above:

$$\mathbf{t} \mapsto \pm \mathbf{t}, \ \dot{\mathbf{t}} \mapsto \dot{\mathbf{t}}, \ \mathbf{n} \mapsto \mathbf{n}, \ \mathbf{b} \mapsto \pm \mathbf{b}, \ \dot{\mathbf{b}} \mapsto \dot{\mathbf{b}}.$$

It follows from Eq. (8) that $\tau \mapsto \tau$, as required.

Just as we did for the curvature in Proposition 2.1, it is possible to give a formula for the torsion of a space curve $\boldsymbol{\gamma}$ in terms of $\boldsymbol{\gamma}$ alone, without requiring a unit-speed reparametrisation:

Proposition 2.3

Let $\boldsymbol{\gamma}(t)$ be a regular curve in $\mathbf{R}^3$ with nowhere vanishing curvature. Then, denoting d/dt by a dot, its torsion is given by

$$\tau = \frac{(\dot{\boldsymbol{\gamma}} \times \ddot{\boldsymbol{\gamma}}).\dddot{\boldsymbol{\gamma}}}{\| \dot{\boldsymbol{\gamma}} \times \ddot{\boldsymbol{\gamma}} \|^2}. \tag{11}$$

Note that this formula shows that $\tau(t)$ is defined at all points $\boldsymbol{\gamma}(t)$ of the curve at which its curvature $\kappa(t)$ is non-zero, since by Proposition 2.1 this is the condition for the denominator on the right-hand side to be non-zero.

Proof 2.3

We could 'derive' Eq. (11) by imitating the proof of Proposition 2.1. But it is easier and clearer to proceed as follows, even though this method has the disadvantage that one must know the formula for τ in Eq. (11) in advance.

We first treat the case in which $\boldsymbol{\gamma}$ is unit-speed. Using Eqs. (7) and (10),

$$\tau = -\mathbf{n}.\dot{\mathbf{b}} = -\mathbf{n}.(\mathbf{t} \times \mathbf{n})^{\cdot} = -\mathbf{n}.(\dot{\mathbf{t}} \times \mathbf{n} + \mathbf{t} \times \dot{\mathbf{n}}) = -\mathbf{n}.(\mathbf{t} \times \dot{\mathbf{n}}).$$

Now, $\mathbf{n} = \frac{1}{\kappa}\dot{\mathbf{t}} = \frac{1}{\kappa}\ddot{\boldsymbol{\gamma}}$, so

$$\begin{aligned}
\tau &= -\frac{1}{\kappa}\ddot{\boldsymbol{\gamma}}.\left(\dot{\boldsymbol{\gamma}} \times \frac{d}{dt}\left(\frac{1}{\kappa}\ddot{\boldsymbol{\gamma}}\right)\right) \\
&= -\frac{1}{\kappa}\ddot{\boldsymbol{\gamma}}.\left(\dot{\boldsymbol{\gamma}} \times \left(\frac{1}{\kappa}\dddot{\boldsymbol{\gamma}} - \frac{\dot{\kappa}}{\kappa^2}\ddot{\boldsymbol{\gamma}}\right)\right) \\
&= \frac{1}{\kappa^2}\dddot{\boldsymbol{\gamma}}.(\dot{\boldsymbol{\gamma}} \times \ddot{\boldsymbol{\gamma}}),
\end{aligned}$$

since $\ddot{\boldsymbol{\gamma}}.(\dot{\boldsymbol{\gamma}} \times \ddot{\boldsymbol{\gamma}}) = 0$ and $\ddot{\boldsymbol{\gamma}}.(\dot{\boldsymbol{\gamma}} \times \dddot{\boldsymbol{\gamma}}) = -\dddot{\boldsymbol{\gamma}}.(\dot{\boldsymbol{\gamma}} \times \ddot{\boldsymbol{\gamma}})$. This agrees with Eq. (11), for, since $\boldsymbol{\gamma}$ is unit-speed, $\dot{\boldsymbol{\gamma}}$ and $\ddot{\boldsymbol{\gamma}}$ are perpendicular, so

$$\| \dot{\boldsymbol{\gamma}} \times \ddot{\boldsymbol{\gamma}} \| = \| \dot{\boldsymbol{\gamma}} \| \| \ddot{\boldsymbol{\gamma}} \| = \| \ddot{\boldsymbol{\gamma}} \| = \kappa.$$

In the general case, let s be arc-length along $\boldsymbol{\gamma}$ and denote d/ds by a dash.

Then,

$$\dot{\boldsymbol{\gamma}} = \frac{ds}{dt}\boldsymbol{\gamma}',$$
$$\ddot{\boldsymbol{\gamma}} = \left(\frac{ds}{dt}\right)^2 \boldsymbol{\gamma}'' + \frac{d^2 s}{dt^2}\boldsymbol{\gamma}',$$
$$\dddot{\boldsymbol{\gamma}} = \left(\frac{ds}{dt}\right)^3 \boldsymbol{\gamma}''' + 3\frac{ds}{dt}\frac{d^2 s}{dt^2}\boldsymbol{\gamma}'' + \frac{d^3 s}{dt^3}\boldsymbol{\gamma}'.$$

Hence,

$$\dot{\boldsymbol{\gamma}} \times \ddot{\boldsymbol{\gamma}} = \left(\frac{ds}{dt}\right)^3 \boldsymbol{\gamma}' \times \boldsymbol{\gamma}'',$$
$$\dddot{\boldsymbol{\gamma}}.(\dot{\boldsymbol{\gamma}} \times \ddot{\boldsymbol{\gamma}}) = \left(\frac{ds}{dt}\right)^6 \boldsymbol{\gamma}'''.(\boldsymbol{\gamma}' \times \boldsymbol{\gamma}''),$$

and so

$$\frac{\dddot{\boldsymbol{\gamma}}.(\dot{\boldsymbol{\gamma}} \times \ddot{\boldsymbol{\gamma}})}{\| \dot{\boldsymbol{\gamma}} \times \ddot{\boldsymbol{\gamma}} \|^2} = \frac{\boldsymbol{\gamma}'''.(\boldsymbol{\gamma}' \times \boldsymbol{\gamma}'')}{\| \boldsymbol{\gamma}' \times \boldsymbol{\gamma}'' \|^2}.$$ □

Example 2.4

We compute the torsion of the circular helix

$$\boldsymbol{\gamma}(\theta) = (a\cos\theta, a\sin\theta, b\theta)$$

studied in Example 2.1. We have

$$\dot{\boldsymbol{\gamma}}(\theta) = (-a\sin\theta, a\cos\theta, b),$$
$$\ddot{\boldsymbol{\gamma}}(\theta) = (-a\cos\theta, -a\sin\theta, 0),$$
$$\dddot{\boldsymbol{\gamma}}(\theta) = (a\sin\theta, -a\cos\theta, 0).$$

Hence,

$$\dot{\boldsymbol{\gamma}} \times \ddot{\boldsymbol{\gamma}} = (ab\sin\theta, -ab\cos\theta, a^2),$$
$$\| \dot{\boldsymbol{\gamma}} \times \ddot{\boldsymbol{\gamma}} \|^2 = a^2(a^2 + b^2),$$
$$(\dot{\boldsymbol{\gamma}} \times \ddot{\boldsymbol{\gamma}}).\dddot{\boldsymbol{\gamma}} = a^2 b,$$

and so the torsion

$$\tau = \frac{(\dot{\boldsymbol{\gamma}} \times \ddot{\boldsymbol{\gamma}}).\dddot{\boldsymbol{\gamma}}}{\| \dot{\boldsymbol{\gamma}} \times \ddot{\boldsymbol{\gamma}} \|^2} = \frac{a^2 b}{a^2(a^2 + b^2)} = \frac{b}{a^2 + b^2}.$$

Note that the torsion of the circular helix in Example 2.4 becomes zero when $b = 0$, in which case the helix is just a circle in the xy-plane. This gives us a clue as to the geometrical interpretation of torsion, contained in

Proposition 2.4

Let $\boldsymbol{\gamma}$ be a regular curve in $\mathbf{R}^3$ with nowhere vanishing curvature (so that the torsion τ of $\boldsymbol{\gamma}$ is defined). Then, the image of $\boldsymbol{\gamma}$ is contained in a plane if and only if τ is zero at every point of the curve.

Proof 2.4

We can assume that $\boldsymbol{\gamma}$ is unit-speed (for this can be achieved by reparametrising $\boldsymbol{\gamma}$, and reparametrising changes neither the torsion nor the fact that $\boldsymbol{\gamma}$ is, or is not, contained in a plane). We denote the parameter of $\boldsymbol{\gamma}$ by s and d/ds by a dot as usual.

Suppose first that the image of $\boldsymbol{\gamma}$ is contained in the plane $\mathbf{r}.\mathbf{a} = d$, where $\mathbf{a}$ is a constant vector and d is a constant scalar ($\mathbf{r}$ is the position vector of an arbitrary point of $\mathbf{R}^3$). We can assume that $\mathbf{a}$ is a unit vector. Differentiating $\boldsymbol{\gamma}.\mathbf{a} = d$ with respect to s, we get

$$\mathbf{t}.\mathbf{a} = 0, \tag{12}$$

$$\therefore \quad \dot{\mathbf{t}}.\mathbf{a} = 0 \quad (\text{since } \dot{\mathbf{a}} = \mathbf{0}),$$

$$\therefore \quad \kappa\mathbf{n}.\mathbf{a} = 0 \quad (\text{since } \dot{\mathbf{t}} = \kappa\mathbf{n}),$$

$$\therefore \quad \mathbf{n}.\mathbf{a} = 0 \quad (\text{since } \kappa \neq 0). \tag{13}$$

Equations (12) and (13) show that $\mathbf{t}$ and $\mathbf{n}$ are perpendicular to $\mathbf{a}$. It follows that $\mathbf{b} = \mathbf{t} \times \mathbf{n}$ is parallel to $\mathbf{a}$. Since $\mathbf{a}$ and $\mathbf{b}$ are both unit vectors, and $\mathbf{b}(s)$ is a smooth (hence continuous) function of s, we must have $\mathbf{b}(s) = \mathbf{a}$ for all s or $\mathbf{b}(s) = -\mathbf{a}$ for all s. In either case, $\mathbf{b}$ is a constant vector. But then $\dot{\mathbf{b}} = \mathbf{0}$, so $\tau = 0$.

Conversely, suppose that $\tau = 0$ everywhere. By Eq. (10), $\dot{\mathbf{b}} = \mathbf{0}$, so $\mathbf{b}$ is a constant vector. The first part of the proof suggests that $\boldsymbol{\gamma}$ should be contained in a plane $\mathbf{r}.\mathbf{b} =$ constant. We therefore consider

$$\frac{d}{ds}(\boldsymbol{\gamma}.\mathbf{b}) = \dot{\boldsymbol{\gamma}}.\mathbf{b} = \mathbf{t}.\mathbf{b} = 0,$$

so $\boldsymbol{\gamma}.\mathbf{b}$ is a constant (scalar), say d. This means that $\boldsymbol{\gamma}$ is indeed contained in the plane $\mathbf{r}.\mathbf{b} = d$. □

There is a gap in our calculations which we would like to fill. Namely, we know that, for a unit-speed curve, we have

$$\dot{\mathbf{t}} = \kappa\mathbf{n} \quad \text{and} \quad \dot{\mathbf{b}} = -\tau\mathbf{n}$$

(these were our definitions of $\mathbf{n}$ and τ, respectively), but we have not computed $\dot{\mathbf{n}}$. This is not difficult. Since $\{\mathbf{t}, \mathbf{n}, \mathbf{b}\}$ is a right-handed orthonormal basis of $\mathbf{R}^3$,

$$\mathbf{t} \times \mathbf{n} = \mathbf{b}, \quad \mathbf{n} \times \mathbf{b} = \mathbf{t}, \quad \mathbf{b} \times \mathbf{t} = \mathbf{n}.$$

Hence,

$$\dot{\mathbf{n}} = \dot{\mathbf{b}} \times \mathbf{t} + \mathbf{b} \times \dot{\mathbf{t}} = -\tau\mathbf{n} \times \mathbf{t} + \kappa\mathbf{b} \times \mathbf{n} = -\kappa\mathbf{t} + \tau\mathbf{b}.$$

Putting all this together, we get

Theorem 2.2

Let $\boldsymbol{\gamma}$ be a unit-speed curve in $\mathbf{R}^3$ with nowhere vanishing curvature. Then,

$$\begin{aligned} \dot{\mathbf{t}} &= \kappa\mathbf{n} \\ \dot{\mathbf{n}} &= -\kappa\mathbf{t} + \tau\mathbf{b} \\ \dot{\mathbf{b}} &= -\tau\mathbf{n}. \end{aligned} \qquad (14)$$

Equations (14) are called the *Frenet–Serret equations* (or sometimes the *Serret–Frenet equations*). Notice that the matrix

$$\begin{pmatrix} 0 & \kappa & 0 \\ -\kappa & 0 & \tau \\ 0 & -\tau & 0 \end{pmatrix}$$

which expresses $\dot{\mathbf{t}}$, $\dot{\mathbf{n}}$ and $\dot{\mathbf{b}}$ in terms of $\mathbf{t}$, $\mathbf{n}$ and $\mathbf{b}$ is *skew-symmetric*, i.e. it is equal to the negative of its transpose. This helps when trying to remember the equations. (The 'reason' for this skew-symmetry can be seen in Exercise 2.22.)

Here is a simple application of Frenet–Serret:

Proposition 2.5

Let $\boldsymbol{\gamma}$ be a unit-speed curve in $\mathbf{R}^3$ with constant curvature and zero torsion. Then, $\boldsymbol{\gamma}$ is (part of) a circle.

Proof 2.5

This result is actually an immediate consequence of Example 2.2 and Proposition 2.4, but the following proof is instructive and gives more information, namely the centre and radius of the circle and the plane in which it lies.

By the proof of Proposition 2.4, the binormal $\mathbf{b}$ is a constant vector and $\boldsymbol{\gamma}$ is contained in a plane perpendicular to $\mathbf{b}$. Now consider

$$\frac{d}{ds}\left(\boldsymbol{\gamma} + \frac{1}{\kappa}\mathbf{n}\right) = \mathbf{t} + \frac{1}{\kappa}\dot{\mathbf{n}} = \mathbf{0},$$

where we have used the fact that the curvature κ is constant and the Frenet–Serret equation

$$\dot{\mathbf{n}} = -\kappa\mathbf{t} + \tau\mathbf{b} = -\kappa\mathbf{t} \quad (\text{since } \tau = 0).$$

Hence $\gamma + \frac{1}{\kappa}\mathbf{n}$ is a constant vector, say $\mathbf{a}$, and we have

$$\gamma + \frac{1}{\kappa}\mathbf{n} = \mathbf{a}, \tag{15}$$

$$\therefore \quad \| \gamma - \mathbf{a} \| = \| -\frac{1}{\kappa}\mathbf{n} \| = \frac{1}{\kappa}.$$

This shows that γ lies on the sphere of centre $\mathbf{a}$ and radius $1/\kappa$. Since the intersection of a plane and a sphere is a circle, this completes the proof. (Note that the plane actually intersects the sphere in a great circle: this is because $\mathbf{n}$ is parallel to the plane, so by Eq. (15) the centre $\mathbf{a}$ of the sphere lies in the plane.) □

We conclude this chapter with the analogue of Theorem 2.1 for space curves. Recall that a rigid motion of $\mathbf{R}^3$ is a rotation about the origin followed by a translation.

Theorem 2.3

Let $\gamma(s)$ and $\tilde{\gamma}(s)$ be two unit-speed curves in $\mathbf{R}^3$ with the same curvature $\kappa(s) > 0$ and the same torsion $\tau(s)$ for all s. Then, there is a rigid motion M of $\mathbf{R}^3$ such that

$$\tilde{\gamma}(s) = M(\gamma(s)) \quad \text{for all } s.$$

Further, if k and t are smooth functions with $k > 0$ everywhere, there is a unit-speed curve in $\mathbf{R}^3$ whose curvature is k and whose torsion is t.

Proof 2.3

Let $\mathbf{t}, \mathbf{n}$ and $\mathbf{b}$ be the tangent vector, principal normal and binormal of γ, and let $\tilde{\mathbf{t}}, \tilde{\mathbf{n}}$ and $\tilde{\mathbf{b}}$ be those of $\tilde{\gamma}$. Let s_0 be a fixed value of the parameter s. Since $\{\mathbf{t}(s_0), \mathbf{n}(s_0), \mathbf{b}(s_0)\}$ and $\{\tilde{\mathbf{t}}(s_0), \tilde{\mathbf{n}}(s_0), \tilde{\mathbf{b}}(s_0)\}$ are both right-handed orthonormal bases of $\mathbf{R}^3$, there is a rotation about the origin of $\mathbf{R}^3$ that takes $\mathbf{t}(s_0), \mathbf{n}(s_0)$ and $\mathbf{b}(s_0)$ to $\tilde{\mathbf{t}}(s_0), \tilde{\mathbf{n}}(s_0)$ and $\tilde{\mathbf{b}}(s_0)$, respectively. Further, there is a translation that takes $\gamma(s_0)$ to $\tilde{\gamma}(s_0)$ (and this has no effect on $\mathbf{t}$, $\mathbf{n}$ and $\mathbf{b}$). By applying the rotation followed by the translation, we can therefore assume that

$$\gamma(s_0) = \tilde{\gamma}(s_0), \ \mathbf{t}(s_0) = \tilde{\mathbf{t}}(s_0), \ \mathbf{n}(s_0) = \tilde{\mathbf{n}}(s_0), \ \mathbf{b}(s_0) = \tilde{\mathbf{b}}(s_0). \tag{16}$$

The trick now is to consider the expression

$$A(s) = \tilde{\mathbf{t}}.\mathbf{t} + \tilde{\mathbf{n}}.\mathbf{n} + \tilde{\mathbf{b}}.\mathbf{b}.$$

In view of Eqs. (16), we have $A(s_0) = 3$. On the other hand, since $\tilde{\mathbf{t}}$ and $\mathbf{t}$ are unit vectors, $\tilde{\mathbf{t}}.\mathbf{t} \leq 1$, with equality holding if and only if $\tilde{\mathbf{t}} = \mathbf{t}$; and similarly for $\tilde{\mathbf{n}}.\mathbf{n}$ and $\tilde{\mathbf{b}}.\mathbf{b}$. It follows that $A(s) \leq 3$, with equality holding if and only if $\tilde{\mathbf{t}} = \mathbf{t}, \tilde{\mathbf{n}} = \mathbf{n}$ and $\tilde{\mathbf{b}} = \mathbf{b}$. Thus, if we can prove that A is constant, it will follow in particular that $\tilde{\mathbf{t}} = \mathbf{t}$, i.e. that $\dot{\tilde{\boldsymbol{\gamma}}} = \dot{\boldsymbol{\gamma}}$, and hence that $\tilde{\boldsymbol{\gamma}}(s) - \boldsymbol{\gamma}(s)$ is a constant. But by Eqs. (16) again, this constant vector must be zero, so $\tilde{\boldsymbol{\gamma}} = \boldsymbol{\gamma}$.

For the first part of the theorem, we are therefore reduced to proving that A is constant. But, using the Frenet–Serret equations,

$$\begin{aligned}\dot{A} &= \dot{\tilde{\mathbf{t}}}.\mathbf{t} + \dot{\tilde{\mathbf{n}}}.\mathbf{n} + \dot{\tilde{\mathbf{b}}}.\mathbf{b} + \tilde{\mathbf{t}}.\dot{\mathbf{t}} + \tilde{\mathbf{n}}.\dot{\mathbf{n}} + \tilde{\mathbf{b}}.\dot{\mathbf{b}} \\ &= \kappa\tilde{\mathbf{n}}.\mathbf{t} + (-\kappa\tilde{\mathbf{t}} + \tau\tilde{\mathbf{b}}).\mathbf{n} + (-\tau\tilde{\mathbf{n}}).\mathbf{b} + \tilde{\mathbf{t}}.\kappa\mathbf{n} + \tilde{\mathbf{n}}.(-\kappa\mathbf{t} + \tau\mathbf{b}) + \mathbf{b}.(-\tau\mathbf{n}),\end{aligned}$$

and this vanishes since the terms cancel in pairs.

For the second part of the theorem, we observe first that it follows from the theory of ordinary differential equations that the equations

$$\dot{\mathbf{T}} = k\mathbf{N}, \tag{17}$$

$$\dot{\mathbf{N}} = -k\mathbf{T} + t\mathbf{B}, \tag{18}$$

$$\dot{\mathbf{B}} = -t\mathbf{N} \tag{19}$$

have a unique solution $\mathbf{T}(s), \mathbf{N}(s), \mathbf{B}(s)$ such that $\mathbf{T}(s_0), \mathbf{N}(s_0), \mathbf{B}(s_0)$ are the standard orthonormal vectors $\mathbf{i} = (1,0,0)$, $\mathbf{j} = (0,1,0)$, $\mathbf{k} = (0,0,1)$, respectively. Since the matrix

$$\begin{pmatrix} 0 & k & 0 \\ -k & 0 & t \\ 0 & -t & 0 \end{pmatrix}$$

expressing $\dot{\mathbf{T}}, \dot{\mathbf{N}}$ and $\dot{\mathbf{B}}$ in terms of $\mathbf{T}, \mathbf{N}$ and $\mathbf{B}$ is skew-symmetric, it follows that the vectors $\mathbf{T}$, $\mathbf{N}$ and $\mathbf{B}$ are orthonormal for all values of s (see Exercise 2.22).

Now define

$$\boldsymbol{\gamma}(s) = \int_{s_0}^{s} \mathbf{T}(u)du.$$

Then, $\dot{\boldsymbol{\gamma}} = \mathbf{T}$, so since $\mathbf{T}$ is a unit vector, $\boldsymbol{\gamma}$ is unit-speed. Next, $\dot{\mathbf{T}} = k\mathbf{N}$ by Eq. (17), so since $\mathbf{N}$ is a unit vector, k is the curvature of $\boldsymbol{\gamma}$ and $\mathbf{N}$ is its principal normal. Next, since $\mathbf{B}$ is a unit vector perpendicular to $\mathbf{T}$ and $\mathbf{N}$, $\mathbf{B} = \lambda\mathbf{T} \times \mathbf{N}$ where λ is a smooth function of s that is equal to ± 1 for all s. Since $\mathbf{k} = \mathbf{i} \times \mathbf{j}$, we have $\lambda(s_0) = 1$, so it follows that $\lambda(s) = 1$ for all s. Hence, $\mathbf{B}$ is the binormal of $\boldsymbol{\gamma}$ and by Eq. (19), t is its torsion. □

EXERCISES

2.14 Compute $\kappa, \tau, \mathbf{t}, \mathbf{n}$ and $\mathbf{b}$ for each of the following curves, and verify that the Frenet–Serret equations are satisfied:

(i) $\boldsymbol{\gamma}(t) = \left(\frac{1}{3}(1+t)^{3/2}, \frac{1}{3}(1-t)^{3/2}, \frac{t}{\sqrt{2}}\right)$;

(ii) $\boldsymbol{\gamma}(t) = \left(\frac{4}{5}\cos t, 1 - \sin t, -\frac{3}{5}\cos t\right)$.

2.15 Show that the curve

$$\boldsymbol{\gamma}(t) = \left(\frac{1+t^2}{t}, t+1, \frac{1-t}{t}\right)$$

is planar.

2.16 Show that the curve in Exercise 2.14(ii) is a circle, and find its centre, radius and the plane in which it lies.

2.17 Describe all curves in $\mathbf{R}^3$ which have *constant* curvature $\kappa > 0$ and *constant* torsion τ. (Observe that it is enough to find *one* curve with curvature κ and torsion τ.)

2.18 Show that the torsion of a regular curve $\boldsymbol{\gamma}(t)$ is a smooth function of t whenever it is defined.

2.19 Let $\boldsymbol{\gamma}(t)$ be a unit-speed curve in $\mathbf{R}^3$, and assume that its curvature $\kappa(t)$ is non-zero for all t. Define a new curve $\boldsymbol{\delta}$ by

$$\boldsymbol{\delta}(t) = \frac{d\boldsymbol{\gamma}(t)}{dt}.$$

Show that $\boldsymbol{\delta}$ is regular and that, if s is an arc-length parameter for $\boldsymbol{\delta}$, then

$$\frac{ds}{dt} = \kappa.$$

Prove that the curvature of $\boldsymbol{\delta}$ is

$$\left(1 + \frac{\tau^2}{\kappa^2}\right)^{\frac{1}{2}},$$

and find a formula for the torsion of $\boldsymbol{\delta}$ in terms of κ, τ and their derivatives with respect to t.

2.20 A regular curve $\boldsymbol{\gamma}$ in $\mathbf{R}^3$ with curvature > 0 is called a *general helix* if its tangent vector makes a fixed angle θ with a fixed unit vector $\mathbf{a}$. Show that the torsion τ and curvature κ of $\boldsymbol{\gamma}$ are related by $\tau = \pm\kappa\cot\theta$. (Assume that $\boldsymbol{\gamma}$ is unit-speed and show that $\mathbf{a} = \mathbf{t}\cos\theta + \mathbf{b}\sin\theta$.)

Show conversely that, if the torsion and curvature of a regular curve are related by $\tau = \lambda\kappa$ where λ is a constant, then the curve is a general helix. (Thus, Examples 2.1 and 2.4 show that a circular helix is a general helix.)

2.21 Let $\boldsymbol{\gamma}(t)$ be a unit-speed curve with $\kappa(t) > 0$ and $\tau(t) \neq 0$ for all t. Show that, if $\boldsymbol{\gamma}$ lies on the surface of a sphere, then

$$\frac{\tau}{\kappa} = \frac{d}{ds}\left(\frac{\dot{\kappa}}{\tau\kappa^2}\right). \tag{20}$$

(If $\boldsymbol{\gamma}$ lies on the sphere of centre $\mathbf{a}$ and radius r, then $(\boldsymbol{\gamma}-\mathbf{a}).(\boldsymbol{\gamma}-\mathbf{a}) = r^2$; now differentiate repeatedly.) Conversely, show that if Eq. (20) holds, then

$$\rho^2 + (\dot{\rho}\sigma)^2 = r^2$$

for some (positive) constant r, where $\rho = 1/\kappa$ and $\sigma = 1/\tau$, and deduce that $\boldsymbol{\gamma}$ lies on a sphere of radius r. (Consider $\boldsymbol{\gamma} + \rho\mathbf{n} + \dot{\rho}\sigma\mathbf{b}$.) Verify that Eq. (20) holds for Viviani's curve (Exercise 1.9).

2.22 Let (a_{ij}) be a skew-symmetric 3×3 matrix (i.e. $a_{ij} = -a_{ji}$ for all i, j). Let $\mathbf{v}_1, \mathbf{v}_2$ and $\mathbf{v}_3$ be smooth functions of a parameter s satisfying the differential equations

$$\dot{\mathbf{v}}_i = \sum_{j=1}^{3} a_{ij}\mathbf{v}_j,$$

for $i = 1, 2$ and 3, and suppose that for some parameter value s_0 the vectors $\mathbf{v}_1(s_0), \mathbf{v}_2(s_0)$ and $\mathbf{v}_3(s_0)$ are orthonormal. Show that the vectors $\mathbf{v}_1(s), \mathbf{v}_2(s)$ and $\mathbf{v}_3(s)$ are orthonormal for all values of s. (Find a system of first order differential equations satisfied by the dot products $\mathbf{v}_i.\mathbf{v}_j$, and use the fact that such a system has a unique solution with given initial conditions.)

For the remainder of this book,
all parametrised curves will be assumed to be regular.

3
Global Properties of Curves

All the properties of curves that we have discussed so far are 'local': they depend only on the behaviour of a curve near a given point, and not on the 'global' shape of the curve. In this chapter, we discuss some global results about curves. The most famous, and perhaps the oldest, of these is the 'isoperimetric inequality', which relates the length of certain 'closed' curves to the area they contain.

3.1. Simple Closed Curves

Our first task is to describe the kind of curves that we shall be considering in this chapter, namely 'simple closed curves'. Intuitively, these are curves that 'join up', but do not otherwise self-intersect. A precise definition is as follows:

Definition 3.1

Let $a \in \mathbf{R}$ be a positive constant. A *simple closed curve* in $\mathbf{R}^2$ with *period* a is a (regular) curve $\boldsymbol{\gamma} : \mathbf{R} \to \mathbf{R}^2$ such that

$$\boldsymbol{\gamma}(t) = \boldsymbol{\gamma}(t') \text{ if and only if } t' - t = ka \text{ for some integer } k.$$

Thus, the point $\boldsymbol{\gamma}(t)$ returns to its starting point when t increases by a, but not before that.

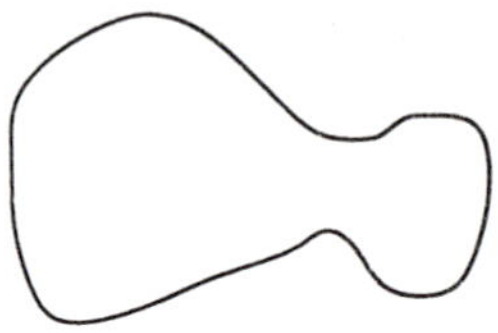

simple closed curve

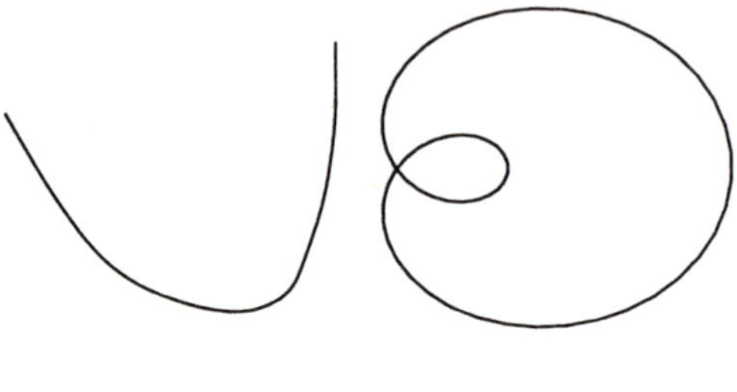

non-simple closed curves

It is a standard, but highly non-trivial, result of the topology of $\mathbf{R}^2$, called the *Jordan Curve Theorem*, that any simple closed curve in the plane has an 'interior' and an 'exterior': more precisely, the set of points of $\mathbf{R}^2$ that are *not* on the curve $\boldsymbol{\gamma}$ is the disjoint union of two subsets of $\mathbf{R}^2$, denoted by $\text{int}(\boldsymbol{\gamma})$ and $\text{ext}(\boldsymbol{\gamma})$, with the following properties:

(i) $\text{int}(\boldsymbol{\gamma})$ is *bounded*, i.e. it is contained inside a circle of sufficiently large radius;

(ii) $\text{ext}(\boldsymbol{\gamma})$ is unbounded;

(iii) both of the regions $\text{int}(\boldsymbol{\gamma})$ and $\text{ext}(\boldsymbol{\gamma})$ are *connected*, i.e. they have the property that any two points in the same region can be joined by a curve contained entirely in the region (but any curve joining a point of $\text{int}(\boldsymbol{\gamma})$ to a point of $\text{ext}(\boldsymbol{\gamma})$ must cross the curve $\boldsymbol{\gamma}$).

Example 3.1

The parametrised circle

$$\boldsymbol{\gamma}(t) = \left(\cos\left(\frac{2\pi t}{a}\right), \sin\left(\frac{2\pi t}{a}\right)\right)$$

is a simple closed curve with period a. The interior and exterior of $\boldsymbol{\gamma}$ are, of course, given by $\{(x,y) \in \mathbf{R}^2 \,|\, x^2 + y^2 < 1\}$ and $\{(x,y) \in \mathbf{R}^2 \,|\, x^2 + y^2 > 1\}$, respectively.

Not all examples of simple closed curves have such an obvious interior and exterior, however. Is the point P in the interior or the exterior of the simple closed curve shown at the top of the next page?

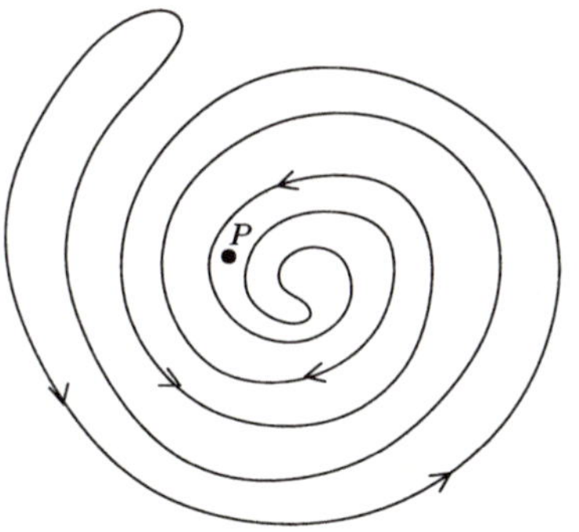

Since every point in the image of a simple closed curve $\boldsymbol{\gamma}$ of period a is traced out as the parameter t of $\boldsymbol{\gamma}$ varies through any interval of length a, e.g. $0 \leq t \leq a$, it is reasonable to define the *length of* $\boldsymbol{\gamma}$ to be

$$\ell(\boldsymbol{\gamma}) = \int_0^a \| \dot{\boldsymbol{\gamma}}(t) \| \, dt, \tag{1}$$

where a dot denotes the derivative with respect to the parameter of the curve $\boldsymbol{\gamma}$. Since $\boldsymbol{\gamma}$ is regular, it has a unit-speed reparametrisation $\tilde{\boldsymbol{\gamma}}$ with the arc-length

$$s = \int_0^t \| \dot{\boldsymbol{\gamma}}(u) \| \, du$$

of $\boldsymbol{\gamma}$ as its parameter (so that $\tilde{\boldsymbol{\gamma}}(s) = \boldsymbol{\gamma}(t)$). Note that

$$s(t+a) = \int_0^{t+a} \| \dot{\boldsymbol{\gamma}}(u) \| \, du = \int_0^a \| \dot{\boldsymbol{\gamma}}(u) \| \, du + \int_a^{t+a} \| \dot{\boldsymbol{\gamma}}(u) \| \, du = \ell(\boldsymbol{\gamma}) + s(t),$$

since, putting $v = u - a$ and using $\boldsymbol{\gamma}(u-a) = \boldsymbol{\gamma}(u)$, we get

$$\int_a^{t+a} \| \dot{\boldsymbol{\gamma}}(u) \| \, du = \int_0^t \| \dot{\boldsymbol{\gamma}}(v) \| \, dv = s(t).$$

Hence,

$$\tilde{\boldsymbol{\gamma}}(s(t)) = \tilde{\boldsymbol{\gamma}}(s(t')) \iff \boldsymbol{\gamma}(t) = \boldsymbol{\gamma}(t') \iff t' - t = ka \iff s(t') - s(t) = k\ell(\boldsymbol{\gamma}),$$

where k is an integer. This shows that $\tilde{\boldsymbol{\gamma}}$ is a simple closed curve with period $\ell(\boldsymbol{\gamma})$. Note that, since $\tilde{\boldsymbol{\gamma}}$ is unit-speed, this is also the length of $\tilde{\boldsymbol{\gamma}}$. In short, *we can always assume that a simple closed curve is unit-speed and that its period is equal to its length.*

We shall usually assume that our simple closed curves $\boldsymbol{\gamma}$ are *positively-oriented.* This means that the signed unit normal $\mathbf{n}_s$ of $\boldsymbol{\gamma}$ (see Section 2.2) points into $\mathrm{int}(\boldsymbol{\gamma})$ at every point of $\boldsymbol{\gamma}$. This can always be achieved by replacing the parameter t of $\boldsymbol{\gamma}$ by $-t$, if necessary.

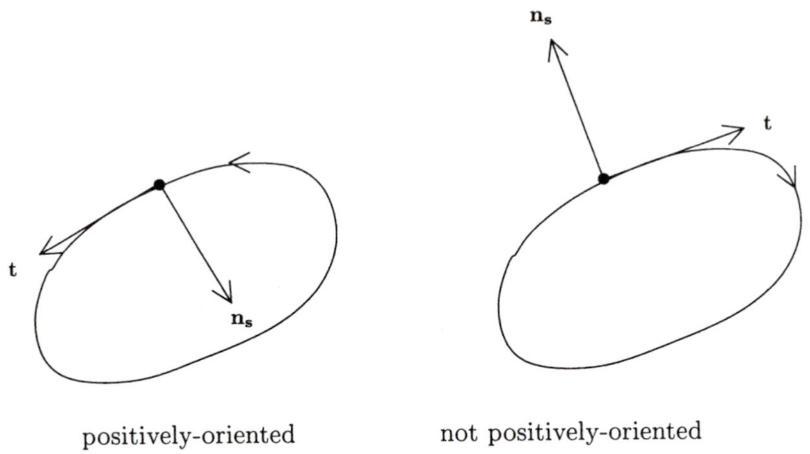

In the above diagrams, the arrow indicates the direction of increasing parameter. Is the simple closed curve shown at the top of the previous page positively-oriented?

In the next section, we shall be interested in the area contained by a simple closed curve $\boldsymbol{\gamma}$, i.e.

$$\mathcal{A}(\mathrm{int}(\boldsymbol{\gamma})) = \int\int_{\mathrm{int}(\boldsymbol{\gamma})} dxdy. \tag{2}$$

This can be computed by using *Green's Theorem*, which says that, for all smooth functions $f(x,y)$ and $g(x,y)$ (i.e. functions with continuous partial derivatives of all orders),

$$\int\int_{\mathrm{int}(\boldsymbol{\gamma})} \left(\frac{\partial g}{\partial x} - \frac{\partial f}{\partial y}\right) dxdy = \int_{\boldsymbol{\gamma}} f(x,y)dx + g(x,y)dy,$$

if $\boldsymbol{\gamma}$ is a positively-oriented simple closed curve.

Proposition 3.1

If $\boldsymbol{\gamma}(t) = (x(t), y(t))$ is a positively-oriented simple closed curve in $\mathbf{R}^2$ with period a, then

$$\mathcal{A}(\mathrm{int}(\boldsymbol{\gamma})) = \frac{1}{2}\int_0^a (x\dot{y} - y\dot{x})dt. \tag{3}$$

Proof 3.1

Taking $f = -\frac{1}{2}y$, $g = \frac{1}{2}x$ in Green's theorem, we get

$$\mathcal{A}(\mathrm{int}(\boldsymbol{\gamma})) = \frac{1}{2}\int_{\boldsymbol{\gamma}} xdy - ydx,$$

which gives Eq. (3) immediately. □

Note that, although the formula in Eq. (3) involves the parameter t of $\boldsymbol{\gamma}$, it is clear from the definition (2) of $\mathcal{A}(\mathrm{int}(\boldsymbol{\gamma}))$ that it is unchanged if $\boldsymbol{\gamma}$ is reparametrised.

EXERCISES

3.1 Show that the length $\ell(\boldsymbol{\gamma})$ and the area $\mathcal{A}(\mathrm{int}(\boldsymbol{\gamma}))$ are unchanged by applying a rigid motion to $\boldsymbol{\gamma}$ (see Section 2.2).

3.2 Show that the ellipse

$$\boldsymbol{\gamma}(t) = (a\cos t, b\sin t),$$

where a and b are positive constants, is a simple closed curve and compute the area of its interior.

3.3 Show that the limaçon

$$\boldsymbol{\gamma}(t) = ((1+2\cos t)\cos t, (1+2\cos t)\sin t)$$

is a (regular) curve such that $\boldsymbol{\gamma}(t+2\pi) = \boldsymbol{\gamma}(t)$ for all values of t, but that $\boldsymbol{\gamma}$ is *not* a simple closed curve.

3.4 Show that, if $\boldsymbol{\gamma}(t)$ is a simple closed curve of period a, and $\mathbf{t}$, $\mathbf{n}_s$ and κ_s are its unit tangent vector, signed unit normal and signed curvature, respectively, then

$$\mathbf{t}(t+a) = \mathbf{t}(t), \quad \mathbf{n}_s(t+a) = \mathbf{n}_s(t), \quad \kappa_s(t+a) = \kappa_s(t).$$

(Differentiate the equation $\boldsymbol{\gamma}(t+a) = \boldsymbol{\gamma}(t)$.)

3.2. The Isoperimetric Inequality

The most important global result about plane curves is

Theorem 3.1 (The Isoperimetric Inequality)

Let $\boldsymbol{\gamma}$ be a simple closed curve, let $\ell(\boldsymbol{\gamma})$ be its length and let $\mathcal{A}(\mathrm{int}(\boldsymbol{\gamma}))$ be the area of its interior. Then,

$$\mathcal{A}(\mathrm{int}(\boldsymbol{\gamma})) \leq \frac{1}{4\pi}\ell(\boldsymbol{\gamma})^2,$$

with equality holding if and only if $\boldsymbol{\gamma}$ is a circle.

Of course, it is obvious that equality holds when $\boldsymbol{\gamma}$ is a circle, since in that case $\ell(\boldsymbol{\gamma}) = 2\pi R$ and $\mathcal{A}(\mathrm{int}(\boldsymbol{\gamma})) = \pi R^2$, where R is the radius of the circle.

To prove this theorem, we need a result from analysis called *Wirtinger's Inequality*:

Proposition 3.2

Let $F : [0, \pi] \to \mathbf{R}$ be a smooth function such that $F(0) = F(\pi) = 0$. Then,

$$\int_0^\pi \left(\frac{dF}{dt}\right)^2 dt \geq \int_0^\pi F(t)^2 dt,$$

with equality holding if and only if $F(t) = A \sin t$ for all $t \in [0, \pi]$, where A is a constant.

Assuming this result for the moment, we show how to deduce the isoperimetric inequality from it.

Proof 3.1

We start by making some assumptions about $\boldsymbol{\gamma}$ that will simplify the proof. First, we can if we wish assume that $\boldsymbol{\gamma}$ is parametrised by arc-length s. However, because of the π that appears in Theorem 3.1, it turns out to be more convenient to assume that the period of $\boldsymbol{\gamma}$ is π. If we change the parameter of $\boldsymbol{\gamma}$ from s to

$$t = \pi s / \ell(\boldsymbol{\gamma}), \tag{4}$$

the resulting curve is still simple closed, and has period π because when s increases by $\ell(\boldsymbol{\gamma})$, t increases by π. We shall therefore assume that $\boldsymbol{\gamma}$ is parametrised using the parameter t in Eq. (4) from now on.

For the second simplification, we note that both $\ell(\boldsymbol{\gamma})$ and $\mathcal{A}(\boldsymbol{\gamma})$ are unchanged if $\boldsymbol{\gamma}$ is subjected to a translation $\boldsymbol{\gamma}(t) \mapsto \boldsymbol{\gamma}(t) + \mathbf{b}$, where $\mathbf{b}$ is any constant vector (see Exercise 3.1). Taking $\mathbf{b} = -\boldsymbol{\gamma}(0)$, we might as well assume that $\boldsymbol{\gamma}(0) = \mathbf{0}$ to begin with, i.e. we assume that $\boldsymbol{\gamma}$ begins and ends at the origin.

To prove Theorem 3.1, we shall calculate $\ell(\boldsymbol{\gamma})$ and $\mathcal{A}(\mathrm{int}(\boldsymbol{\gamma}))$ by using polar coordinates

$$x = r\cos\theta, \quad y = r\sin\theta.$$

Using the chain rule, it is easy to show that

$$\dot{x}^2 + \dot{y}^2 = \dot{r}^2 + r^2\dot{\theta}^2, \quad x\dot{y} - y\dot{x} = r^2\dot{\theta},$$

with d/dt denoted by a dot. Then, using Eq. (4),

$$\dot{r}^2 + r^2\dot{\theta}^2 = \left(\frac{dx}{dt}\right)^2 + \left(\frac{dy}{dt}\right)^2 = \left(\left(\frac{dx}{ds}\right)^2 + \left(\frac{dy}{ds}\right)^2\right)\left(\frac{ds}{dt}\right)^2 = \frac{\ell(\boldsymbol{\gamma})^2}{\pi^2}, \tag{5}$$

since $(dx/ds)^2 + (dy/ds)^2 = 1$. Further, by Eq. (3), we have

$$\mathcal{A}(\mathrm{int}(\boldsymbol{\gamma})) = \frac{1}{2}\int_0^\pi (x\dot{y} - y\dot{x})dt = \frac{1}{2}\int_0^\pi r^2\dot{\theta}dt. \tag{6}$$

To prove Theorem 3.1, we have to show that

$$\frac{\ell(\boldsymbol{\gamma})^2}{4\pi} - \mathcal{A}(\mathrm{int}(\boldsymbol{\gamma})) \geq 0,$$

with equality holding if and only if $\boldsymbol{\gamma}$ is a circle. By Eq. (5),

$$\int_0^\pi (\dot{r}^2 + r^2\dot{\theta}^2)dt = \frac{\ell(\boldsymbol{\gamma})^2}{\pi}.$$

Hence, using Eq. (6),

$$\frac{\ell(\boldsymbol{\gamma})^2}{4\pi} - \mathcal{A}(\mathrm{int}(\boldsymbol{\gamma})) = \frac{1}{4}\int_0^\pi (\dot{r}^2 + r^2\dot{\theta}^2)dt - \frac{1}{2}\int_0^\pi r^2\dot{\theta}dt = \frac{1}{4}\mathcal{I},$$

where

$$\mathcal{I} = \int_0^\pi (\dot{r}^2 + r^2\dot{\theta}^2 - 2r^2\dot{\theta})dt. \tag{7}$$

Thus, to prove Theorem 3.1, we have to show that $\mathcal{I} \geq 0$, and that $\mathcal{I} = 0$ if and only if $\boldsymbol{\gamma}$ is a circle.

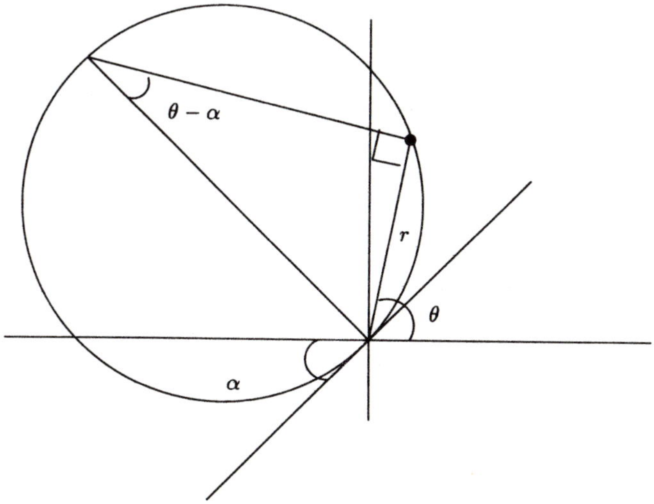

By simple algebra,

$$\mathcal{I} = \int_0^\pi r^2(\dot{\theta} - 1)^2 dt + \int_0^\pi (\dot{r}^2 - r^2)dt. \tag{8}$$

The first integral on the right-hand side of Eq. (8) is obviously ≥ 0, and the second integral is ≥ 0 by Wirtinger's inequality (we are taking $F = r$: note that $r(0) = r(\pi) = 0$ since $\boldsymbol{\gamma}(0) = \boldsymbol{\gamma}(\pi) = \mathbf{0}$). Hence, $\mathcal{I} \geq 0$. Further, since both integrals on the right-hand side of Eq. (8) are ≥ 0, their sum $\mathcal{I}$ is zero if and only if both of these integrals are zero. But the first integral is zero only if $\dot{\theta} = 1$ for all t, and the second is zero only if $r = A \sin t$ for some constant A (by Wirtinger again). So $\theta = t + \alpha$, where α is a constant, and hence $r = A\sin(\theta - \alpha)$. It is easy to see that this is the polar equation of a circle of diameter A, thus completing the proof of Theorem 3.1 (see the diagram above). □

We now prove Wirtinger's inequality.

Let $G(t) = F(t)/\sin t$. Then, denoting d/dt by a dot as usual,

$$\int_0^\pi \dot{F}^2 dt = \int_0^\pi (\dot{G}\sin t + G\cos t)^2 dt$$
$$= \int_0^\pi \dot{G}^2 \sin^2 t\, dt + 2\int_0^\pi G\dot{G}\sin t \cos t\, dt + \int_0^\pi G^2\cos^2 t\, dt.$$

Integrating by parts:

$$2\int_0^\pi G\dot{G}\sin t\cos t\, dt = G^2 \sin t \cos t\big|_0^\pi - \int_0^\pi G^2(\cos^2 t - \sin^2 t)dt$$
$$= \int_0^\pi G^2(\sin^2 t - \cos^2 t)dt.$$

Hence,

$$\int_0^\pi \dot{F}^2 dt = \int_0^\pi \dot{G}^2\sin^2 t\, dt + \int_0^\pi G^2(\sin^2 t - \cos^2 t)dt + \int_0^\pi G^2 \cos^2 t\, dt$$
$$= \int_0^\pi (G^2 + \dot{G}^2)\sin^2 t\, dt = \int_0^\pi F^2 dt + \int_0^\pi \dot{G}^2 \sin^2 t\, dt,$$

and so

$$\int_0^\pi \dot{F}^2 dt - \int_0^\pi F^2 dt = \int_0^\pi \dot{G}^2 \sin^2 t\, dt.$$

The integral on the right-hand side is obviously ≥ 0, and it is zero if and only if $\dot{G} = 0$ for all t, i.e. if and only if $G(t)$ is equal to a constant, say A, for all t. Then, $F(t) = A\sin t$, as required. □

EXERCISES

3.5 By applying the isoperimetric inequality to the ellipse

$$\frac{x^2}{a^2} + \frac{y^2}{b^2} = 1$$

(where a and b are positive constants), prove that

$$\int_0^{2\pi} \sqrt{a^2 \sin^2 t + b^2 \cos^2 t}\, dt \geq 2\pi\sqrt{ab},$$

with equality holding if and only if $a = b$ (see Exercise 3.2).

3.3. The Four Vertex Theorem

We conclude this chapter with a famous result about convex curves in the plane. A simple closed curve $\boldsymbol{\gamma}$ is called *convex* if its interior $\mathrm{int}(\boldsymbol{\gamma})$ is convex, in the usual sense that the straight line segment joining any two points of $\mathrm{int}(\boldsymbol{\gamma})$ is contained entirely in $\mathrm{int}(\boldsymbol{\gamma})$.

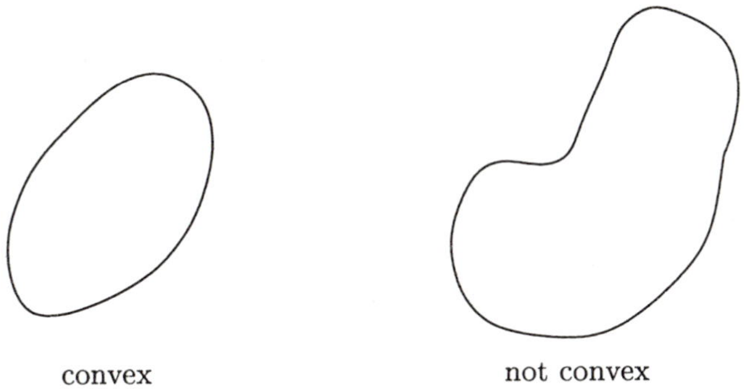

Definition 3.2

A *vertex* of a curve $\boldsymbol{\gamma}(t)$ in $\mathbf{R}^2$ is a point where its signed curvature κ_s has a stationary point, i.e. where $d\kappa_s/dt = 0$.

It is easy to see that this definition is independent of the parametrisation of $\boldsymbol{\gamma}$ (see Exercise 3.7).

Example 3.2

The ellipse $\boldsymbol{\gamma}(t) = (a\cos t, b\sin t)$, where a and b are positive constants, is a convex simple closed curve with period 2π (see Exercises 3.2 and 3.6). Its signed curvature is easily found to be

$$\kappa_s(t) = \frac{ab}{(a^2 \sin^2 t + b^2 \cos^2 t)^{3/2}}.$$

Then,

$$\frac{d\kappa_s}{dt} = \frac{3ab(b^2 - a^2)\sin t \cos t}{(a^2 \sin^2 t + b^2 \cos^2 t)^{5/2}}$$

vanishes at exactly four points of the ellipse, namely the points with $t = 0, \pi/2, \pi$ and $3\pi/2$, which are the ends of the two axes of the ellipse.

The following theorem says that this is the smallest number of vertices a convex simple closed curve can have.

Theorem 3.2 (Four Vertex Theorem)

Every convex simple closed curve in $\mathbf{R}^2$ has at least four vertices.

It is actually the case that this theorem remains true without the assumption of convexity, but the proof is then more difficult than the one we are about to give.

Proof 3.2

We might as well assume that the curve $\boldsymbol{\gamma}(t)$ is unit-speed, so that its period is the length ℓ of $\boldsymbol{\gamma}$. We consider the integral

$$\int_0^\ell \dot{\kappa}_s(t)\boldsymbol{\gamma}(t)\,dt,$$

where a dot denotes d/dt. (Recall from Exercise 2.4 that κ_s is a smooth function of t.) Integrating by parts, and using the equation $\dot{\mathbf{n}}_s = -\kappa_s \mathbf{t}$ (see Exercise 2.3), we get

$$\int_0^\ell \dot{\kappa}_s \boldsymbol{\gamma}\,dt = -\int_0^\ell \kappa_s \dot{\boldsymbol{\gamma}}\,dt = -\int_0^\ell \kappa_s \mathbf{t}\,dt = \int_0^\ell \dot{\mathbf{n}}_s\,dt = \mathbf{n}_s(\ell) - \mathbf{n}_s(0) = \mathbf{0}. \quad (9)$$

Now, κ_s attains all of its values on the closed interval $[0,\ell]$, so κ_s must attain its maximum and minimum values at some points P and Q of $\boldsymbol{\gamma}$, say. We can assume that $P \neq Q$, since otherwise κ_s would be constant, $\boldsymbol{\gamma}$ would be a circle (by Example 2.2), and every point of $\boldsymbol{\gamma}$ would be a vertex. Let $\mathbf{a}$ be a unit vector parallel to the vector $\mathbf{PQ}$, and let $\mathbf{b}$ be the vector obtained by rotating $\mathbf{a}$ anti-clockwise by $\pi/2$. Taking the dot product of the integral in Eq. (9) with the constant vector $\mathbf{b}$ gives

$$\int_0^\ell \dot{\kappa}_s(\boldsymbol{\gamma}.\mathbf{b})\,dt = 0. \quad (10)$$

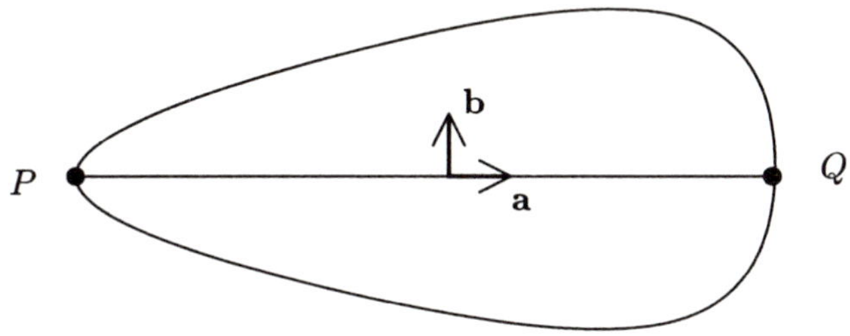

Suppose that P and Q are the only vertices of $\boldsymbol{\gamma}$. Since $\boldsymbol{\gamma}$ is convex, the straight line joining P and Q divides $\boldsymbol{\gamma}$ into two segments, and since there are no other vertices, we must have $\dot{\kappa}_s > 0$ on one segment and $\dot{\kappa}_s < 0$ on the other. But then the integrand on the left-hand side of Eq. (10) is either always > 0 or always < 0 (except at P and Q where it vanishes), so the integral is definitely > 0 or < 0, a contradiction.

Hence, there must be at least one more vertex, say R. If there are no other vertices, the points P, Q and R divide $\boldsymbol{\gamma}$ into three segments, on each of which $\dot{\kappa}_s$ is either always > 0 or always < 0. But then $\dot{\kappa}_s$ must have the same sign on two adjacent segments. Hence, there is a straight line that divides $\boldsymbol{\gamma}$ into two segments, on one of which $\dot{\kappa}_s$ is always positive, and on the other $\dot{\kappa}_s$ is always < 0. The argument in the preceding paragraph shows that this is impossible. So there must be a fourth vertex. □

EXERCISES

3.6 Show that the ellipse in Exercise 3.2 is convex. (You may need to use the inequality $2x_1x_2 \leq x_1^2 + x_2^2$.)

3.7 Show that the definition of a vertex of a plane curve is independent of its parametrisation.

3.8 Show that the limacon in Exercise 3.3 has only two vertices.

4
Surfaces in Three Dimensions

In this chapter, we introduce several different ways to formulate mathematically the notion of a surface. Although the simplest of these, that of a surface patch, is all that is needed for most of the book, it does not describe adequately most of the objects that we would want to call surfaces. For example, a sphere is not a surface patch, but it can be described by gluing two surface patches together suitably. The idea behind this gluing procedure is simple enough, but making it precise turns out to be a little complicated. We have tried to minimise the trauma by collecting the most demanding proofs in a section at the end of the chapter; this section is not used anywhere else in the book and can safely be omitted if desired. In fact, surfaces (as opposed to surface patches) will be used in a serious way on only a few occasions in this book.

4.1. What is a Surface?

A surface is a subset of $\mathbf{R}^3$ that looks like a piece of $\mathbf{R}^2$ in the vicinity of any given point, just as the surface of the Earth, although actually nearly spherical, appears to be a flat plane to an observer on the surface who sees only to the horizon. To make the phrases 'looks like' and 'in the vicinity' precise, we must first introduce some preliminary material. We describe this for $\mathbf{R}^n$ for any $n \geq 1$, although we shall need it only for $n = 1, 2$ or 3.

First, a subset U of $\mathbf{R}^n$ is called *open* if, whenever $\mathbf{a}$ is a point in U, there is a positive number ϵ such that every point $\mathbf{u} \in \mathbf{R}^n$ within a distance ϵ of $\mathbf{a}$ is also in U:

$$\mathbf{a} \in U \text{ and } \| \mathbf{u} - \mathbf{a} \| < \epsilon \quad \Longrightarrow \quad \mathbf{u} \in U.$$

For example, the whole of $\mathbf{R}^n$ is an open set, as is

$$\mathcal{D}_r(\mathbf{a}) = \{\mathbf{u} \in \mathbf{R}^n \mid \| \mathbf{u} - \mathbf{a} \| < r\},$$

the *open ball* with centre $\mathbf{a}$ and radius $r > 0$. (If $n = 1$, an open ball is called an open interval; if $n = 2$ it is called an open disc.) However,

$$\overline{\mathcal{D}}_r(\mathbf{a}) = \{\mathbf{u} \in \mathbf{R}^n \mid \| \mathbf{u} - \mathbf{a} \| \leq r\}$$

is *not* open, because however small the positive number ϵ is, there is a point within a distance ϵ of the point $(a_1 + r, a_2, \ldots, a_n) \in \overline{\mathcal{D}}^n$ (say) that is not in $\overline{\mathcal{D}}_r(\mathbf{a})$ (e.g. the point $(a_1 + r + \frac{\epsilon}{2}, a_2, \ldots, a_n)$).

Next, if X and Y are subsets of $\mathbf{R}^m$ and $\mathbf{R}^n$, respectively, a map $f : X \to Y$ is said to be continuous at a point $\mathbf{a} \in X$ if points in X near $\mathbf{a}$ are mapped by f onto points in Y near $f(\mathbf{a})$. More precisely, f is *continuous* at $\mathbf{a}$ if, given any number $\epsilon > 0$, there is a number $\delta > 0$ such that

$$\mathbf{u} \in X \quad \text{and} \quad \| \mathbf{u} - \mathbf{a} \| < \delta \Longrightarrow \| f(\mathbf{u}) - f(\mathbf{a}) \| < \epsilon.$$

Then f is said to be *continuous* if it is continuous at every point of X. Composites of continuous maps are continuous.

In view of the definition of an open set, this is equivalent to the following: f is continuous if and only if, for any open set V of $\mathbf{R}^n$, there is an open set U of $\mathbf{R}^m$ such that f maps $U \cap X$ into $V \cap Y$.

If $f : X \to Y$ is continuous and bijective, and if its inverse map $f^{-1} : Y \to X$ is also continuous, then f is called a *homeomorphism* and X and Y are said to be *homeomorphic*.

We are now in a position to make our first attempt at defining the notion of a surface in $\mathbf{R}^3$.

Definition 4.1

A subset $\mathcal{S}$ of $\mathbf{R}^3$ is a *surface* if, for every point $P \in \mathcal{S}$, there is an open set U in $\mathbf{R}^2$ and an open set W in $\mathbf{R}^3$ containing P such that $\mathcal{S} \cap W$ is homeomorphic to U.

Thus, a surface comes equipped with a collection of homeomorphisms $\boldsymbol{\sigma} : U \to \mathcal{S} \cap W$, which we call *surface patches* or *parametrisations*. The collection of all these surface patches is called the *atlas* of $\mathcal{S}$. Every point of $\mathcal{S}$ lies in the image of at least one surface patch in the atlas of $\mathcal{S}$. The reason for this terminology will become clear from the following examples.

Example 4.1

Every plane in $\mathbf{R}^3$ is a surface with an atlas containing a single surface patch. In fact, let $\mathbf{a}$ be a point on the plane, and let $\mathbf{p}$ and $\mathbf{q}$ be two unit vectors that are parallel to the plane and perpendicular to each other. Then, any vector parallel to the plane is a linear combination of $\mathbf{p}$ and $\mathbf{q}$, say $u\mathbf{p} + v\mathbf{q}$ for some scalars u and v. If $\mathbf{r}$ is the position vector of any point of the plane, $\mathbf{r} - \mathbf{a}$ is parallel to the plane, so

$$\mathbf{r} - \mathbf{a} = u\mathbf{p} + v\mathbf{q},$$
$$\therefore \quad \mathbf{r} = \mathbf{a} + u\mathbf{p} + v\mathbf{q},$$

for some scalars u and v. Thus, the desired surface patch is

$$\boldsymbol{\sigma}(u, v) = \mathbf{a} + u\mathbf{p} + v\mathbf{q},$$

and its inverse map is

$$\boldsymbol{\sigma}^{-1}(\mathbf{r}) = ((\mathbf{r} - \mathbf{a}).\mathbf{p}, (\mathbf{r} - \mathbf{a}).\mathbf{q}).$$

These formulas make it clear that $\boldsymbol{\sigma}$ and $\boldsymbol{\sigma}^{-1}$ are continuous, and hence that $\boldsymbol{\sigma}$ is a homeomorphism. (We shall not verify this in detail.)

The next example shows why we have to consider surfaces, and not just surface patches.

Example 4.2

The *unit sphere*

$$S^2 = \{(x, y, z) \in \mathbf{R}^3 \mid x^2 + y^2 + z^2 = 1\}$$

is a surface. The most obvious parametrisation is probably that given by latitude θ and longitude φ:

$$\boldsymbol{\sigma}(\theta, \varphi) = (\cos\theta \cos\varphi, \cos\theta \sin\varphi, \sin\theta).$$

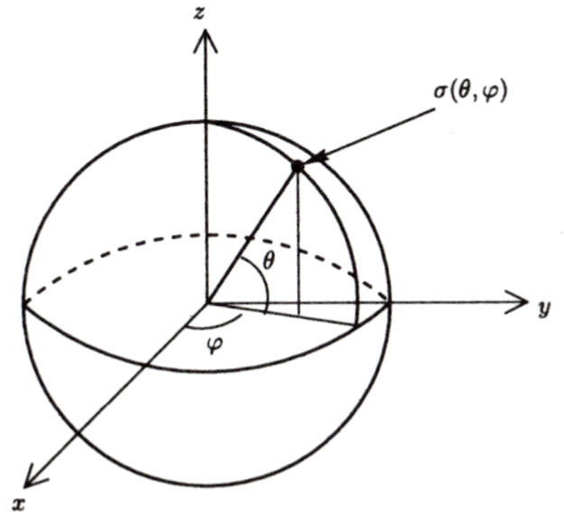

Without some restriction on (θ, φ), $\boldsymbol{\sigma}$ is not injective (and so is not a homeomorphism). To cover the whole sphere, it is clearly sufficient to take

$$-\frac{\pi}{2} \leq \theta \leq \frac{\pi}{2}, \quad 0 \leq \varphi \leq 2\pi.$$

However, the set of points (θ, φ) satisfying these inequalities is not an open subset of $\mathbf{R}^2$, and so cannot be used as the domain of a surface patch. The largest open set consistent with the above inequalities is

$$U = \left\{ (\theta, \varphi) \mid -\frac{\pi}{2} < \theta < \frac{\pi}{2}, \ 0 < \varphi < 2\pi \right\},$$

but now the image of $\boldsymbol{\sigma} : U \to \mathbf{R}^3$ is not the whole of the sphere, but rather the complement of the great semi-circle $\mathcal{C}$ consisting of the points of the sphere of the form $(x, 0, z)$ with $x \geq 0$. Hence, $\boldsymbol{\sigma} : U \to \mathbf{R}^3$ covers only a 'patch' of the sphere. Again, we shall not verify in detail that $\boldsymbol{\sigma}$ is a homeomorphism from U to the intersection of the sphere with the open set

$$W = \{(x, y, z) \in \mathbf{R}^3 \mid x < 0 \text{ or } y \neq 0\}.$$

To show that the sphere is a surface, we must therefore produce at least one more surface patch covering the part of the sphere omitted by $\boldsymbol{\sigma}$. For example, let $\tilde{\boldsymbol{\sigma}}$ be the patch obtained by first rotating $\boldsymbol{\sigma}$ by π about the z-axis and then by $\pi/2$ about the x-axis. Explicitly, $\tilde{\boldsymbol{\sigma}} : U \to \mathbf{R}^3$ is given by

$$\tilde{\boldsymbol{\sigma}}(\theta, \varphi) = (-\cos\theta\cos\varphi, -\sin\theta, -\cos\theta\sin\varphi)$$

(the open set U is the same as for $\boldsymbol{\sigma}$). The image of $\tilde{\boldsymbol{\sigma}}$ is the complement of the great semi-circle $\tilde{\mathcal{C}}$ consisting of the points of the sphere of the form $(x, y, 0)$ with $x \leq 0$ (see the diagram at the top of the next page).

It is clear that $\mathcal{C}$ and $\tilde{\mathcal{C}}$ do not intersect, so the union of the images of $\boldsymbol{\sigma}$ and $\tilde{\boldsymbol{\sigma}}$ is the whole sphere. Note that most points of the sphere are in the images of *both* surface patches.

It is intuitively obvious, although not quite trivial to prove, that the sphere *cannot* be covered by a single surface patch (see Exercise 4.5).

Our last example (for the moment) is a subset of $\mathbf{R}^3$ that is nearly, but not quite, a surface.

Example 4.3

We consider the *double cone*

$$\mathcal{S} = \{(x, y, z) \in \mathbf{R}^3 \mid x^2 + y^2 = z^2\}.$$

To see that this is *not* a surface, suppose that $\boldsymbol{\sigma} : U \to \mathcal{S} \cap W$ is a surface patch containing the vertex $(0, 0, 0)$ of the cone, and let $\mathbf{a} \in U$ correspond to the vertex. We can assume that U is an open ball with centre $\mathbf{a}$, since any open set U containing $\mathbf{a}$ must contain such an open ball. The open set W must obviously contain a point $\mathbf{p}$ in the lower half $\mathcal{S}_-$ of $\mathcal{S}$ where $z < 0$ and a point $\mathbf{q}$ in the upper half $\mathcal{S}_+$ where $z > 0$; let $\mathbf{b}$ and $\mathbf{c}$ be the corresponding points in U. It is clear that there is a curve $\boldsymbol{\pi}$ in U passing through $\mathbf{b}$ and $\mathbf{c}$, but not passing through $\mathbf{a}$. This is mapped by $\boldsymbol{\sigma}$ into the curve $\boldsymbol{\gamma} = \boldsymbol{\sigma} \circ \boldsymbol{\pi}$ lying entirely in $\mathcal{S}$, passing through $\mathbf{p}$ and $\mathbf{q}$, and not passing through the vertex. (It is true that $\boldsymbol{\gamma}$ will in general only be continuous, and not smooth, but this does not affect the argument.) This is clearly impossible. (The reader versed in point set topology will be able to make this heuristic argument rigorous.)

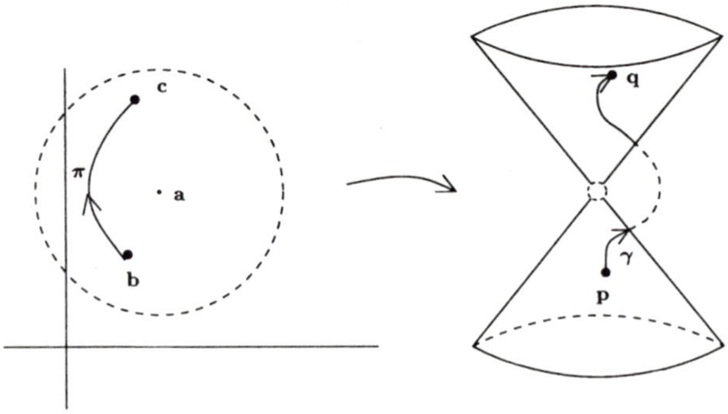

If we remove the vertex, however, we do get a surface $\mathcal{S}_- \cup \mathcal{S}_+$. It has an atlas consisting of the two surface patches $\boldsymbol{\sigma}_\pm : U \to \mathbf{R}^3$, where $U = \mathbf{R}^2 \backslash \{(0,0)\}$, given by the inverse of projection onto the xy plane:

$$\boldsymbol{\sigma}_\pm(u,v) = (u, v, \pm\sqrt{u^2+v^2}).$$

As the example of the sphere shows, a point $\mathbf{a}$ of a surface $\mathcal{S}$ will generally lie in the image of more than one surface patch. Suppose then that $\boldsymbol{\sigma} : U \to \mathcal{S} \cap W$

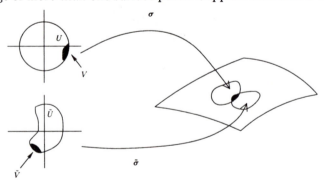

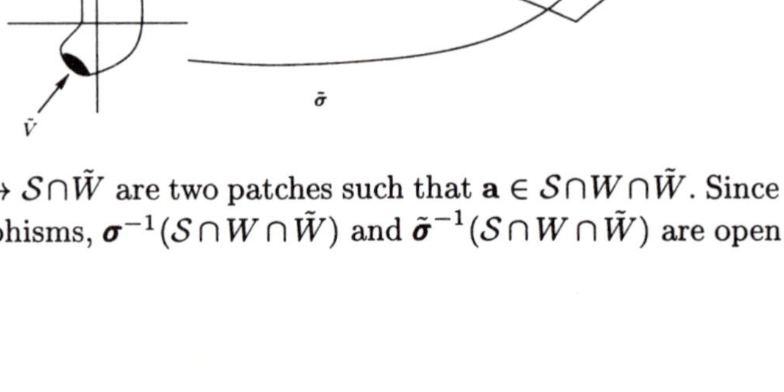

and $\tilde{\boldsymbol{\sigma}} : \tilde{U} \to \mathcal{S} \cap \tilde{W}$ are two patches such that $\mathbf{a} \in \mathcal{S} \cap W \cap \tilde{W}$. Since $\boldsymbol{\sigma}$ and $\tilde{\boldsymbol{\sigma}}$ are homeomorphisms, $\boldsymbol{\sigma}^{-1}(\mathcal{S} \cap W \cap \tilde{W})$ and $\tilde{\boldsymbol{\sigma}}^{-1}(\mathcal{S} \cap W \cap \tilde{W})$ are open sets $V \subseteq U$

and $\tilde{V} \subseteq \tilde{U}$, respectively. The composite homeomorphism $\boldsymbol{\sigma}^{-1} \circ \tilde{\boldsymbol{\sigma}} : \tilde{V} \to V$ is called the *transition map* from $\boldsymbol{\sigma}$ to $\tilde{\boldsymbol{\sigma}}$. If we denote this map by Φ, we have

$$\tilde{\boldsymbol{\sigma}}(\tilde{u}, \tilde{v}) = \boldsymbol{\sigma}(\Phi(\tilde{u}, \tilde{v}))$$

for all $(\tilde{u}, \tilde{v}) \in \tilde{V}$.

EXERCISES

4.1 Show that an open disc in the xy-plane is a surface.

4.2 Show that the *circular cylinder*

$$\mathcal{S} = \{(x, y, z) \in \mathbf{R}^3 \mid x^2 + y^2 = 1\}$$

can be covered by a single surface patch, and so is a surface. (Take U to be an annulus.)

4.3 Define surface patches $\boldsymbol{\sigma}^x_\pm : U \to \mathbf{R}^3$ for the unit sphere by solving the equation $x^2 + y^2 + z^2 = 1$ for x in terms of y and z:

$$\boldsymbol{\sigma}^x_\pm(u, v) = (\pm\sqrt{1 - u^2 - v^2}, u, v),$$

defined on the open set $U = \{(u, v) \in \mathbf{R}^2 \mid u^2 + v^2 < 1\}$. Define $\boldsymbol{\sigma}^y_\pm$ and $\boldsymbol{\sigma}^z_\pm$ similarly (with the same U) by solving for y and z, respectively. Show that these six patches give the sphere the structure of a surface.

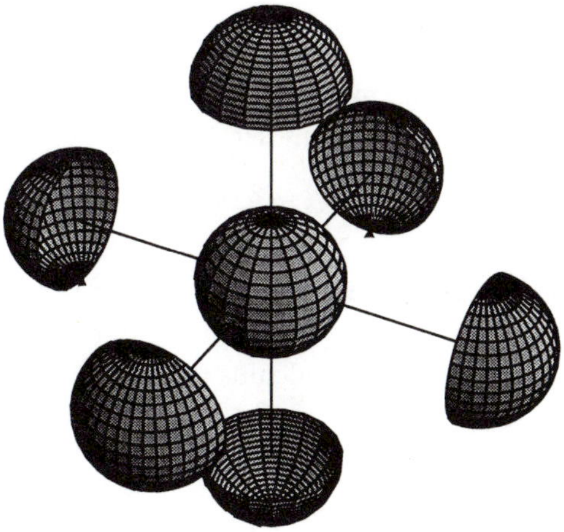

4.4 The *hyperboloid of one sheet* is

$$\mathcal{S} = \{(x, y, z) \in \mathbf{R}^3 \mid x^2 + y^2 - z^2 = 1\}.$$

Show that, for every θ, the straight line

$$(x - z)\cos\theta = (1 - y)\sin\theta, \quad (x + z)\sin\theta = (1 + y)\cos\theta$$

is contained in $\mathcal{S}$, and that every point of the hyperboloid lies on one of these lines. Deduce that $\mathcal{S}$ can be covered by a single surface patch, and hence is a surface. (Compare the case of the cylinder in Exercise 4.2.)

Find a second family of straight lines on $\mathcal{S}$, and show that no two lines of the same family intersect, while every line of the first family intersects every line of the second family with one exception. One says that the surface $\mathcal{S}$ is *doubly ruled.*

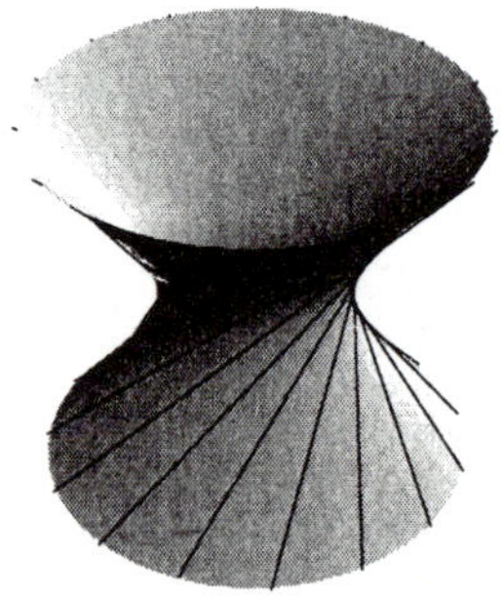

4.5 Show that the unit sphere cannot be covered by a single surface patch. (This requires some point set topology.)

4.2. Smooth Surfaces

In Differential Geometry we use calculus to analyse surfaces (and other geometric objects). We must be able to make sense of the statement that a function on a surface is differentiable, for example. For this, we have to consider surfaces with some extra structure.

First, if U is an open subset of $\mathbf{R}^m$, we say that a map $\mathbf{f} : U \to \mathbf{R}^n$ is smooth if each of the n components of $\mathbf{f}$, which are functions $U \to \mathbf{R}$, have continuous partial derivatives of all orders. The partial derivatives of $\mathbf{f}$ are then computed componentwise. For example, if $m = 2$ and $n = 3$, and

$$\mathbf{f}(u,v) = (f_1(u,v), f_2(u,v), f_3(u,v)),$$

then

$$\frac{\partial \mathbf{f}}{\partial u} = \left(\frac{\partial f_1}{\partial u}, \frac{\partial f_2}{\partial u}, \frac{\partial f_3}{\partial u}\right), \quad \frac{\partial \mathbf{f}}{\partial v} = \left(\frac{\partial f_1}{\partial v}, \frac{\partial f_2}{\partial v}, \frac{\partial f_3}{\partial v}\right),$$

and similarly for higher derivatives. We often use the following abbreviations:

$$\frac{\partial \mathbf{f}}{\partial u} = \mathbf{f}_u, \quad \frac{\partial \mathbf{f}}{\partial v} = \mathbf{f}_v,$$

$$\frac{\partial^2 \mathbf{f}}{\partial u^2} = \mathbf{f}_{uu}, \quad \frac{\partial^2 \mathbf{f}}{\partial u \partial v} = \mathbf{f}_{uv}, \quad \frac{\partial^2 \mathbf{f}}{\partial v \partial u} = \mathbf{f}_{vu}, \quad \frac{\partial^2 \mathbf{f}}{\partial v^2} = \mathbf{f}_{vv},$$

and so on. Note that $\mathbf{f}_{uv} = \mathbf{f}_{vu}$, because all the partial derivatives of the components of $\mathbf{f}$ are continuous.

It now makes sense to say that the surface patches $\boldsymbol{\sigma} : U \to \mathbf{R}^3$ in the atlas of a surface S are smooth. But we shall require one further condition.

Definition 4.2

A surface patch $\boldsymbol{\sigma} : U \to \mathbf{R}^3$ is called *regular* if it is smooth and the vectors $\boldsymbol{\sigma}_u$ and $\boldsymbol{\sigma}_v$ are linearly independent at all points $(u,v) \in U$. Equivalently, $\boldsymbol{\sigma}$ should be smooth and the vector product $\boldsymbol{\sigma}_u \times \boldsymbol{\sigma}_v$ should be non-zero at every point of U.

We can finally define the class of surfaces to be studied in this book.

Definition 4.3

A *smooth surface* is a surface $\boldsymbol{\sigma}$ whose atlas consists of regular surface patches.

Example 4.4

The plane in Example 4.1 is a smooth surface. For

$$\boldsymbol{\sigma}(u,v) = \mathbf{a} + u\mathbf{p} + v\mathbf{q}$$

is clearly smooth and $\boldsymbol{\sigma}_u = \mathbf{p}$ and $\boldsymbol{\sigma}_v = \mathbf{q}$ are linearly independent because $\mathbf{p}$ and $\mathbf{q}$ were chosen to be perpendicular unit vectors.

Example 4.5

For the unit sphere S^2 in Example 4.2, it is again clear that $\boldsymbol{\sigma}$ and $\tilde{\boldsymbol{\sigma}}$ are smooth. As for regularity, we compute

$$\boldsymbol{\sigma}_\theta = (-\sin\theta\cos\varphi, -\sin\theta\sin\varphi, \cos\theta), \quad \boldsymbol{\sigma}_\varphi = (-\cos\theta\sin\varphi, \cos\theta\cos\varphi, 0),$$

which gives

$$\boldsymbol{\sigma}_\theta \times \boldsymbol{\sigma}_\varphi = (-\cos^2\theta\cos\varphi, -\cos^2\theta\sin\varphi, -\sin\theta\cos\theta)$$

and hence $\| \boldsymbol{\sigma}_\theta \times \boldsymbol{\sigma}_\varphi \| = |\cos\theta|$. But if $(\theta, \varphi) \in U$, then $-\pi/2 < \theta < \pi/2$ so $\cos\theta \neq 0$. Similarly, one checks that $\tilde{\boldsymbol{\sigma}}$ is regular.

In Exercise 4.3 we gave another family of surface patches covering the unit sphere S^2, and it is easy to check that they are regular (see Exercise 4.7). Together with Example 4.5, this gives two atlases for S^2 consisting of regular surface patches, and an obvious question is: which atlas should we use to study the sphere? The answer is that we can use either, or both. For the eight patches in Exercise 4.3 and Example 4.5 together form a third atlas. In most situations (although not in all – see Definition 4.5), one might as well use the *maximal atlas* for a given surface $\mathcal{S}$ consisting of *all* the regular surface patches $\boldsymbol{\sigma} : U \to \mathcal{S} \cap W$, with U and W being open subsets of $\mathbf{R}^2$ and $\mathbf{R}^3$, respectively. Such surface patches are called *allowable* surface patches for $\mathcal{S}$. The maximal atlas is independent of any arbitrary choices.

Although not at first sight very interesting, the next two results are very important for what is to follow.

Proposition 4.1

The transition maps of a smooth surface are smooth.

The proof of this will be given in Section 4.7. The next result is a kind of converse.

Proposition 4.2

Let U and $\tilde{U}$ be open subsets of $\mathbf{R}^2$ and let $\boldsymbol{\sigma} : U \to \mathbf{R}^3$ be a regular surface patch. Let $\Phi : \tilde{U} \to U$ be a bijective smooth map with smooth inverse map $\Phi^{-1} : U \to \tilde{U}$. Then, $\tilde{\boldsymbol{\sigma}} = \boldsymbol{\sigma} \circ \Phi : \tilde{U} \to U$ is a regular surface patch.

Proof 4.2

The patch $\tilde{\boldsymbol{\sigma}}$ is smooth because any composite of smooth maps is smooth. As for regularity, let $(u, v) = \Phi(\tilde{u}, \tilde{v})$. By the chain rule,

$$\tilde{\boldsymbol{\sigma}}_{\tilde{u}} = \frac{\partial u}{\partial \tilde{u}}\boldsymbol{\sigma}_u + \frac{\partial v}{\partial \tilde{u}}\boldsymbol{\sigma}_v, \quad \tilde{\boldsymbol{\sigma}}_{\tilde{v}} = \frac{\partial u}{\partial \tilde{v}}\boldsymbol{\sigma}_u + \frac{\partial v}{\partial \tilde{v}}\boldsymbol{\sigma}_v,$$

so

$$\tilde{\boldsymbol{\sigma}}_{\tilde{u}} \times \tilde{\boldsymbol{\sigma}}_{\tilde{v}} = \left(\frac{\partial u}{\partial \tilde{u}}\frac{\partial v}{\partial \tilde{v}} - \frac{\partial u}{\partial \tilde{v}}\frac{\partial v}{\partial \tilde{u}}\right)\boldsymbol{\sigma}_u \times \boldsymbol{\sigma}_v. \tag{1}$$

The scalar on the right-hand side of this equation is the determinant of the *jacobian matrix*

$$J(\Phi) = \begin{pmatrix} \frac{\partial u}{\partial \tilde{u}} & \frac{\partial u}{\partial \tilde{v}} \\ \frac{\partial v}{\partial \tilde{u}} & \frac{\partial v}{\partial \tilde{v}} \end{pmatrix}$$

of Φ. We recall from calculus that, if Ψ and $\tilde{\Psi}$ are two maps between open sets in $\mathbf{R}^2$,

$$J(\tilde{\Psi} \circ \Psi) = J(\tilde{\Psi})J(\Psi).$$

(In fact, this is equivalent to the chain rule that expresses the first partial derivatives of $\tilde{\Psi} \circ \Psi$ in terms of those of $\tilde{\Psi}$ and Ψ.) Taking $\Psi = \Phi$ and $\tilde{\Psi} = \Phi^{-1}$, we see that $J(\Phi^{-1}) = J(\Phi)^{-1}$. In particular, $J(\Phi)$ is invertible, so its determinant is non-zero and Eq. (1) shows that $\tilde{\boldsymbol{\sigma}}$ is regular. □

If regular surface patches $\boldsymbol{\sigma}$ and $\tilde{\boldsymbol{\sigma}}$ are related as in this proposition, we say that $\tilde{\boldsymbol{\sigma}}$ is a *reparametrisation* of $\boldsymbol{\sigma}$, and that Φ is a *reparametrisation map*. Note that $\boldsymbol{\sigma}$ is then a reparametrisation of $\tilde{\boldsymbol{\sigma}}$, since $\boldsymbol{\sigma} = \tilde{\boldsymbol{\sigma}} \circ \Phi^{-1}$.

Note also that, if $\boldsymbol{\sigma} : U \to \mathcal{S} \cap W$ and $\tilde{\boldsymbol{\sigma}} : \tilde{U} \to \mathcal{S} \cap \tilde{W}$ are two allowable surface patches of a smooth surface $\mathcal{S}$, and if $V \subseteq U$ and $\tilde{V} \subseteq \tilde{U}$ are the open subsets such that $\boldsymbol{\sigma}(V) = \tilde{\boldsymbol{\sigma}}(\tilde{V}) = \mathcal{S} \cap W \cap \tilde{W}$, then $\Phi = \boldsymbol{\sigma}^{-1} \circ \tilde{\boldsymbol{\sigma}} : \tilde{V} \to V$ is bijective, smooth and has a smooth inverse by Proposition 4.1. Thus, $\tilde{\boldsymbol{\sigma}}$ *is a reparametrisation of* $\boldsymbol{\sigma}$ *where they are both defined.*

These observations give rise to a very important principle that we shall use throughout the book. The principle is that *we can define a property of any smooth surface provided we can define it for any regular surface patch in such a way that it is unchanged when the patch is reparametrised.*

As an illustration of this principle, let us define what is meant by a *smooth map* $f : \mathcal{S}_1 \to \mathcal{S}_2$, where $\mathcal{S}_1$ and $\mathcal{S}_2$ are smooth surfaces. By our general principle, we can assume that $\mathcal{S}_1$ and $\mathcal{S}_2$ are covered by single surface patches $\boldsymbol{\sigma}_1 : U_1 \to \mathbf{R}^3$ and $\boldsymbol{\sigma}_2 : U_2 \to \mathbf{R}^3$ provided we verify that the definition we give is unaffected by a reparametrisation of $\boldsymbol{\sigma}_1$ and $\boldsymbol{\sigma}_2$. Since $\boldsymbol{\sigma}_1$ and $\boldsymbol{\sigma}_2$ are bijective, any map $f : \mathcal{S}_1 \to \mathcal{S}_2$ gives rise to the map $\boldsymbol{\sigma}_2^{-1} \circ f \circ \boldsymbol{\sigma}_1 : U_1 \to U_2$, and we say that f is smooth if this map is smooth (we already know what it means for a map between open subsets of $\mathbf{R}^2$ to be smooth). Now suppose that $\tilde{\boldsymbol{\sigma}}_1 : \tilde{U}_1 \to \mathbf{R}^3$ and $\tilde{\boldsymbol{\sigma}}_2 : \tilde{U}_2 \to \mathbf{R}^3$ are reparametrisations of $\boldsymbol{\sigma}_1$ and $\boldsymbol{\sigma}_2$, with reparametrisation maps $\Phi_1 : \tilde{U}_1 \to U_1$ and $\Phi_2 : \tilde{U}_2 \to U_2$, respectively. We have to show that the corresponding map $\tilde{\boldsymbol{\sigma}}_2^{-1} \circ f \circ \tilde{\boldsymbol{\sigma}}_1 : \tilde{U}_1 \to \tilde{U}_2$ is smooth if $\boldsymbol{\sigma}_2^{-1} \circ f \circ \boldsymbol{\sigma}_1 : U_1 \to U_2$ is smooth. But this is true, since

$$\begin{aligned} \tilde{\boldsymbol{\sigma}}_2^{-1} \circ f \circ \tilde{\boldsymbol{\sigma}}_1 &= \tilde{\boldsymbol{\sigma}}_2^{-1} \circ (\boldsymbol{\sigma}_2 \circ \boldsymbol{\sigma}_2^{-1}) \circ f \circ (\boldsymbol{\sigma}_1 \circ \boldsymbol{\sigma}_1^{-1}) \circ \tilde{\boldsymbol{\sigma}}_1 \\ &= (\tilde{\boldsymbol{\sigma}}_2^{-1} \circ \boldsymbol{\sigma}_2) \circ (\boldsymbol{\sigma}_2^{-1} \circ f \circ \boldsymbol{\sigma}_1) \circ (\boldsymbol{\sigma}_1^{-1} \circ \tilde{\boldsymbol{\sigma}}_1) \\ &= \Phi_2^{-1} \circ (\boldsymbol{\sigma}_2^{-1} \circ f \circ \boldsymbol{\sigma}_1) \circ \Phi_1, \end{aligned}$$

and Φ_1 and Φ_2^{-1} are smooth maps (between open subsets of $\mathbf{R}^2$). The reader should check that composites of smooth maps between surfaces are smooth.

We shall be especially interested in smooth maps $f : \mathcal{S}_1 \to \mathcal{S}_2$ which are bijective and whose inverse map $f^{-1} : \mathcal{S}_2 \to \mathcal{S}_1$ is smooth. Such maps are called *diffeomorphisms*, and $\mathcal{S}_1$ and $\mathcal{S}_2$ are said to be *diffeomorphic* if there is a diffeomorphism between them. The following observation will be useful.

Proposition 4.3

Let $f : \mathcal{S}_1 \to \mathcal{S}_2$ be a diffeomorphism. If $\boldsymbol{\sigma}_1$ is an allowable surface patch on $\mathcal{S}_1$, then $f \circ \boldsymbol{\sigma}_1$ is an allowable surface patch on $\mathcal{S}_2$.

Proof 4.3

We can assume that $\mathcal{S}_1$ and $\mathcal{S}_2$ are covered by single allowable patches $\boldsymbol{\sigma}_1 : U_1 \to \mathbf{R}^3$ and $\boldsymbol{\sigma}_2 : U_2 \to \mathbf{R}^3$, respectively. Since f is a diffeomorphism, $f(\boldsymbol{\sigma}_1(u,v)) = \boldsymbol{\sigma}_2(F(u,v))$, where $F : U_1 \to U_2$ is bijective, smooth and F^{-1} is smooth. The result now follows from Proposition 4.2. □

Example 4.6

We consider the map which wraps the plane onto the circular cylinder of radius 1 and axis the z-axis, which we parametrise by $\boldsymbol{\sigma}_2 : U \to \mathbf{R}^3$, where

$$\boldsymbol{\sigma}_2(u,v) = (\cos u, \sin u, v), \quad U = \{(u,v) \in \mathbf{R}^2 \mid 0 < u < 2\pi\}.$$

If we wrap the whole plane onto the cylinder, we shall not get a bijective map, since the plane will wrap around infinitely many times. So we consider the infinite strip in the yz-plane of width 2π parametrised by $\boldsymbol{\sigma}_1 : U \to \mathbf{R}^3$, where

$$\boldsymbol{\sigma}_1(u,v) = (0, u, v).$$

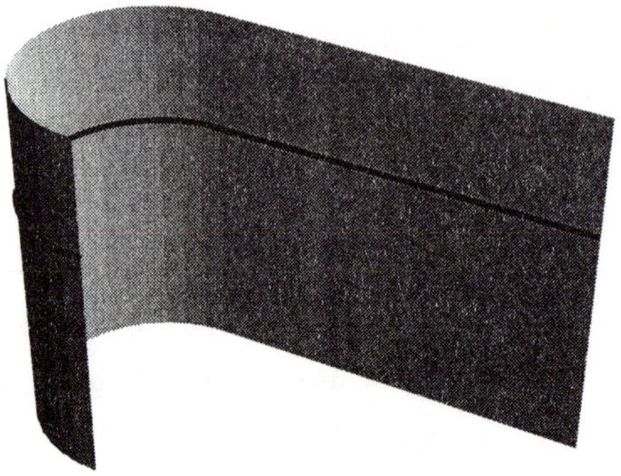

We wrap the strip around the cylinder by wrapping the line $z = v$ parallel to the y-axis around the 'waist' of the cylinder at height v above the xy-plane. Since the width of the strip is equal to the circumference of the cylinder, a point in the strip with y-coordinate u will go to a point on the cylinder with polar angle u. Putting this together, we see that the wrapping map takes the point $(0, u, v)$ of the strip to the point $(\cos u, \sin u, v)$ of the cylinder, i.e. in the notation above,

$$f(0, u, v) = (\cos u, \sin u, v).$$

From this, it is clear that the expression for f in terms of parameters is simply

$$F(u, v) = (u, v),$$

since

$$f(\boldsymbol{\sigma}_1(u, v)) = \boldsymbol{\sigma}_2(u, v).$$

We conclude this section with a general construction of smooth surfaces. In fact, it can be shown that every smooth surface arises in this way.

As we have seen in the above examples, surfaces are often given to us as *level surfaces* $\{(x, y, z) \in \mathbf{R}^3 \mid f(x, y, z) = c\}$, where f is a smooth function and c is a constant (of course, we can always assume that $c = 0$ by replacing f by $f - c$). For example, the unit sphere S^2 is the level surface $x^2 + y^2 + z^2 = 1$. In these examples, we constructed atlases by fairly ad hoc methods. The following result gives conditions under which a level surface is a smooth surface. In fact, it deals with a slightly more general situation in which different regions of a surface may be defined by different functions.

Theorem 4.1

Let $\mathcal{S}$ be a subset of $\mathbf{R}^3$ with the following property: for each point $P \in \mathcal{S}$, there is an open subset W of $\mathbf{R}^3$ containing P and a smooth function $f : W \to \mathbf{R}$ such that
(i) $\mathcal{S} \cap W = \{(x, y, z) \in W \mid f(x, y, z) = 0\}$;
(ii) the partial derivatives f_x, f_y and f_z do not all vanish at P.
Then, $\mathcal{S}$ is a smooth surface.

We postpone the proof to Section 4.7.

Example 4.7

For the unit sphere S^2, we can take $W = \mathbf{R}^3$ and use the single function $f(x, y, z) = x^2 + y^2 + z^2 - 1$. Then, $(f_x, f_y, f_z) = (2x, 2y, 2z)$ so $\| (f_x, f_y, f_z) \|$

$= 2$ at all points of S^2. In particular, (f_x, f_y, f_z) is non-zero everywhere on S^2. Hence, the theorem tells us that S^2 is a smooth surface.

Example 4.8

For the double cone of Example 4.3, $f(x, y, z) = x^2 + y^2 - z^2$. Hence, $(f_x, f_y, f_z) = (2x, 2y, -2z)$, and this vanishes only at the vertex $(0, 0, 0)$. Hence, removing the vertex gives a smooth surface, as we have already seen.

For the rest of this book, by a surface we shall mean a smooth surface, and by a surface patch we shall mean a regular smooth surface patch (or equivalently an allowable surface patch).

Unless we indicate otherwise, we shall also assume that all surfaces we consider are *connected*, which means that any two points of $\mathcal{S}$ can be joined by a curve lying entirely in $\mathcal{S}$. This is not a serious restriction, for it is not difficult to prove that any surface is a disjoint union of connected surfaces, and so can be studied by studying each of its connected parts separately. All the surfaces we have encountered so far are connected except the double cone of Example 4.3, which breaks into the union of two disjoint half cones $\mathcal{S}_\pm$ when the vertex is removed, as it must be to have a surface.

EXERCISES

4.6 Show that, if $f(x, y)$ is a smooth function, its *graph*

$$\{(x, y, z) \in \mathbf{R}^3 \mid z = f(x, y)\}$$

is a smooth surface with atlas consisting of the single regular surface patch

$$\boldsymbol{\sigma}(u, v) = (u, v, f(u, v)).$$

4.7 Verify that the six surface patches for the unit sphere in Exercise 4.3 are regular. Calculate the transition maps between them and verify that these maps are smooth.

4.8 Show that

$$\boldsymbol{\sigma}(r, \theta) = (r \cosh\theta, r \sinh\theta, r^2)$$

is a parametrisation of the part $z > 0$ of the *hyperbolic paraboloid* $z = x^2 - y^2$. (A picture of the hyperbolic paraboloid can be found in Proposition 4.6.) Use Exercise 4.6 to find another parametrisation $\tilde{\boldsymbol{\sigma}}$ of the same part, and verify that $\tilde{\boldsymbol{\sigma}}$ is a reparametrisation of $\boldsymbol{\sigma}$. Find two analogous parametrisations for the part $z < 0$ of the hyperbolic paraboloid.

4.9 Show that the level surface

$$\frac{x^2}{a^2} + \frac{y^2}{b^2} + \frac{z^2}{c^2} = 1,$$

where a, b and c are non-zero constants, is a smooth surface (called an *ellipsoid*). (A picture of an ellipsoid can be found in Proposition 4.6.)

4.10 A *torus* is obtained by rotating a circle $\mathcal{C}$ in a plane Π around a straight line $\mathcal{L}$ in Π that does not intersect $\mathcal{C}$. Take Π to be the xz-plane, $\mathcal{L}$ to be the z-axis, $a > 0$ the distance of the centre of $\mathcal{C}$ from $\mathcal{L}$, and $b < a$ the radius of $\mathcal{C}$. Show that the torus is a smooth surface

(i) by showing that it has an atlas consisting of surface patches

$$\boldsymbol{\sigma}(\theta, \varphi) = ((a + b\cos\theta)\cos\varphi, (a + b\cos\theta)\sin\varphi, b\sin\theta),$$

with (θ, φ) in suitable open subsets of $\mathbf{R}^2$;

(ii) by showing that it is the level surface given by

$$(x^2 + y^2 + z^2 + a^2 - b^2)^2 = 4a^2(x^2 + y^2).$$

4.11 If $\mathcal{S}$ is a smooth surface, define the notion of a *smooth function* $\mathcal{S} \to \mathbf{R}$. Show that, if $\mathcal{S}$ is a smooth surface, each component of the inclusion map $\mathcal{S} \to \mathbf{R}^3$ is a smooth function $\mathcal{S} \to \mathbf{R}$.

4.12 Show that translations and invertible linear transformations of $\mathbf{R}^3$ take smooth surfaces to smooth surfaces.

4.3. Tangents, Normals and Orientability

A natural way to study a surface $\mathcal{S}$ is via the (smooth) curves $\boldsymbol{\gamma}$ that lie in $\mathcal{S}$. If $\boldsymbol{\gamma} : (\alpha, \beta) \to \mathbf{R}^3$ is contained in the image of a surface patch $\boldsymbol{\sigma} : U \to \mathbf{R}^3$ in the atlas of $\mathcal{S}$, there is a map $(\alpha, \beta) \to U$, say $t \mapsto (u(t), v(t))$, such that

$$\boldsymbol{\gamma}(t) = \boldsymbol{\sigma}(u(t), v(t)). \tag{2}$$

The functions u and v are necessarily smooth (see Exercise 4.30). Conversely, it is obvious that if $t \mapsto (u(t), v(t))$ is smooth, then Eq. (2) defines a curve lying in $\mathcal{S}$. In general, if $\boldsymbol{\gamma}$ is a curve in $\mathcal{S}$ and some point $\boldsymbol{\gamma}(t_0)$ of $\boldsymbol{\gamma}$ lies in a surface patch $\boldsymbol{\sigma}$ of $\mathcal{S}$, then Eq. (2) will hold for all t in some open interval containing t_0. Thus, we may restrict ourselves to curves of the form (2).

Definition 4.4

The *tangent space* at a point P of a surface $\mathcal{S}$ is the set of tangent vectors at P of all curves in $\mathcal{S}$ passing through P.

Proposition 4.4

Let $\boldsymbol{\sigma} : U \to \mathbf{R}^3$ be a patch of a surface $\mathcal{S}$ containing a point P of $\mathcal{S}$, and let (u, v) be coordinates in U. The tangent space to $\mathcal{S}$ at P is the vector subspace of $\mathbf{R}^3$ spanned by the vectors $\boldsymbol{\sigma}_u$ and $\boldsymbol{\sigma}_v$ (the derivatives are evaluated at the point $(u_0, v_0) \in U$ such that $\boldsymbol{\sigma}(u_0, v_0) = P$).

Proof 4.4

Let $\boldsymbol{\gamma}$ be a smooth curve in $\mathcal{S}$, say

$$\boldsymbol{\gamma}(t) = \boldsymbol{\sigma}(u(t), v(t)).$$

Denoting d/dt by a dot, we have, by the chain rule,

$$\dot{\boldsymbol{\gamma}} = \boldsymbol{\sigma}_u \dot{u} + \boldsymbol{\sigma}_v \dot{v}.$$

Thus, $\dot{\boldsymbol{\gamma}}$ is a linear combination of $\boldsymbol{\sigma}_u$ and $\boldsymbol{\sigma}_v$.

Conversely, any vector in the vector subspace of $\mathbf{R}^3$ spanned by $\boldsymbol{\sigma}_u$ and $\boldsymbol{\sigma}_v$ is of the form $\xi\boldsymbol{\sigma}_u + \eta\boldsymbol{\sigma}_v$ for some scalars ξ and η. Define

$$\boldsymbol{\gamma}(t) = \boldsymbol{\sigma}(u_0 + \xi t, v_0 + \eta t).$$

Then, $\boldsymbol{\gamma}$ is a smooth curve in $\mathcal{S}$ and at $t = 0$, i.e. at the point P in $\mathcal{S}$, we have

$$\dot{\boldsymbol{\gamma}} = \xi\boldsymbol{\sigma}_u + \eta\boldsymbol{\sigma}_v.$$

This shows that every vector in the span of $\boldsymbol{\sigma}_u$ and $\boldsymbol{\sigma}_v$ is the tangent vector at P of some curve in $\mathcal{S}$. □

Since $\boldsymbol{\sigma}_u$ and $\boldsymbol{\sigma}_v$ are linearly independent, the tangent space is two-dimensional, and will be called the *tangent plane* from now on. Note that Definition 4.4 shows that the tangent plane is independent of the choice of patch containing P, even though this is not immmediately obvious from Proposition 4.4 (see Exercise 4.15).

Since the tangent plane at $P \in \mathcal{S}$ passes through the origin of $\mathbf{R}^3$, it is completely determined by giving a unit vector perpendicular to it, called a *unit normal* to $\mathcal{S}$ at P. There are, of course, two such vectors, but Proposition 4.4 shows that choosing a surface patch $\boldsymbol{\sigma} : U \to \mathbf{R}^3$ containing P leads to a definite choice, namely

$$\mathbf{N}_{\boldsymbol{\sigma}} = \frac{\boldsymbol{\sigma}_u \times \boldsymbol{\sigma}_v}{\| \boldsymbol{\sigma}_u \times \boldsymbol{\sigma}_v \|} \tag{3}$$

(with the derivatives evaluated at the point of U corresponding to P), for this is clearly a unit vector perpendicular to every linear combination of $\boldsymbol{\sigma}_u$ and $\boldsymbol{\sigma}_v$. This is called the *standard unit normal* of the surface patch $\boldsymbol{\sigma}$ at P. Unlike the tangent plane, however, $\mathbf{N}_{\boldsymbol{\sigma}}$ is *not* quite independent of the choice of patch $\boldsymbol{\sigma}$ containing P. In fact, if $\tilde{\boldsymbol{\sigma}} : \tilde{U} \to \mathbf{R}^3$ is another surface patch in the atlas of $\mathcal{S}$ containing P, we showed in the proof of Proposition 4.2 that

$$\tilde{\boldsymbol{\sigma}}_{\tilde{u}} \times \tilde{\boldsymbol{\sigma}}_{\tilde{v}} = \det(J(\Phi))\, \boldsymbol{\sigma}_u \times \boldsymbol{\sigma}_v,$$

where $J(\Phi)$ is the jacobian matrix of the transition map Φ from $\boldsymbol{\sigma}$ to $\tilde{\boldsymbol{\sigma}}$. So the standard unit normal of $\tilde{\boldsymbol{\sigma}}$ is

$$\mathbf{N}_{\tilde{\boldsymbol{\sigma}}} = \frac{\tilde{\boldsymbol{\sigma}}_{\tilde{u}} \times \tilde{\boldsymbol{\sigma}}_{\tilde{v}}}{\| \tilde{\boldsymbol{\sigma}}_{\tilde{u}} \times \tilde{\boldsymbol{\sigma}}_{\tilde{v}} \|} = \pm \frac{\boldsymbol{\sigma}_u \times \boldsymbol{\sigma}_v}{\| \boldsymbol{\sigma}_u \times \boldsymbol{\sigma}_v \|} = \pm \mathbf{N}_{\boldsymbol{\sigma}},$$

where the sign is that of the determinant of $J(\Phi)$. This leads to the following definition.

Definition 4.5

An *orientable* surface is a surface with an atlas having the property that, if Φ is the transition map between any two surface patches in the atlas, then $\det(J(\Phi)) > 0$ where Φ is defined.

The preceding discussion gives

Proposition 4.5

An orientable surface $\mathcal{S}$ has a canonical choice of unit normal at each point, obtained by taking the standard unit normal of each surface patch in the atlas of $\mathcal{S}$. □

In fact, the converse is also true: if a surface $\mathcal{S}$ has a unit normal $\mathbf{N}$ defined at each point $P \in \mathcal{S}$ and depending smoothly on P, then $\mathcal{S}$ is orientable. To see this, start with the maximal atlas of $\mathcal{S}$ and retain a patch $\boldsymbol{\sigma}(u,v)$ if $\boldsymbol{\sigma}_u \times \boldsymbol{\sigma}_v$ is a positive multiple of $\mathbf{N}$ at all points in the image of $\boldsymbol{\sigma}$, otherwise discard it. The patches that remain form an atlas satisfying the condition in Proposition 4.5. We leave the details of this to the interested reader (the argument is similar to that used in the next example).

Most of the surfaces we shall discuss are orientable (see Exercise 4.16). Here is one that is not.

Example 4.9

The ***Möbius band*** is the surface obtained by rotating a straight line segment $\mathcal{L}$ around its midpoint P at the same time as P moves around a circle $\mathcal{C}$, in such a way that as P moves once around $\mathcal{C}$, $\mathcal{L}$ makes a half-turn about P. If we take $\mathcal{C}$ to be the circle $x^2+y^2=1$ in the xy-plane, and $\mathcal{L}$ to be a segment of length 1 that is initially parallel to the z-axis with its midpoint P at $(1,0,0)$, then after P has rotated by an angle θ around the z-axis, $\mathcal{L}$ should have rotated by $\theta/2$ around P in the plane containing P and the z-axis. The point of $\mathcal{L}$ initially at $(1,0,t)$ is then at the point

$$\boldsymbol{\sigma}(t,\theta) = \left(\left(1 - t\sin\frac{\theta}{2}\right)\cos\theta,\ \left(1 - t\sin\frac{\theta}{2}\right)\sin\theta,\ t\cos\frac{\theta}{2}\right).$$

We take the domain of definition of $\boldsymbol{\sigma}$ to be

$$U = \{(t,\theta) \in \mathbf{R}^2 \mid -1/2 < t < 1/2,\ 0 < \theta < 2\pi\}.$$

We can define a second patch $\tilde{\boldsymbol{\sigma}}$ by the same formula as $\boldsymbol{\sigma}$ but with domain of definition $\tilde{U} = \{(t,\theta) \in \mathbf{R}^2 \mid -1/2 < t < 1/2,\ -\pi < \theta < \pi\}$. It can be checked that these two patches form an atlas for the Möbius band consisting of regular surface patches, making the Möbius band into a smooth surface $\mathcal{S}$.

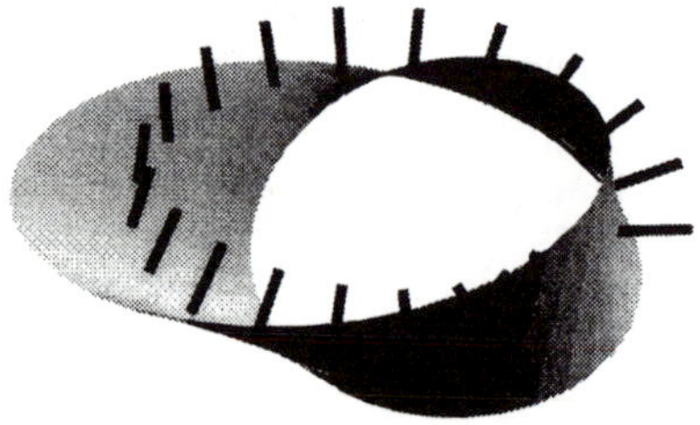

We compute the standard unit normal $\mathbf{N}_{\boldsymbol{\sigma}}$ at points on the median circle (where $t = 0$). At such points, we have

$$\boldsymbol{\sigma}_t = \left(-\sin\frac{\theta}{2}\cos\theta, -\sin\frac{\theta}{2}\sin\theta, \cos\frac{\theta}{2}\right), \quad \boldsymbol{\sigma}_\theta = (-\sin\theta, \cos\theta, 0),$$

so

$$\boldsymbol{\sigma}_t \times \boldsymbol{\sigma}_\theta = \left(-\cos\theta\cos\frac{\theta}{2}, -\sin\theta\cos\frac{\theta}{2}, -\sin\frac{\theta}{2}\right).$$

This is a unit vector, so it is equal to $\mathbf{N}_{\boldsymbol{\sigma}}$.

If the Möbius band were orientable, there would be a well defined unit normal $\mathbf{N}$ defined at every point of $\mathcal{S}$ and varying smoothly over $\mathcal{S}$. At a point $\boldsymbol{\sigma}(0, \theta)$ on the median circle, we would have

$$\mathbf{N} = \lambda(\theta)\mathbf{N}_{\boldsymbol{\sigma}},$$

where $\lambda : (0, 2\pi) \to \mathbf{R}$ is smooth and $\lambda(\theta) = \pm 1$ for all θ. It follows that either $\lambda(\theta) = +1$ for all $\theta \in (0, 2\pi)$, or $\lambda(\theta) = -1$ for all $\theta \in (0, 2\pi)$. Replacing $\mathbf{N}$ by $-\mathbf{N}$ if necessary, we can assume that $\lambda = 1$. At the point $\boldsymbol{\sigma}(0, 0) = \boldsymbol{\sigma}(0, 2\pi)$, we must have (since $\mathbf{N}$ is smooth)

$$\mathbf{N} = \lim_{\theta \downarrow 0} \mathbf{N}_{\boldsymbol{\sigma}} = (-1, 0, 0)$$

and also

$$\mathbf{N} = \lim_{\theta \uparrow 2\pi} \mathbf{N}_{\boldsymbol{\sigma}} = (1, 0, 0).$$

This contradiction shows that the Möbius band is not orientable.

EXERCISES

4.13 Find the equation of the tangent plane of the following surface patches at the indicated points:
(i) $\boldsymbol{\sigma}(u, v) = (u, v, u^2 - v^2)$, $(1, 1, 0)$;
(ii) $\boldsymbol{\sigma}(r, \theta) = (r\cosh\theta, r\sinh\theta, r^2)$, $(1, 0, 1)$.

4.14 A *helicoid* is a surface swept out by an aeroplane propeller, when both the aeroplane and its propeller move at constant speed. (A picture of the helicoid can be found in Example 9.3.) If the aeroplane is flying along the z-axis, show that the helicoid can be parametrised as

$$\boldsymbol{\sigma}(u, v) = (v\cos u, v\sin u, \lambda u),$$

where λ is a constant. Show that the cotangent of the angle that the standard unit normal of $\boldsymbol{\sigma}$ at a point P makes with the z-axis is proportional to the distance of P from the axis.

4.15 If $\boldsymbol{\sigma}(u,v)$ is a surface patch, show that the set of linear combinations of $\boldsymbol{\sigma}_u$ and $\boldsymbol{\sigma}_v$ is unchanged when $\boldsymbol{\sigma}$ is reparametrised.

4.16 Consider the surface $\mathcal{S}$ defined by $f(x,y,z)=0$, where f is a smooth function such that f_x, f_y and f_z do not all vanish at any point of $\mathcal{S}$. Show that the vector

$$\nabla f = (f_x, f_y, f_z)$$

is perpendicular to the tangent plane at every point of $\mathcal{S}$, and deduce that $\mathcal{S}$ is orientable. (Compare Exercise 1.17.)

4.17 Let $\mathcal{S}$ be a surface and let $F : \mathcal{S} \to \mathbf{R}$ be a smooth function (see Exercise 4.11). Show that, at each point $P \in \mathcal{S}$, there is a unique vector $\nabla_{\mathcal{S}} F$ in the tangent plane at P such that

$$(\nabla_{\mathcal{S}} F).\dot{\boldsymbol{\gamma}}(0) = \left.\frac{d}{dt}\right|_{t=0} F(\boldsymbol{\gamma}(t))$$

for all curves $\boldsymbol{\gamma}$ in $\mathcal{S}$ with $\boldsymbol{\gamma}(0) = P$. Deduce that $\nabla_{\mathcal{S}} F = \mathbf{0}$ if F has a local maximum or a local minimum at P.
Show that, if $\mathcal{S}$ is the surface in Exercise 4.16, then $\nabla_{\mathcal{S}} F$ is the perpendicular projection of ∇F onto the tangent plane to $\mathcal{S}$, and deduce that, if F has a local maximum or a local minimum at P, then $\nabla F = \lambda \nabla f$ for some scalar λ. (This is called *Lagrange's Method of Undetermined Multipliers.*)

4.4. Examples of Surfaces

We now describe some of the simplest classes of surfaces. Others will be introduced later in the book.

Example 4.10

A *(generalised) cylinder* is the surface $\mathcal{S}$ obtained by translating a curve. If the curve is $\boldsymbol{\gamma} : (\alpha, \beta) \to \mathbf{R}^3$ and $\mathbf{a}$ is a unit vector in the direction of translation, the point obtained by translating the point $\boldsymbol{\gamma}(u)$ of $\boldsymbol{\gamma}$ by the vector $v\mathbf{a}$ parallel to $\mathbf{a}$ is

$$\boldsymbol{\sigma}(u,v) = \boldsymbol{\gamma}(u) + v\mathbf{a}.$$

Then, $\boldsymbol{\sigma} : U \to \mathbf{R}^3$, where $U = \{(u,v) \in \mathbf{R}^2 \mid \alpha < u < \beta\}$, and clearly $\boldsymbol{\sigma}$ is smooth. Since

$$\boldsymbol{\sigma}(u,v) = \boldsymbol{\sigma}(u',v') \Longleftrightarrow \boldsymbol{\gamma}(u) - \boldsymbol{\gamma}(u') = (v'-v)\mathbf{a},$$

for $\boldsymbol{\sigma}$ to be a surface patch (and hence injective), no straight line parallel to $\mathbf{a}$ should meet $\boldsymbol{\gamma}$ in more than one point. Finally, $\boldsymbol{\sigma}_u = \dot{\boldsymbol{\gamma}}$, $\boldsymbol{\sigma}_v = \mathbf{a}$ (with a dot denoting d/du), so $\boldsymbol{\sigma}$ is regular if and only if the tangent vector of $\boldsymbol{\gamma}$ is never parallel to $\mathbf{a}$.

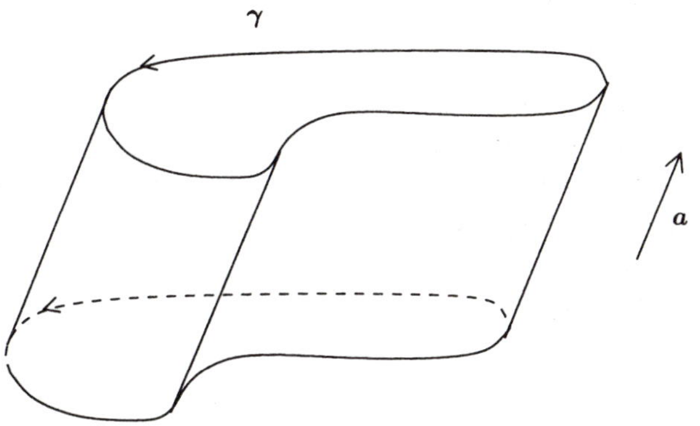

The parametrisation is simplest when $\boldsymbol{\gamma}$ lies in a plane perpendicular to $\mathbf{a}$. In fact, this can always be achieved by replacing $\boldsymbol{\gamma}$ by its projection onto such a plane (see Exercise 4.22). The regularity condition is then clearly satisfied provided $\dot{\boldsymbol{\gamma}}$ is never zero, i.e. provided $\boldsymbol{\gamma}$ is regular. We might as well take the plane to be the xy-plane and $\mathbf{a} = (0, 0, 1)$ to be parallel to the z-axis. Then, $\boldsymbol{\gamma}(u) = (f(u), g(u), 0)$ for some smooth functions f and g, and the parametrisation becomes

$$\boldsymbol{\sigma}(u, v) = (f(u), g(u), v).$$

As an example, starting with a circle, we get an ordinary (circular) cylinder. Taking the circle to have centre the origin, radius 1 and to lie in the xy-plane, it can be parametrised by

$$\boldsymbol{\gamma}(u) = (\cos u, \sin u, 0),$$

defined for $0 < u < 2\pi$ and $-\pi < u < \pi$, say. This gives an atlas for the cylinder consisting of two patches, both given by

$$\boldsymbol{\sigma}(u, v) = (\cos u, \sin u, v),$$

and defined on the open sets

$$\{(u, v) \in \mathbf{R}^2 \mid 0 < u < 2\pi\}, \quad \{(u, v) \in \mathbf{R}^2 \mid -\pi < u < \pi\}.$$

Example 4.11

A *(generalised) cone* is the union of the straight lines passing through a fixed point and the points of a curve. If $\mathbf{p}$ is the fixed point and $\boldsymbol{\gamma} : (\alpha, \beta) \to \mathbf{R}^3$ is the curve, the most general point on the straight line passing through $\mathbf{p}$ and a point $\boldsymbol{\gamma}(u)$ of the curve is

$$\boldsymbol{\sigma}(u,v) = (1-v)\mathbf{p} + v\boldsymbol{\gamma}(u).$$

Then, $\boldsymbol{\sigma}$ is clearly smooth. Now,

$$\boldsymbol{\sigma}(u,v) = \boldsymbol{\sigma}(u',v') \iff v\boldsymbol{\gamma}(u) - v'\boldsymbol{\gamma}(u') + (v'-v)\mathbf{p} = \mathbf{0},$$

which says that the points $\mathbf{p}$, $\boldsymbol{\gamma}(u)$ and $\boldsymbol{\gamma}(u')$ are collinear. So, for $\boldsymbol{\sigma}$ to be a surface patch, no straight line passing through $\mathbf{p}$ should pass through more than one point of $\boldsymbol{\gamma}$ (in particular, $\boldsymbol{\gamma}$ should not pass through $\mathbf{p}$). Finally, we have $\boldsymbol{\sigma}_u = v\dot{\boldsymbol{\gamma}}$, $\boldsymbol{\sigma}_v = \boldsymbol{\gamma} - \mathbf{p}$ (with d/du denoted by a dot), so $\boldsymbol{\sigma}$ is regular provided $v \neq 0$, i.e. the vertex of the cone is omitted (cf. Exercise 4.3), and none of the straight lines forming the cone is tangent to $\boldsymbol{\gamma}$.

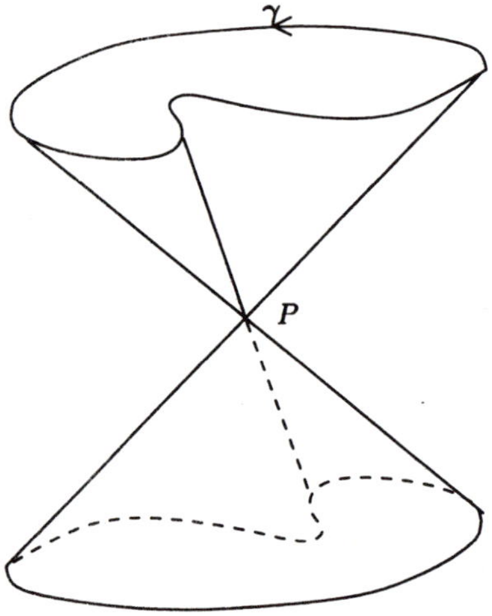

The parametrisation is simplest when $\boldsymbol{\gamma}$ lies in a plane. If this plane contains $\mathbf{p}$, the cone is simply part of that plane. Otherwise, we can take $\mathbf{p}$ to be the origin and the plane to be $z = 1$. Then, $\boldsymbol{\gamma}(u) = (f(u), g(u), 1)$ for some smooth functions f and g, and the parametrisation takes the form

$$\boldsymbol{\sigma}(u,v) = v(f(u), g(u), 1).$$

Examples 4.10 and 4.11 are both special cases of the next class of surfaces.

Example 4.12

A *ruled surface* is a surface that is a union of straight lines, called the *rulings* of the surface. Suppose that C is a curve in $\mathbf{R}^3$ that meets each of these lines. Any point P of the surface lies on one of the given straight lines which intersects $\mathcal{C}$ at Q, say. If $\boldsymbol{\gamma}$ is a parametrisation of $\mathcal{C}$ with $\boldsymbol{\gamma}(u) = Q$, and if $\boldsymbol{\delta}(u)$ is a non-zero vector in the direction of the line passing through $\boldsymbol{\gamma}(u)$, P has position vector of the form

$$\boldsymbol{\sigma}(u, v) = \boldsymbol{\gamma}(u) + v\boldsymbol{\delta}(u),$$

for some scalar v.

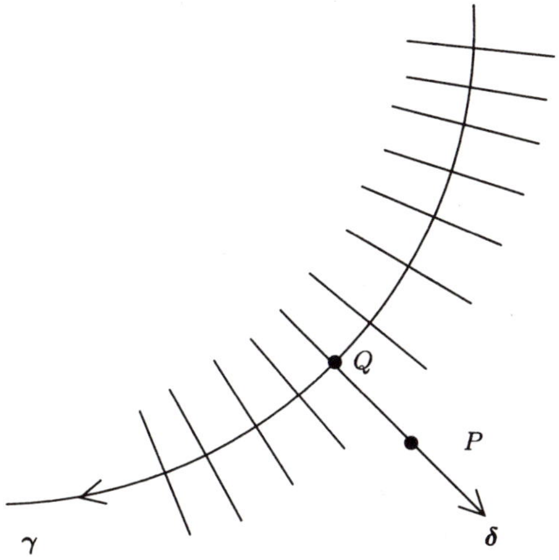

We have, with d/du denoted by a dot,

$$\boldsymbol{\sigma}_u = \dot{\boldsymbol{\gamma}} + v\dot{\boldsymbol{\delta}}, \quad \boldsymbol{\sigma}_v = \boldsymbol{\delta}.$$

Thus, $\boldsymbol{\sigma}$ is regular if $\dot{\boldsymbol{\gamma}} + v\dot{\boldsymbol{\delta}}$ and $\boldsymbol{\delta}$ are linearly independent. This will be true, for example, if $\dot{\boldsymbol{\gamma}}$ and $\boldsymbol{\delta}$ are linearly independent and v is sufficiently small. Thus, to get a surface, the curve $\mathcal{C}$ must never be tangent to the rulings.

Example 4.13

A *surface of revolution* is the surface obtained by rotating a plane curve, called the *profile curve*, around a straight line in the plane. The circles obtained by rotating a fixed point on the profile curve around the axis of rotation are called the *parallels* of the surface, and the curves on the surface obtained by rotating the profile curve through a fixed angle are called its *meridians*. (This agrees with the use of these terms in geography, if we think of the earth as the surface obtained by rotating a great circle passing through the poles about the polar axis and we take u and v to be latitude and longitude, respectively.)

Let us take the axis of rotation to be the z-axis and the plane to be the xz-plane. Any point P of the surface is obtained by rotating some point Q of the profile curve through an angle v (say) around the z-axis. If

$$\gamma(u) = (f(u), 0, g(u))$$

is a parametrisation of the profile curve containing Q, P has position vector of the form

$$\boldsymbol{\sigma}(u, v) = (f(u)\cos v, f(u)\sin v, g(u)).$$

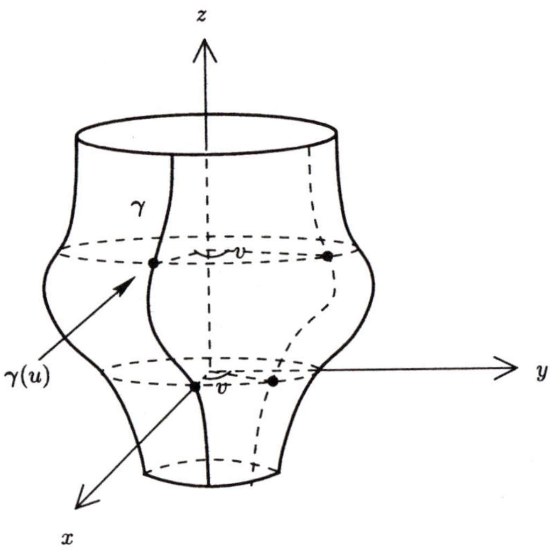

To check regularity, we compute (with a dot denoting d/du):

$$\boldsymbol{\sigma}_u = (\dot{f}\cos v, \dot{f}\sin v, \dot{g}), \quad \boldsymbol{\sigma}_v = (-f\sin v, f\cos v, 0),$$
$$\therefore \quad \boldsymbol{\sigma}_u \times \boldsymbol{\sigma}_v = (f\dot{g}\cos v, -f\dot{g}\sin v, f\dot{f}),$$
$$\therefore \quad \| \boldsymbol{\sigma}_u \times \boldsymbol{\sigma}_v \|^2 = f^2(\dot{f}^2 + \dot{g}^2).$$

Thus, $\boldsymbol{\sigma}_u \times \boldsymbol{\sigma}_v$ will be non-vanishing if $f(u)$ is never zero, i.e. if $\boldsymbol{\gamma}$ does not intersect the z-axis, and if $\dot{f}$ and $\dot{g}$ are never zero simultaneously, i.e. if $\boldsymbol{\gamma}$ is regular. In this case, we might as well assume that $f(u) > 0$, so that $f(u)$ is the distance of $\boldsymbol{\sigma}(u, v)$ from the axis of rotation. Then, $\boldsymbol{\sigma}$ is injective provided that $\boldsymbol{\gamma}$ does not self-intersect and the angle of rotation v is restricted to lie in an open interval of length $\leq 2\pi$. Under these conditions, surface patches of the form $\boldsymbol{\sigma}$ give the surface of revolution the structure of a smooth surface.

EXERCISES

4.18 The surface obtained by rotating the curve $x = \cosh z$ in the xz-plane around the z-axis is called a *catenoid*. Describe an atlas for this surface. (A picture of the catenoid can be found in Section 9.2.)

4.19 Show that

$$\boldsymbol{\sigma}(u,v) = (\operatorname{sech} u \cos v, \operatorname{sech} u \sin v, \tanh u)$$

is a regular surface patch for the unit sphere (it is called *Mercator's projection*). Show that meridians and parallels on the sphere correspond under $\boldsymbol{\sigma}$ to perpendicular straight lines in the plane.

4.20 A *loxodrome* is a curve on the unit sphere that intersects the meridians at a fixed angle, say α. Show that, in the Mercator surface patch $\boldsymbol{\sigma}$ (see Exercise 4.19), a unit-speed loxodrome satisfies

$$\dot{u} = \cos\alpha \cosh u, \quad \dot{v} = \pm \sin\alpha \cosh u$$

(a dot denoting differentiation with respect to the parameter of the loxodrome). Deduce that loxodromes correspond under $\boldsymbol{\sigma}$ to straight lines in the uv-plane.

4.21 A *right conoid* is a ruled surface whose rulings are parallel to a given plane Π and pass through a given straight line $\mathcal{L}$ perpendicular to Π. If Π is the xy-plane and $\mathcal{L}$ is the z-axis, show that

$$\boldsymbol{\sigma}(u,v) = (v\cos\theta(u), v\sin\theta(u), u)$$

is a regular surface patch for the conoid, where $\theta(u)$ is the angle that the ruling through $(0,0,u)$ makes with the z-axis ($\theta(u)$ is assumed to be a smooth function of u). Taking $\theta(u) = u$ gives a helicoid (Exercise 4.14).

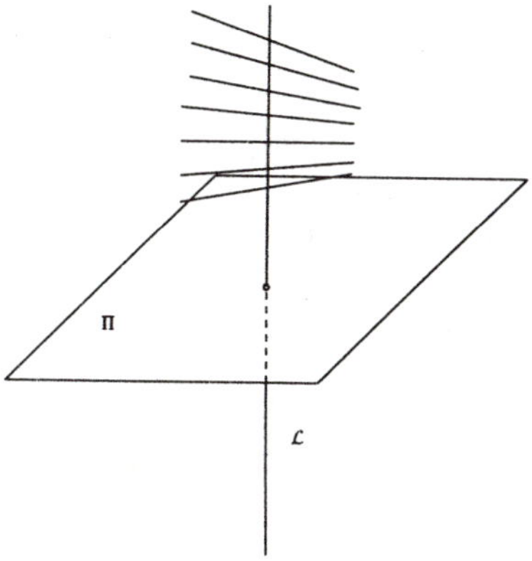

4.22 Show that, if $\boldsymbol{\sigma}(u,v)$ is the (generalised) cylinder in Example 4.10:

(i) the curve $\tilde{\boldsymbol{\gamma}}(u) = \boldsymbol{\gamma}(u) - (\boldsymbol{\gamma}(u).\mathbf{a})\mathbf{a}$ is contained in a plane perpendicular to $\mathbf{a}$;

(ii) $\boldsymbol{\sigma}(u,v) = \tilde{\boldsymbol{\gamma}}(u) + \tilde{v}\mathbf{a}$, where $\tilde{v} = v + \boldsymbol{\gamma}(u).\mathbf{a}$;

(iii) $\tilde{\boldsymbol{\sigma}}(u,\tilde{v}) = \tilde{\boldsymbol{\gamma}}(u) + \tilde{v}\mathbf{a}$ is a reparametrisation of $\boldsymbol{\sigma}(u,v)$.

4.5. Quadric Surfaces

The simplest surfaces, namely planes, have cartesian equations that are linear in x, y and z. From this point of view, the next simplest surfaces should be those whose cartesian equations are given by quadratic expressions in x, y and z. This leads to the following definition.

Definition 4.6

A *quadric* is the subset of $\mathbf{R}^3$ defined by an equation of the form

$$(\mathbf{r}A).\mathbf{r} + \mathbf{b}.\mathbf{r} + c = 0,$$

where $\mathbf{r} = (x, y, z)$, A is a constant symmetric 3×3 matrix, $\mathbf{b} \in \mathbf{R}^3$ is a constant vector, and c is a constant scalar.

To see this more explicitly, let

$$A = \begin{pmatrix} a_1 & a_4 & a_6 \\ a_4 & a_2 & a_5 \\ a_6 & a_5 & a_3 \end{pmatrix}, \quad \mathbf{b} = (b_1, b_2, b_3).$$

Then, the equation of the quadric is

$$a_1x^2 + a_2y^2 + a_3z^2 + 2a_4xy + 2a_5yz + 2a_6xz + b_1x + b_2y + b_3z + c = 0. \quad (4)$$

A quadric is not necessarily a surface. For example, the quadric with equation $x^2 + y^2 + z^2 = 0$ is a single point, and that with equation $x^2 + y^2 = 0$ is a straight line. A more interesting example is the quadric $xy = 0$, which is the union of two intersecting planes, which is also not a surface. (Intuitively, it has a 'corner' along the line of intersection of the planes.) The following theorem shows that it is sufficient to consider quadrics whose equations take on a particularly simple form.

Proposition 4.6

By applying a rigid motion of $\mathbf{R}^3$, every non-empty quadric (4) in which the coefficients are not all zero can be transformed into one whose cartesian equation *is one of the following:*

(i) *ellipsoid:* $\frac{x^2}{p^2} + \frac{y^2}{q^2} + \frac{z^2}{r^2} = 1$

(ii) *hyperboloid of one sheet:* $\frac{x^2}{p^2} + \frac{y^2}{q^2} - \frac{z^2}{r^2} = 1$

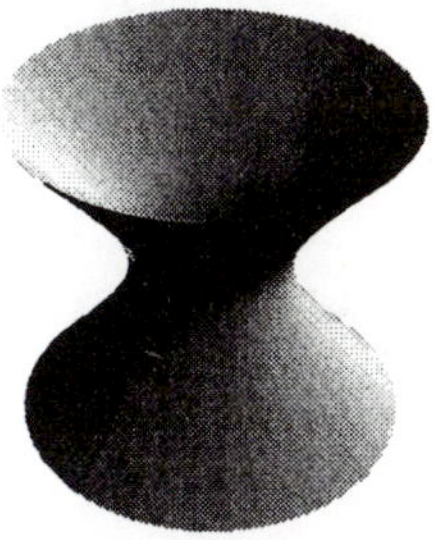

(iii) *hyperboloid of two sheets:* $\frac{z^2}{r^2} - \frac{x^2}{p^2} - \frac{y^2}{q^2} = 1$

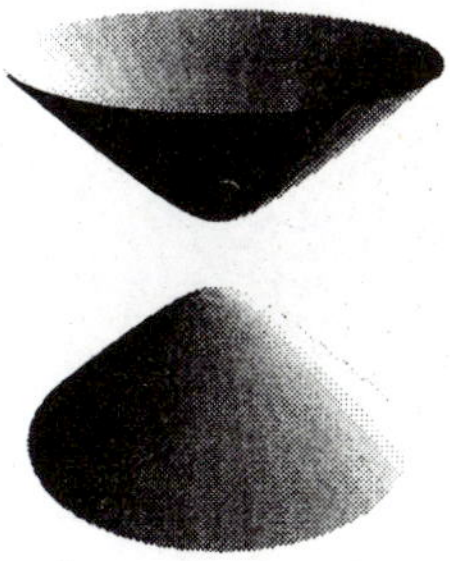

(iv) *elliptic paraboloid:* $\frac{x^2}{p^2} + \frac{y^2}{q^2} = z$

(v) *hyperbolic paraboloid:* $\frac{x^2}{p^2} - \frac{y^2}{q^2} = z$

(vi) *quadric cone:* $\frac{x^2}{p^2} + \frac{y^2}{q^2} - \frac{z^2}{r^2} = 0$

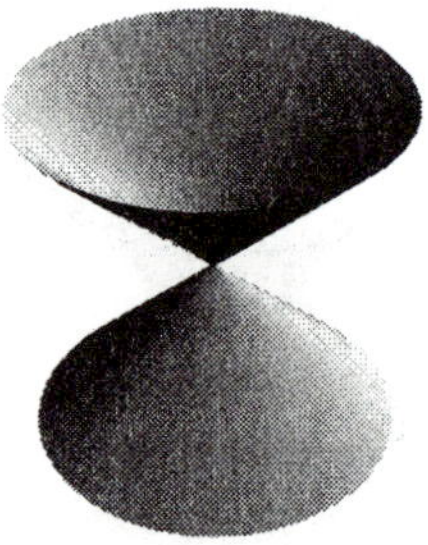

(vii) *elliptic cylinder:* $\frac{x^2}{p^2} + \frac{y^2}{q^2} = 1$

(viii) *hyperbolic cylinder:* $\frac{x^2}{p^2} - \frac{y^2}{q^2} = 1$

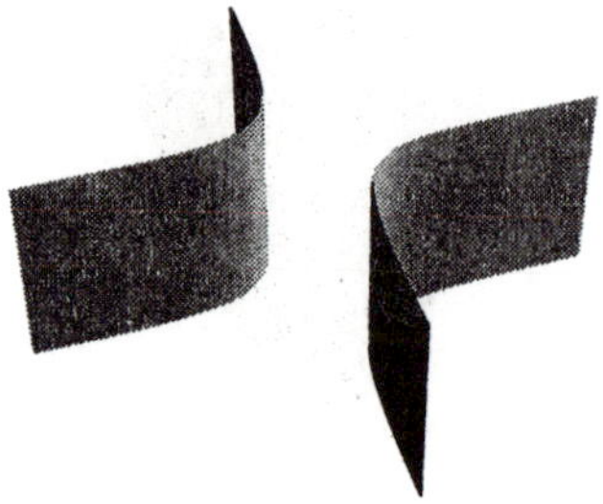

(ix) *parabolic cylinder:* $\frac{x^2}{p^2} = y$

(x) *plane:* $x = 0$
(xi) *two parallel planes:* $x^2 = p^2$
(xii) *two intersecting planes:* $\frac{x^2}{p^2} - \frac{y^2}{q^2} = 0$
(xiii) *straight line:* $\frac{x^2}{p^2} + \frac{y^2}{q^2} = 0$
(xiv) *single point:* $\frac{x^2}{p^2} + \frac{y^2}{q^2} + \frac{z^2}{r^2} = 0.$

In each case, p, q and r are non-zero constants.

Proof 4.6

The proof depends on the following fact. If A is a real symmetric matrix, there is a matrix P with $P^tP = I$ and $\det(P) = 1$ such that P^tAP is a *diagonal* matrix A' (P^t denotes the transpose of P). The diagonal entries of A' are the eigenvalues of A, and the rows of P are the corresponding eigenvectors.

With A as in Definition 4.6, we define $\mathbf{r}' = (x', y', z')$, $\mathbf{b}' = (b_1', b_2', b_3')$, where

$$(x'\,y'\,z') = (x\,y\,z)P, \qquad (b_1'\,b_2'\,b_3') = (b_1\,b_2\,b_3)P.$$

Writing the equation of the quadric as

$$(x\,y\,z)A(x\,y\,z)^t + (b_1\,b_2\,b_3)(x\,y\,z)^t + c = 0$$

and noting that

$$(x\,y\,z) = (x'\,y'\,z')P^t, \quad (b_1\,b_2\,b_3) = (b_1'\,b_2'\,b_3')P^t,$$

we get

$$(x'\,y'\,z')A'(x'\,y'\,z')^t + (b_1'\,b_2'\,b_3')(x'\,y'\,z')^t + c = 0,$$

$$\therefore \quad (\mathbf{r}'A').\mathbf{r}' + \mathbf{b}'.\mathbf{r}' + c = 0,$$

$$\therefore \quad a_1'x'^2 + a_2'y'^2 + a_3'z'^2 + b_1'x' + b_2'y' + b_3'z' + c = 0,$$

where a_1', a_2' and a_3' are the diagonal entries of A', i.e. the eigenvalues of A. Since any 3×3 matrix P with $P^tP = I$ and $\det(P) = 1$ represents a rotation of $\mathbf{R}^3$, this new quadric is obtained from the given one by a rigid motion. Hence, we might as well consider the quadric in (4), but assume that $a_4 = a_5 = a_6 = 0$, i.e.

$$a_1x^2 + a_2y^2 + a_3z^2 + b_1x + b_2y + b_3z + c = 0. \tag{5}$$

Suppose now that, in Eq. (5), $a_1 \neq 0$. If we define $x' = x + b_1/2a_1$, corresponding to a translation of $\mathbf{R}^3$, the equation becomes

$$a_1x'^2 + a_2y^2 + a_3z^2 + b_2y + b_3z + c' = 0,$$

where c' is a constant. In other words, if $a_1 \neq 0$, we can assume that $b_1 = 0$, and similarly for a_2 and a_3, of course.

If a_1, a_2 and a_3 in Eq. (5) are all non-zero, we may therefore reduce to the form

$$a_1x^2 + a_2y^2 + a_3z^2 + c = 0.$$

If $c \neq 0$, we get cases (i), (ii) and (iii), depending on the signs of a_1, a_2, a_3 and c, and if $c = 0$ we get cases (vi) and (xiv).

If exactly one of a_1, a_2 and a_3 is zero, say $a_3 = 0$, we are reduced to the form

$$a_1x^2 + a_2y^2 + b_3z + c = 0. \tag{6}$$

If $b_3 \neq 0$, we may define $z' = z + c/b_3$. Thus, by a translation (and by dividing by b_3), we are reduced to the case

$$a_1x^2 + a_2y^2 + z = 0.$$

This gives cases (iv) and (v).

If $b_3 = 0$ in Eq. (6), we have

$$a_1x^2 + a_2y^2 + c = 0.$$

If $c = 0$ we get cases (xii) and (xiii). If $c \neq 0$, dividing through by it leads to cases (vii) and (viii).

Suppose now that $a_2 = a_3 = 0$, but $a_1 \neq 0$. Then we have

$$a_1x^2 + b_2y + b_3z + c = 0. \tag{7}$$

If b_2 and b_3 are not both zero, by rotating the yz-plane so that the y-axis becomes parallel to the vector (b_2, b_3), we can arrive at the situation $b_2 \neq 0, b_3 = 0$, and then by a translation along the y-axis we can arrange that $c = 0$. This leads to the equation

$$a_1x^2 + y = 0,$$

which gives case (ix). If $b_2 = b_3 = 0$ in Eq. (7), then $c = 0$ gives case (x) and $c \neq 0$ gives case (xi).

Finally, if $a_1 = a_2 = a_3 = 0$, (5) is the equation of a plane, so after applying a rigid motion we are in case (x) again. □

Example 4.14

Consider the quadric

$$x^2 + 2y^2 + 6x - 4y + 3z = 7.$$

Setting $x' = x + 3, y' = y - 1$ (a translation), we get

$$x'^2 + 2y'^2 + 3z = 18.$$

Setting $z' = z - 6$ (another translation) gives

$$x'^2 + 2y'^2 + 3z' = 0.$$

Finally, setting $x'' = x', y'' = -y', z'' = -z'$ (a rotation by π about the x-axis) gives

$$\frac{1}{3}x''^2 + \frac{2}{3}y''^2 = z'',$$

which is an elliptic paraboloid. It can be parametrised by setting $x'' = u, y'' = v, z'' = \frac{1}{3}u^2 + \frac{2}{3}v^2$. This corresponds to $x = u - 3, y = 1 - v, z = 6 - \frac{1}{3}u^2 - \frac{2}{3}v^2$, and shows that the given quadric is a smooth surface with an atlas consisting of the single surface patch

$$\boldsymbol{\sigma}(u, v) = \left(u - 3, 1 - v, 6 - \frac{1}{3}u^2 - \frac{2}{3}v^2\right).$$

EXERCISES

4.23 Write down parametrisations of each of the quadrics in parts (i)–(xi) of Proposition 4.6 (in case (vi) one must remove the origin).

4.24 Which quadric surfaces are
  (a) generalised cylinders;
  (b) generalised cones;
  (c) ruled surfaces;
  (d) surfaces of revolution?

4.25 By setting

$$u = \frac{x}{p} - \frac{y}{q}, \qquad v = \frac{x}{p} + \frac{y}{q},$$

find a surface patch covering the hyperbolic paraboloid

$$\frac{x^2}{p^2} - \frac{y^2}{q^2} = z.$$

Deduce that the hyperbolic paraboloid is doubly ruled.

4.26 Show that, if a quadric contains three points on a straight line, it contains the whole line. (Parametrise the line by $\boldsymbol{\gamma}(t) = \mathbf{a} + \mathbf{b}t$, and note that substituting into Eq. (4) gives a quadratic equation for t.) Deduce that, if L_1, L_2 and L_3 are non-intersecting straight lines in $\mathbf{R}^3$, there is a quadric containing all three lines. (Take three points on each line and show that there is a quadric passing through all nine points.)

4.27 Use Exercise 4.26 to show that any doubly ruled surface is (part of) a quadric surface. (A surface is doubly ruled if it is the union of each of two families of straight lines such that no two lines of the same family intersect, but every line of the first family intersects every line of the second family, with at most a finite number of exceptions.) Which quadric surfaces are doubly ruled?

4.6. Triply Orthogonal Systems

We conclude this section by showing how quadric surfaces furnish some beautiful examples of *triply orthogonal systems*. By such a system, we mean three families of surfaces, each depending on a single parameter, with the property that, if P is a point that is on one surface of each family, the tangent planes of these surfaces at P are mutually perpendicular. The simplest example, of course, consists of the planes parallel to one of the three coordinate planes. Other examples, almost as obvious, are given in Exercise 4.28. But a more interesting example can be constructed in which the three families consist of ellipsoids, hyperboloids of one sheet and hyperboloids of two sheets, respectively.

To see this, let p, q and r be constants, and assume that $0 < p^2 < q^2 < r^2$. For $(x, y, z) \in \mathbf{R}^3$, $t \neq p^2, q^2$ or r^2, let

$$F_t(x, y, z) = \frac{x^2}{p^2 - t} + \frac{y^2}{q^2 - t} + \frac{z^2}{r^2 - t}.$$

Fix a point $(a, b, c) \in \mathbf{R}^3$ with a, b and c all non-zero. The following properties are clear:

(i) $F_t(a, b, c)$ is a continuous function of t in each of the open intervals $(-\infty, p^2)$, (p^2, q^2), (q^2, r^2) and (r^2, ∞);

(ii) $F_t(a, b, c) \to 0$ as $t \to \pm\infty$;

(iii) $F_t(a, b, c) \to \infty$ as t approaches p^2, q^2 or r^2 from the left, and $F_t(a, b, c) \to -\infty$ as t approaches p^2, q^2 or r^2 from the right.

It follows from these properties and the Intermediate Value Theorem that there is at least one value of t in each open interval $(-\infty, p^2)$, (p^2, q^2) and (q^2, r^2) such that $F_t(a,b,c) = 1$. On the other hand, the equation $F_t(a,b,c) = 1$ is equivalent to the cubic equation $G_t(a,b,c) = 0$, where

$$\begin{aligned} G_t(a,b,c) = a^2(q^2-t)(r^2-t) + b^2(p^2-t)(r^2-t) + c^2(p^2-t)(q^2-t) \\ - (p^2-t)(q^2-t)(r^2-t), \end{aligned} \tag{8}$$

and so has at most three real roots. It follows that there are unique numbers $u \in (-\infty, p^2)$, $v \in (p^2, q^2)$ and $w \in (q^2, r^2)$ (depending on (a,b,c), of course) such that

$$F_u(a,b,c) = 1, \quad F_v(a,b,c) = 1, \quad F_w(a,b,c) = 1. \tag{9}$$

The three quadrics $F_u(x,y,z) = 1$, $F_v(x,y,z) = 1$ and $F_w(x,y,z) = 1$ are ellipsoids, hyperboloids of one sheet and hyperboloids of two sheets, respectively, and we have shown that there is one of each passing through each point $(a,b,c) \in \mathbf{R}^3$ that does not lie on any of the coordinate planes. We show that they form a triply orthogonal system.

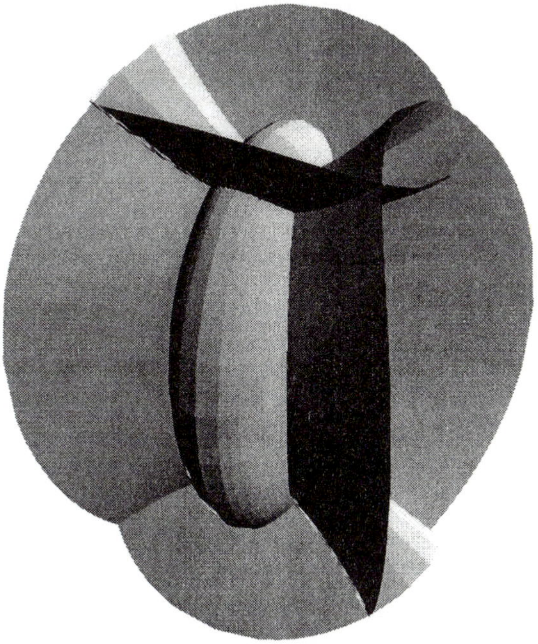

Indeed, the vector

$$\left(\frac{x}{p^2-t}, \frac{y}{q^2-t}, \frac{z}{r^2-t}\right)$$

is perpendicular to the tangent plane of the surface $F_t(x,y,z) = 1$ at (x,y,z) (see Exercise 4.16). Thus, to show that the first two surfaces in (9) are perpendicular at (a,b,c), for example, we have to show that

$$\frac{a^2}{(p^2-u)(p^2-v)} + \frac{b^2}{(q^2-u)(q^2-v)} + \frac{c^2}{(r^2-u)(r^2-v)} = 0.$$

But the left-hand side of this equation is

$$\frac{F_u(a,b,c) - F_v(a,b,c)}{u-v} = \frac{1-1}{u-v} = 0.$$

We can also construct a simultaneous parametrisation of the three families. Note that the cubic $G_t(a,b,c)$ is equal to $(t-u)(t-v)(t-w)$, since it is divisible by this product and the coefficients of t^3 agree. Putting $t = p^2, q^2$ and r^2 and solving the resulting equations for a^2, b^2 and c^2, we find that

$$\begin{aligned} a &= \pm\sqrt{\frac{(p^2-u)(p^2-v)(p^2-w)}{(r^2-p^2)(q^2-p^2)}}, \\ b &= \pm\sqrt{\frac{(q^2-u)(q^2-v)(q^2-w)}{(p^2-q^2)(r^2-q^2)}}, \\ c &= \pm\sqrt{\frac{(r^2-u)(r^2-v)(r^2-w)}{(p^2-r^2)(q^2-r^2)}}. \end{aligned} \tag{10}$$

Define $\boldsymbol{\sigma}(u,v,w) = (x,y,z)$, where x, y and z are the right-hand sides of the three equations in (10), respectively, with any combination of signs. For fixed u (resp. fixed v, fixed w), this gives eight surface patches for the corresponding ellipsoid $F_u(x,y,z) = 1$ (resp. hyperboloid of one sheet $F_v(x,y,z) = 1$, hyperboloid of two sheets $F_w(x,y,z) = 1$).

EXERCISES

4.28 Show that the following are triply orthogonal systems:

(i) the spheres with centre the origin, the planes containing the z-axis, and the circular cones with axis the z-axis;

(ii) the planes parallel to the xy-plane, the planes containing the z-axis and the circular cylinders with axis the z-axis.

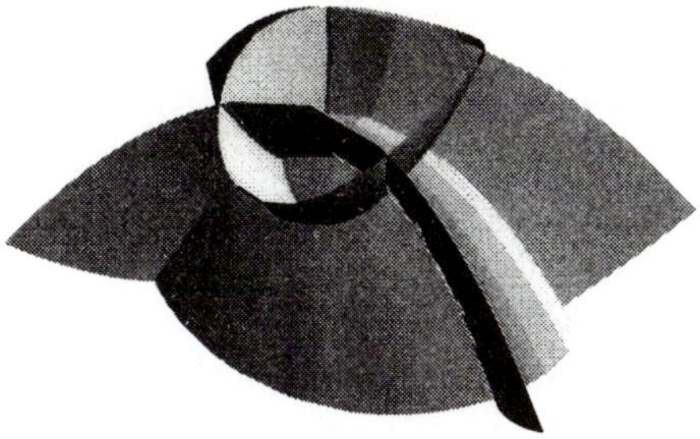

4.29 By considering the quadric surface $F_t(x,y,z) = 0$, where

$$F_t(x,y,z) = \frac{x^2}{p^2 - t} + \frac{y^2}{q^2 - t} - 2z + t,$$

construct a triply orthogonal system consisting of two families of elliptic paraboloids and one family of hyperbolic paraboloids (see above). Find a parametrisation of these surfaces analogous to (10).

4.7. Applications of the Inverse Function Theorem

In this section we give the proofs of Proposition 4.1 and Theorem 4.1.

Suppose first that $f : U \to \mathbf{R}^n$ is a smooth map, where U is an open subset of $\mathbf{R}^m$. If we write $(\tilde{u}_1, \ldots, \tilde{u}_n) = f(u_1, \ldots, u_m)$, the jacobian matrix of f is

$$J(f) = \begin{pmatrix} \frac{\partial \tilde{u}_1}{\partial u_1} & \frac{\partial \tilde{u}_1}{\partial u_2} & \cdots & \frac{\partial \tilde{u}_1}{\partial u_m} \\ \frac{\partial \tilde{u}_2}{\partial u_1} & \frac{\partial \tilde{u}_2}{\partial u_2} & \cdots & \frac{\partial \tilde{u}_2}{\partial u_m} \\ \vdots & \vdots & \ddots & \vdots \\ \frac{\partial \tilde{u}_n}{\partial u_1} & \frac{\partial \tilde{u}_n}{\partial u_2} & \cdots & \frac{\partial \tilde{u}_n}{\partial u_m} \end{pmatrix}.$$

This has already been used in the case $m = n = 2$ in Section 4.2, but now we shall need it in other cases too.

The main tool that we use is

Theorem 4.2 (Inverse Function Theorem)

Let $f : U \to \mathbf{R}^n$ be a smooth map defined on an open subset U of $\mathbf{R}^n$ ($n \geq 1$). Assume that, at some point $x_0 \in U$, the jacobian matrix $J(f)$ is invertible.

Then, there is an open subset V of $\mathbf{R}^n$ and a smooth map $g : V \to \mathbf{R}^n$ such that
(i) $y_0 = f(x_0) \in V$;
(ii) $g(y_0) = x_0$;
(iii) $g(V) \subseteq U$;
(iv) $g(V)$ *is an open subset of* $\mathbf{R}^n$;
(v) $f(g(y)) = y$ *for all* $y \in V$.
In particular, $g : V \to g(V)$ *and* $f : g(V) \to V$ *are inverse bijections.*

Thus, the inverse function theorem says that, if $J(f)$ is invertible at some point, then f is bijective near that point and its inverse map is smooth. A proof of this theorem can be found in books on multivariable calculus.

We use the inverse function theorem to give the proof of Proposition 4.1. We want to show that, if $\boldsymbol{\sigma} : U \to \mathbf{R}^3$ and $\tilde{\boldsymbol{\sigma}} : \tilde{U} \to \mathbf{R}^3$ are two regular patches in the atlas of a surface $\mathcal{S}$, the transition map from $\boldsymbol{\sigma}$ to $\tilde{\boldsymbol{\sigma}}$ is smooth where it is defined.

Suppose that a point P lies in both patches, say $\boldsymbol{\sigma}(u_0, v_0) = \tilde{\boldsymbol{\sigma}}(\tilde{u}_0, \tilde{v}_0) = P$. Write

$$\boldsymbol{\sigma}(u, v) = (f(u, v), g(u, v), h(u, v)).$$

Since $\boldsymbol{\sigma}_u$ and $\boldsymbol{\sigma}_v$ are linearly independent, the jacobian matrix

$$\begin{pmatrix} f_u & f_v \\ g_u & g_v \\ h_u & h_v \end{pmatrix}$$

of $\boldsymbol{\sigma}$ has rank 2 everywhere. Hence, at least one of its three 2×2 submatrices is invertible at each point. Suppose that the submatrix

$$\begin{pmatrix} f_u & f_v \\ g_u & g_v \end{pmatrix}$$

is invertible at P. (The proof is similar in the other two cases.) By the inverse function theorem applied to the map $F : U \to \mathbf{R}^2$ given by

$$F(u, v) = (f(u, v), g(u, v)),$$

there is an open subset V of $\mathbf{R}^2$ containing $F(u_0, v_0)$ and an open subset W of U containing (u_0, v_0) such that $F : W \to V$ is bijective with a smooth inverse $F^{-1} : V \to W$. Since $\boldsymbol{\sigma} : W \to \boldsymbol{\sigma}(W)$ is bijective, the projection $\pi : \boldsymbol{\sigma}(W) \to V$ given by $\pi(x, y, z) = (x, y)$ is also bijective, since $\pi = F \circ \boldsymbol{\sigma}^{-1}$ on $\boldsymbol{\sigma}(W)$. It follows that $\tilde{W} = \tilde{\boldsymbol{\sigma}}^{-1}(\boldsymbol{\sigma}(W))$ is an open subset of $\tilde{U}$ and that

$$\boldsymbol{\sigma}^{-1} \circ \tilde{\boldsymbol{\sigma}} = F^{-1} \circ \tilde{F}$$

on $\tilde{W}$, where $\tilde{F} = \pi \circ \tilde{\boldsymbol{\sigma}}$. Since F^{-1} and $\tilde{F}$ are smooth on $\tilde{W}$, so is the transition map $\boldsymbol{\sigma}^{-1} \circ \tilde{\boldsymbol{\sigma}}$. Since $\boldsymbol{\sigma}^{-1} \circ \tilde{\boldsymbol{\sigma}}$ is smooth on an open set containing any point (u_0, v_0) where it is defined, it is smooth. □

We now give the proof of Theorem 4.1. Let P, W and f be as in the statement of the theorem, and suppose that $P = (x_0, y_0, z_0)$ and that $f_z \neq 0$ at P. (The proof is similar in the other two cases.) Consider the map $F : W \to \mathbf{R}^3$ defined by

$$F(x, y, z) = (x, y, f(x, y, z)).$$

The jacobian matrix of F is

$$\begin{pmatrix} 1 & 0 & 0 \\ 0 & 1 & 0 \\ f_x & f_y & f_z \end{pmatrix},$$

and is clearly invertible at P since $f_z \neq 0$. By the inverse function theorem, there is an open subset V of $\mathbf{R}^3$ containing $F(x_0, y_0, z_0) = (x_0, y_0, 0)$ and a smooth map $G : V \to W$ such that $\tilde{W} = G(V)$ is open and $F : \tilde{W} \to V$ and $G : V \to \tilde{W}$ are inverse bijections.

Since V is open, there are open subsets U_1 of $\mathbf{R}^2$ containing (x_0, y_0) and U_2 of $\mathbf{R}$ containing 0 such that V contains the open set $U_1 \times U_2$ of all points (x, y, w) with $(x, y) \in U_1$ and $w \in U_2$. Hence, we might as well assume that $V = U_1 \times U_2$. The fact that F and G are inverse bijections means that

$$G(x, y, w) = (x, y, g(x, y, w))$$

for some smooth map $g : U_1 \times U_2 \to \mathbf{R}$, and

$$f(x, y, g(x, y, w)) = w$$

for all $(x, y) \in U_1$, $w \in U_2$.

Define $\boldsymbol{\sigma} : U_1 \to \mathbf{R}^3$ by

$$\boldsymbol{\sigma}(x, y) = (x, y, g(x, y, 0)).$$

Then $\boldsymbol{\sigma}$ is a homeomorphism from U_1 to $\mathcal{S} \cap \tilde{W}$ (whose inverse is the restriction to $\mathcal{S} \cap \tilde{W}$ of the projection $\pi(x, y, z) = (x, y)$). It is obvious that $\boldsymbol{\sigma}$ is smooth, and it is regular because

$$\boldsymbol{\sigma}_x \times \boldsymbol{\sigma}_y = (-g_x, -g_y, 1)$$

is nowhere zero. So $\boldsymbol{\sigma}$ is a regular surface patch on $\mathcal{S}$ containing the given point P. Since P was an arbitrary point of $\mathcal{S}$, we have constructed an atlas for $\mathcal{S}$ making it into a (smooth) surface. □

EXERCISES

4.30 Show that, if $\boldsymbol{\gamma} : (\alpha, \beta) \to \mathbf{R}^3$ is a curve whose image is contained in a surface patch $\boldsymbol{\sigma} : U \to \mathbf{R}^3$, then $\boldsymbol{\gamma}(t) = \boldsymbol{\sigma}(u(t), v(t))$ for some smooth map $(\alpha, \beta) \to U$, $t \mapsto (u(t), v(t))$. (Imitate the proof of Proposition 4.1.)

4.31 Prove Theorem 1.1 and its generalization to level curves in $\mathbf{R}^3$ (Exercise 1.17).

5
The First Fundamental Form

Perhaps the first thing that a geometrically inclined bug living on a surface might wish to do is to measure the distance between two points of the surface. Of course, this will usually be different from the distance between these points as measured by an inhabitant of the ambient three dimensional space, since the straight line segment which furnishes the shortest path between the points in $\mathbf{R}^3$ will generally not be contained in the surface. The object which allows one to compute lengths on a surface, and also angles and areas, is the first fundamental form of the surface.

5.1. Lengths of Curves on Surfaces

If $\boldsymbol{\gamma}(t) = \boldsymbol{\sigma}(u(t), v(t))$ is a curve in a surface patch $\boldsymbol{\sigma}$, its arc-length starting at a point $\boldsymbol{\gamma}(t_0)$ is given by

$$s = \int_{t_0}^{t} \parallel \dot{\boldsymbol{\gamma}}(u) \parallel du.$$

By the chain rule, $\dot{\boldsymbol{\gamma}} = \boldsymbol{\sigma}_u \dot{u} + \boldsymbol{\sigma}_v \dot{v}$, so

$$\begin{aligned} \parallel \dot{\boldsymbol{\gamma}} \parallel^2 &= (\boldsymbol{\sigma}_u \dot{u} + \boldsymbol{\sigma}_v \dot{v}).(\boldsymbol{\sigma}_u \dot{u} + \boldsymbol{\sigma}_v \dot{v}) \\ &= (\boldsymbol{\sigma}_u.\boldsymbol{\sigma}_u)\dot{u}^2 + (\boldsymbol{\sigma}_u.\boldsymbol{\sigma}_v)\dot{u}\dot{v} + (\boldsymbol{\sigma}_v.\boldsymbol{\sigma}_u)\dot{v}\dot{u} + (\boldsymbol{\sigma}_v.\boldsymbol{\sigma}_v)\dot{v}^2 \\ &= (\boldsymbol{\sigma}_u.\boldsymbol{\sigma}_u)\dot{u}^2 + 2(\boldsymbol{\sigma}_u.\boldsymbol{\sigma}_v)\dot{u}\dot{v} + (\boldsymbol{\sigma}_v.\boldsymbol{\sigma}_v)\dot{v}^2 \\ &= E\dot{u}^2 + 2F\dot{u}\dot{v} + G\dot{v}^2, \end{aligned}$$

where

$$E = \| \boldsymbol{\sigma}_u \|^2, \quad F = \boldsymbol{\sigma}_u.\boldsymbol{\sigma}_v, \quad G = \| \boldsymbol{\sigma}_v \|^2 .$$

So

$$s = \int_{t_0}^{t} (E\dot{u}^2 + 2F\dot{u}\dot{v} + G\dot{v}^2)^{1/2} dt. \tag{1}$$

If we bring the dt inside the square root and write $\left(\frac{du}{dt}\right)^2 (dt)^2 = du^2$, etc. (!), we see that s is the integral of the square root of the expression

$$Edu^2 + 2Fdudv + Gdv^2. \tag{2}$$

This is called the *first fundamental form* of $\boldsymbol{\sigma}$. Since

$$s = \int \sqrt{ds^2}, \quad (!)$$

Eq. (1) is sometimes written as

$$ds^2 = Edu^2 + 2Fdudv + Gdv^2.$$

We shall not attempt to justify these apparently dubious manipulations. We simply regard the expression in (2) as a way of keeping track of the functions E, F and G from which the length of any curve in the surface patch can be computed using Eq. (1).

If now $\boldsymbol{\gamma}$ is a curve on an arbitrary surface $\mathcal{S}$, its length can be computed by breaking $\boldsymbol{\gamma}$ into segments, each of which lies in a surface patch, and using Eq. (1) to compute the length of each segment. The first fundamental form will change when the surface patch is changed, in the manner described in Exercise 5.4.

Example 5.1

For the plane

$$\boldsymbol{\sigma}(u, v) = \mathbf{a} + u\mathbf{p} + v\mathbf{q}$$

(see Example 4.1) with $\mathbf{p}$ and $\mathbf{q}$ being perpendicular unit vectors, we have $\boldsymbol{\sigma}_u = \mathbf{p}$, $\boldsymbol{\sigma}_v = \mathbf{q}$, so $E = \| \boldsymbol{\sigma}_u \|^2 = \| \mathbf{p} \|^2 = 1$, $F = \boldsymbol{\sigma}_u.\boldsymbol{\sigma}_v = \mathbf{p}.\mathbf{q} = 0$, $G = \| \boldsymbol{\sigma}_v \|^2 = \| \mathbf{q} \|^2 = 1$, and the first fundamental form is simply

$$du^2 + dv^2.$$

Example 5.2

For the sphere in latitude longitude coordinates

$$\boldsymbol{\sigma}(\theta, \varphi) = (\cos\theta \cos\varphi, \cos\theta \sin\varphi, \sin\theta)$$

(see Example 4.2), we have

$$\boldsymbol{\sigma}_\theta = (-\sin\theta\cos\varphi, -\sin\theta\sin\varphi, \cos\theta), \quad \boldsymbol{\sigma}_\varphi = (-\cos\theta\sin\varphi, \cos\theta\cos\varphi, 0),$$

$$\therefore \quad E = \| \boldsymbol{\sigma}_\theta \|^2 = 1, \quad F = \boldsymbol{\sigma}_\theta . \boldsymbol{\sigma}_\varphi = 0, \quad G = \| \boldsymbol{\sigma}_\varphi \|^2 = \cos^2\theta,$$

and so the first fundamental form is

$$d\theta^2 + \cos^2\theta\, d\varphi^2.$$

Example 5.3

We consider a (generalised) cylinder

$$\boldsymbol{\sigma}(u,v) = \boldsymbol{\gamma}(u) + v\mathbf{a}$$

defined in Example 4.10. As we saw in that example we can assume that $\boldsymbol{\gamma}$ is unit-speed, that $\mathbf{a}$ is a unit vector, and that $\boldsymbol{\gamma}$ is contained in a plane perpendicular to $\mathbf{a}$. Then, denoting d/du by a dot,

$$\boldsymbol{\sigma}_u = \dot{\boldsymbol{\gamma}}, \qquad \boldsymbol{\sigma}_v = \mathbf{a},$$

so $E = \| \boldsymbol{\sigma}_u \|^2 = \| \dot{\boldsymbol{\gamma}} \|^2 = 1$, $F = \boldsymbol{\sigma}_u . \boldsymbol{\sigma}_v = \dot{\boldsymbol{\gamma}}.\mathbf{a} = 0$, $G = \| \boldsymbol{\sigma}_v \|^2 = \| \mathbf{a} \|^2 = 1$, and the first fundamental form of $\boldsymbol{\sigma}$ is

$$du^2 + dv^2.$$

Note that this is the *same* as the first fundamental form of the plane (see Example 5.1). The geometrical reason for this coincidence will be revealed in the next section.

Example 5.4

We consider a (generalised) cone

$$\boldsymbol{\sigma}(u,v) = (1-v)\mathbf{p} + v\boldsymbol{\gamma}(u)$$

(see Example 4.11). Before computing its first fundamental form, we make some simplifications to $\boldsymbol{\sigma}$.

First, translating the surface by $\mathbf{p}$ (which does not change its first fundamental form by Exercise 5.3), we get the surface patch $\boldsymbol{\sigma}_1 = \boldsymbol{\sigma} - \mathbf{p} = v(\boldsymbol{\gamma} - \mathbf{p})$, so if we replace $\boldsymbol{\gamma}$ by $\boldsymbol{\gamma}_1 = \boldsymbol{\gamma} - \mathbf{p}$ we get $\boldsymbol{\sigma}_1 = v\boldsymbol{\gamma}_1$. This means that we might as well assume that $\mathbf{p} = \mathbf{0}$ to begin with. Next, we saw in Example 4.11 that for $\boldsymbol{\sigma}$ to be a regular surface patch, $\boldsymbol{\gamma}$ must not pass through the origin, so we can define a new curve $\tilde{\boldsymbol{\gamma}}$ by $\tilde{\boldsymbol{\gamma}}(u) = \boldsymbol{\gamma}(u)/ \| \boldsymbol{\gamma}(u) \|$. Setting $\tilde{u} = u$, $\tilde{v} = v/ \| \boldsymbol{\gamma}(u) \|$, we get a reparametrisation $\tilde{\boldsymbol{\sigma}}(\tilde{u},\tilde{v}) = \tilde{v}\tilde{\boldsymbol{\gamma}}(\tilde{u})$ of $\boldsymbol{\sigma}$ with $\| \tilde{\boldsymbol{\gamma}} \| = 1$. We can therefore assume to begin with that $\boldsymbol{\sigma}(u,v) = v\boldsymbol{\gamma}(u)$ with $\| \boldsymbol{\gamma}(u) \| = 1$ for all values of u (geometrically, this means that we can replace $\boldsymbol{\gamma}$ by the intersection of the cone with the unit sphere). Finally, reparametrising again, we can assume that $\boldsymbol{\gamma}$ is unit-speed, for we saw in Example 4.11 that for $\boldsymbol{\sigma}$ to be regular, $\boldsymbol{\gamma}$ must be regular.

With these assumptions, and with a dot denoting d/du, we have

$$\boldsymbol{\sigma}_u = v\dot{\boldsymbol{\gamma}}, \quad \boldsymbol{\sigma}_v = \boldsymbol{\gamma},$$

so $E = \| v\dot{\boldsymbol{\gamma}} \|^2 = v^2 \| \dot{\boldsymbol{\gamma}} \|^2 = v^2$, $F = v\dot{\boldsymbol{\gamma}}.\boldsymbol{\gamma} = 0$ (since $\| \boldsymbol{\gamma} \| = 1$), $G = \| \boldsymbol{\gamma} \|^2 = 1$, and the first fundamental form is

$$v^2du^2 + dv^2.$$

Note that, as for the generalised cylinder in Example 5.3, there is no trace of the curve $\boldsymbol{\gamma}$ in the first fundamental form.

EXERCISES

5.1 Calculate the first fundamental forms of the following surfaces:

(i) $\boldsymbol{\sigma}(u, v) = (\sinh u \sinh v, \sinh u \cosh v, \sinh u)$;

(ii) $\boldsymbol{\sigma}(u, v) = (u - v, u + v, u^2 + v^2)$;

(iii) $\boldsymbol{\sigma}(u, v) = (\cosh u, \sinh u, v)$;

(iv) $\boldsymbol{\sigma}(u, v) = (u, v, u^2 + v^2)$.

What kind of surfaces are these?

5.2 Sketch the curve on the cone

$$\boldsymbol{\sigma}(u, v) = (u \cos v, u \sin v, u)$$

given by $u = e^{\lambda t}$, $v = t$, where λ is a constant. Find the length of the part of the curve with $0 \leq t \leq \pi$.

5.3 Show that applying a rigid motion to a surface does not change its first fundamental form.

5.4 Let a surface $\tilde{\boldsymbol{\sigma}}(\tilde{u}, \tilde{v})$ be a reparametrisation of a surface $\boldsymbol{\sigma}(u, v)$, and let

$$\tilde{E}d\tilde{u}^2 + 2\tilde{F}d\tilde{u}d\tilde{v} + \tilde{G}d\tilde{v}^2 \quad \text{and} \quad Edu^2 + 2Fdudv + Gdv^2$$

be their first fundamental forms. Let

$$J = \begin{pmatrix} \frac{\partial u}{\partial \tilde{u}} & \frac{\partial u}{\partial \tilde{v}} \\ \frac{\partial v}{\partial \tilde{u}} & \frac{\partial v}{\partial \tilde{v}} \end{pmatrix}$$

be the jacobian matrix of the reparametrisation map $(\tilde{u}, \tilde{v}) \mapsto (u, v)$, and let J^t be the transpose of J. Prove that

$$\begin{pmatrix} \tilde{E} & \tilde{F} \\ \tilde{F} & \tilde{G} \end{pmatrix} = J^t \begin{pmatrix} E & F \\ F & G \end{pmatrix} J.$$

5.2. Isometries of Surfaces

We observed in Example 5.3 that a plane and a (generalised) cylinder, when suitably parametrised, have the *same* first fundamental form. The geometric reason for this is not hard to see. A plane piece of paper can be 'wrapped' on a cylinder in the obvious way without crumpling the paper (we take the case of a circular cylinder for simplicity):

If we draw a curve on the plane, then after wrapping it becomes a curve on the cylinder. Because there is no crumpling, the lengths of these two curves will be the same. Since the lengths are computed as the integral of the (square root of the) first fundamental form, it is plausible that the first fundamental forms of the two surfaces should be the same. Experiment suggests, on the other hand, that it is impossible to wrap a plane sheet of paper around a sphere without crumpling. Thus, we expect that a plane and a sphere do not have the same first fundamental form (whatever the choice of parametrisation).

The following definition makes precise what it means to wrap one surface onto another without crumpling.

Definition 5.1

If $\mathcal{S}_1$ and $\mathcal{S}_2$ are surfaces, a diffeomorphism $f : \mathcal{S}_1 \to \mathcal{S}_2$ is called an *isometry* if it takes curves in $\mathcal{S}_1$ to curves of the same length in $\mathcal{S}_2$. If an isometry $f : \mathcal{S}_1 \to \mathcal{S}_2$ exists, we say that $\mathcal{S}_1$ and $\mathcal{S}_2$ are *isometric*.

Theorem 5.1

A diffeomorphism $f : \mathcal{S}_1 \to \mathcal{S}_2$ is an isometry if and only if, for any surface *patch* $\boldsymbol{\sigma}_1$ *of* $\mathcal{S}_1$*, the patches* $\boldsymbol{\sigma}_1$ *and* $f \circ \boldsymbol{\sigma}_1$ *of* $\mathcal{S}_1$ *and* $\mathcal{S}_2$*, respectively, have the* same *first fundamental form.*

Proof 5.1

Since the length of any curve can be computed as the sum of the lengths of curves each lying in a single surface patch, we can assume that $\mathcal{S}_1$ and $\mathcal{S}_2$ are covered by single surface patches. Moreover, since f is a diffeomorphism, Proposition 4.3 shows that we can assume that these patches are of the form $\boldsymbol{\sigma}_1 : U \to \mathbf{R}^3$ (for $\mathcal{S}_1$) and $f \circ \boldsymbol{\sigma}_1 = \boldsymbol{\sigma}_2$ (for $\mathcal{S}_2$). We have to show that f is an isometry if and only if $\boldsymbol{\sigma}_1$ and $\boldsymbol{\sigma}_2$ have the same first fundamental form.

Suppose first that $\boldsymbol{\sigma}_1$ and $\boldsymbol{\sigma}_2$ have the same first fundamental form. If $t \mapsto (u(t), v(t))$ is any curve in U, and $\boldsymbol{\gamma}_1(t) = \boldsymbol{\sigma}_1(u(t), v(t))$ and $\boldsymbol{\gamma}_2(t) = \boldsymbol{\sigma}_2(u(t), v(t))$ are the corresponding curves in $\mathcal{S}_1$ and $\mathcal{S}_2$, then f takes $\boldsymbol{\gamma}_1$ to $\boldsymbol{\gamma}_2$, since

$$f(\boldsymbol{\gamma}_1(t)) = f(\boldsymbol{\sigma}_1(u(t), v(t))) = \boldsymbol{\sigma}_2(u(t), v(t)) = \boldsymbol{\gamma}_2(t).$$

It is clear that $\boldsymbol{\gamma}_1$ and $\boldsymbol{\gamma}_2$ have the same length, since both lengths are found by integrating the expression $(E\dot{u}^2 + 2F\dot{u}\dot{v} + G\dot{v}^2)^{1/2}$, where $Edu^2 + 2Fdudv + Gdv^2$ is the (common) first fundamental form of $\boldsymbol{\sigma}_1$ and $\boldsymbol{\sigma}_2$.

Conversely, suppose that f is an isometry. If $t \mapsto (u(t), v(t))$ is any curve in U, defined for $t \in (\alpha, \beta)$, say, the curves $\boldsymbol{\gamma}_1(t) = \boldsymbol{\sigma}_1(u(t), v(t))$ and $\boldsymbol{\gamma}_2(t) = \boldsymbol{\sigma}_2(u(t), v(t))$ have the same length. Hence,

$$\int_{t_0}^{t_1} (E_1\dot{u}^2 + 2F_1\dot{u}\dot{v} + G_1\dot{v}^2)^{1/2} dt = \int_{t_0}^{t_1} (E_2\dot{u}^2 + 2F_2\dot{u}\dot{v} + G_2\dot{v}^2)^{1/2} dt,$$

for all $t_0, t_1 \in (\alpha, \beta)$, where E_1, F_1 and G_1 are the coefficients of the first fundamental form of $\boldsymbol{\sigma}_1$, and E_2, F_2 and G_2 those of $\boldsymbol{\sigma}_2$. This implies that the two integrands are the same, and hence that

$$E_1\dot{u}^2 + 2F_1\dot{u}\dot{v} + G_1\dot{v}^2 = E_2\dot{u}^2 + 2F_2\dot{u}\dot{v} + G_2\dot{v}^2. \tag{3}$$

Fix $t_0 \in (\alpha, \beta)$, and let $u_0 = u(t_0)$, $v_0 = v(t_0)$. We now apply Eq. (3) for the following three choices of the curve $t \mapsto (u(t), v(t))$ in U:

(i) $u = u_0 + t - t_0, v = v_0$: this gives $E_1 = E_2$;

(ii) $u = u_0, v = v_0 + t - t_0$: this gives $G_1 = G_2$;

(iii) $u = u_0 + t - t_0, v = v_0 + t - t_0$: this gives

$$E_1 + 2F_1 + G_1 = E_2 + 2F_2 + G_2,$$

and hence (in view of (i) and (ii)) $F_1 = F_2$. □

Example 5.5

Let $\mathcal{S}_1$ be the infinite strip in the xy-plane given by $0 < x < 2\pi$, and let $\mathcal{S}_2$ be the circular cylinder $x^2 + y^2 = 1$ with the ruling given by $x = 1, y = 0$ removed. Then, $\mathcal{S}_1$ is covered by the single patch $\boldsymbol{\sigma}_1(u, v) = (u, v, 0)$, and $\mathcal{S}_2$ by the patch $\boldsymbol{\sigma}_2(u, v) = (\cos u, \sin u, v)$, with $0 < u < 2\pi$ in both cases. The map $f : \mathcal{S}_1 \to \mathcal{S}_2$ that takes $\boldsymbol{\sigma}_1(u, v)$ to $\boldsymbol{\sigma}_2(u, v)$ is an isometry since, by Example 5.3, $\boldsymbol{\sigma}_1$ and $\boldsymbol{\sigma}_2$ have the same first fundamental form.

A similar argument shows that a suitable subset of a circular cone is isometric to part of a plane (see Exercise 5.5), and this even extends to generalised cylinders and cones (see Exercise 5.7). It turns out that there is another class of surfaces that are isometric to the plane, called *tangent developables*. (In older works, a 'development' of one surface on another was the term used for an isometry.) A tangent developable is the union of the tangent lines to a curve in $\mathbf{R}^3$ – the tangent line at a point $\boldsymbol{\gamma}(u)$ to a curve $\boldsymbol{\gamma}$ is the straight line passing through $\boldsymbol{\gamma}(u)$ and parallel to the tangent vector $\dot{\boldsymbol{\gamma}}(u)$.

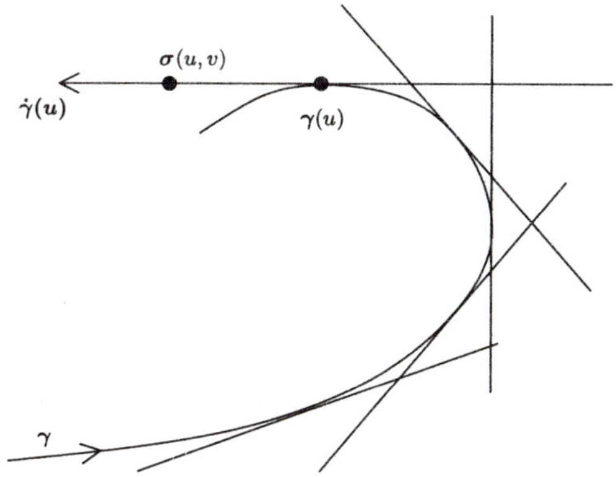

We might as well assume that $\boldsymbol{\gamma}$ is unit-speed. The most general point on the tangent line at $\boldsymbol{\gamma}(u)$ is

$$\boldsymbol{\sigma}(u, v) = \boldsymbol{\gamma}(u) + v\dot{\boldsymbol{\gamma}}(u),$$

for some scalar v. Now

$$\boldsymbol{\sigma}_u \times \boldsymbol{\sigma}_v = (\dot{\boldsymbol{\gamma}} + v\ddot{\boldsymbol{\gamma}}) \times \dot{\boldsymbol{\gamma}} = v\ddot{\boldsymbol{\gamma}} \times \dot{\boldsymbol{\gamma}}.$$

For $\boldsymbol{\sigma}$ to be regular, it is thus necessary that $\ddot{\boldsymbol{\gamma}}$ is never zero, or in other words, the curvature $\kappa = \| \ddot{\boldsymbol{\gamma}} \|$ is > 0 at all points of $\boldsymbol{\gamma}$. Now, $\dot{\boldsymbol{\gamma}} = \mathbf{t}$, the unit tangent vector of $\boldsymbol{\gamma}$, and $\ddot{\boldsymbol{\gamma}} = \dot{\mathbf{t}} = \kappa\mathbf{n}$, where $\mathbf{n}$ is the principal normal to $\boldsymbol{\gamma}$, so

$$\boldsymbol{\sigma}_u \times \boldsymbol{\sigma}_v = \kappa v\mathbf{n} \times \mathbf{t} = -\kappa v\mathbf{b},$$

where $\mathbf{b}$ is the binormal of $\boldsymbol{\gamma}$. Thus, $\boldsymbol{\sigma}$ will be regular if $\kappa > 0$ everywhere and $v \neq 0$. The latter condition means that, for regularity, we must exclude the curve $\boldsymbol{\gamma}$ itself from the surface. Typically, the regions $v > 0$ and $v < 0$ of the tangent developable form two sheets which meet along a sharp edge formed by the curve $\boldsymbol{\gamma}$ where $v = 0$, as the following illustration of the tangent developable of a circular helix indicates:

Our interest in tangent developables stems from the following result.

Proposition 5.1

Any tangent developable is isometric to (part of) a plane.

Proof 5.1

We use the above notation, assuming that $\boldsymbol{\gamma}$ is unit-speed and that $\kappa > 0$. Now,

$$\begin{aligned} E &= \| \boldsymbol{\sigma}_u \|^2 = (\dot{\boldsymbol{\gamma}} + v\ddot{\boldsymbol{\gamma}}).(\dot{\boldsymbol{\gamma}} + v\ddot{\boldsymbol{\gamma}}) \\ &= \dot{\boldsymbol{\gamma}}.\dot{\boldsymbol{\gamma}} + 2v\dot{\boldsymbol{\gamma}}.\ddot{\boldsymbol{\gamma}} + v^2\ddot{\boldsymbol{\gamma}}.\ddot{\boldsymbol{\gamma}} = 1 + v^2\kappa^2, \\ F &= \boldsymbol{\sigma}_u.\boldsymbol{\sigma}_v = (\dot{\boldsymbol{\gamma}} + v\ddot{\boldsymbol{\gamma}}).\dot{\boldsymbol{\gamma}} = \dot{\boldsymbol{\gamma}}.\dot{\boldsymbol{\gamma}} + v\dot{\boldsymbol{\gamma}}.\ddot{\boldsymbol{\gamma}} = 1, \\ G &= \| \boldsymbol{\sigma}_v \|^2 = \dot{\boldsymbol{\gamma}}.\dot{\boldsymbol{\gamma}} = 1, \end{aligned}$$

since $\dot{\boldsymbol{\gamma}}.\dot{\boldsymbol{\gamma}} = 1$, $\dot{\boldsymbol{\gamma}}.\ddot{\boldsymbol{\gamma}} = 0$, $\ddot{\boldsymbol{\gamma}}.\ddot{\boldsymbol{\gamma}} = \kappa^2$. So the first fundamental form of the tangent developable is

$$(1 + v^2\kappa^2)du^2 + 2dudv + dv^2. \tag{4}$$

We are going to show that (part of) the plane can be parametrised so that it has the same first fundamental form. This will prove the proposition.

By Theorem 2.1, there is a *plane* unit-speed curve $\tilde{\boldsymbol{\gamma}}$ whose curvature is κ (we can even assume that its signed curvature is κ). By the above calculations, the first fundamental form of the tangent developable of $\tilde{\boldsymbol{\gamma}}$ is also given by (4).

But since $\tilde{\gamma}$ is a plane curve, its tangent lines obviously fill out part of the plane in which $\tilde{\gamma}$ lies. □

There is a converse to Proposition 5.1: any sufficiently small piece of a surface isometric to (part of) a plane *is* (part of) a plane, a (generalised) cylinder, a (generalised) cone, or a tangent developable. The proof of this will be given in Section 7.3.

EXERCISES

5.5 The circular cone

$$\boldsymbol{\sigma}(u,v) = (u\cos v, u\sin v, u), \qquad u > 0,\ 0 < v < 2\pi,$$

can be 'uwrapped', hence is isometric to (part of) the xy-plane (say). Write down this isometry explicitly, and describe exactly which part of the plane $\boldsymbol{\sigma}$ is isometric to. Verify that the map really is an isometry.

5.6 Is the map from the circular half-cone $x^2 + y^2 = z^2$, $z > 0$, to the xy-plane given by $(x, y, z) \mapsto (x, y, 0)$ an isometry?

5.7 Show that every (generalised) cylinder and every (generalised) cone is isometric to (part of) the plane. (See Examples 5.3 and 5.4 and Exercise 5.5.)

$t = 0$

$t = 0.6$

$t = 0.2$

$t = 0.8$

$t = 0.4$

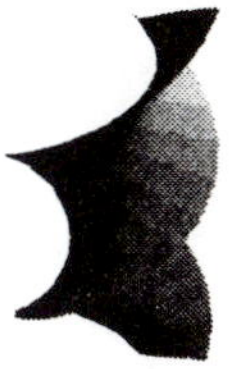

$t = 1$

5.8 Consider the surface patches

$$\sigma(u,v) = (\cosh u \cos v, \cosh u \sin v, u), \quad 0 < v < 2\pi,$$

$$\tilde{\sigma}(u,v) = (u \cos v, u \sin v, v), \quad 0 < v < 2\pi,$$

representing the catenoid (Exercise 4.18) with one meridian removed and the part of the helicoid (Example 4.14) between the planes $z = 0$ and $z = 2\pi$. Show that the map from the catenoid to the helicoid that takes $\sigma(u,v)$ to $\tilde{\sigma}(\sinh u, v)$ is an isometry. Which curves on the helicoid correspond under this isometry to the parallels and meridians of the catenoid?

In fact, there is an *isometric deformation* of the catenoid into the helicoid. For $0 \leq t \leq \pi/2$, define a surface

$$\sigma^t(u,v) = \cos t\, \sigma(u, v+t) + \sin t\, \tilde{\sigma}(\sinh u, v + t - \pi/2),$$

so that $\sigma^0(u,v) = \sigma(u,v)$ and $\sigma^{\pi/2}(u,v) = \tilde{\sigma}(\sinh u, v)$. Show that, for all values of t, the map $\sigma(u,v) \mapsto \sigma^t(u,v)$ is an isometry. The surfaces σ^t are shown above for several values of t.

5.3. Conformal Mappings of Surfaces

Now that we understand how to measure lengths of curves on surfaces, it is natural to ask about angles. Suppose that two curves γ and $\tilde{\gamma}$ on a surface $\mathcal{S}$ intersect at a point P that lies in a surface patch σ of $\mathcal{S}$. Then, $\gamma(t) = \sigma(u(t), v(t))$ and $\tilde{\gamma}(t) = \sigma(\tilde{u}(t), \tilde{v}(t))$ for some smooth functions $u, v, \tilde{u}$ and $\tilde{v}$, and for some parameter values t_0 and $\tilde{t}_0$, we have $\sigma(u(t_0), v(t_0)) = P = \sigma(\tilde{u}(\tilde{t}_0), \tilde{v}(t_0))$.

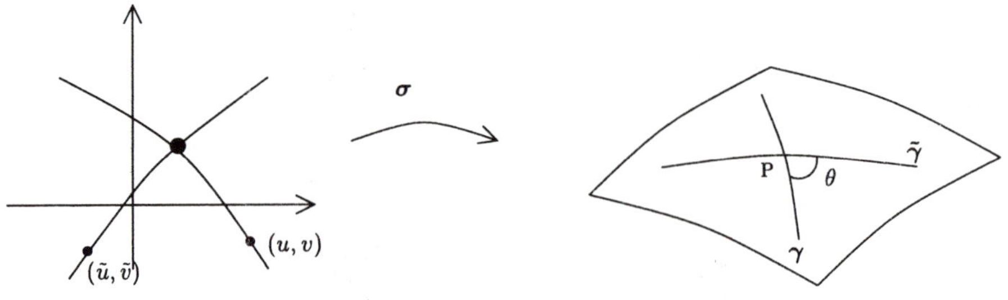

The *angle* θ of intersection of γ and $\tilde{\gamma}$ at P is defined to be the angle between the tangent vectors $\dot{\gamma}$ and $\dot{\tilde{\gamma}}$ (evaluated at $t = t_0$ and $t = \tilde{t}_0$, respectively).

Using the dot product formula for the angle between vectors, we see that θ is given by

$$\cos\theta = \frac{\dot{\boldsymbol{\gamma}}.\dot{\tilde{\boldsymbol{\gamma}}}}{\parallel \dot{\boldsymbol{\gamma}} \parallel \parallel \dot{\tilde{\boldsymbol{\gamma}}} \parallel}.$$

By the chain rule,

$$\dot{\boldsymbol{\gamma}} = \boldsymbol{\sigma}_u \dot{u} + \boldsymbol{\sigma}_v \dot{v}, \quad \dot{\tilde{\boldsymbol{\gamma}}} = \boldsymbol{\sigma}_u \dot{\tilde{u}} + \boldsymbol{\sigma}_v \dot{\tilde{v}},$$

so

$$\begin{aligned}\dot{\boldsymbol{\gamma}}.\dot{\tilde{\boldsymbol{\gamma}}} &= (\boldsymbol{\sigma}_u \dot{u} + \boldsymbol{\sigma}_v \dot{v}).(\boldsymbol{\sigma}_u \dot{\tilde{u}} + \boldsymbol{\sigma}_v \dot{\tilde{v}}) \\ &= (\boldsymbol{\sigma}_u.\boldsymbol{\sigma}_u)\dot{u}\dot{\tilde{u}} + (\boldsymbol{\sigma}_u.\boldsymbol{\sigma}_v)(\dot{u}\dot{\tilde{v}} + \dot{\tilde{u}}\dot{v}) + (\boldsymbol{\sigma}_v.\boldsymbol{\sigma}_v)\dot{v}\dot{\tilde{v}} \\ &= E\dot{u}\dot{\tilde{u}} + F(\dot{u}\dot{\tilde{v}} + \dot{\tilde{u}}\dot{v}) + G\dot{v}\dot{\tilde{v}}.\end{aligned}$$

Replacing $\tilde{\boldsymbol{\gamma}}$ by $\boldsymbol{\gamma}$ (resp. $\boldsymbol{\gamma}$ by $\tilde{\boldsymbol{\gamma}}$) gives similar expressions for $\parallel \dot{\boldsymbol{\gamma}} \parallel^2 = \dot{\boldsymbol{\gamma}}.\dot{\boldsymbol{\gamma}}$ (resp. $\parallel \dot{\tilde{\boldsymbol{\gamma}}} \parallel^2$), which finally give the formula

$$\cos\theta = \frac{E\dot{u}\dot{\tilde{u}} + F(\dot{u}\dot{\tilde{v}} + \dot{\tilde{u}}\dot{v}) + G\dot{v}\dot{\tilde{v}}}{(E\dot{u}^2 + 2F\dot{u}\dot{v} + G\dot{v}^2)^{1/2}(E\dot{\tilde{u}}^2 + 2F\dot{\tilde{u}}\dot{\tilde{v}} + G\dot{\tilde{v}}^2)^{1/2}}. \tag{5}$$

Example 5.6

The *parameter curves* on a surface patch $\boldsymbol{\sigma}(u, v)$ can be parametrised by

$$\boldsymbol{\gamma}(t) = \boldsymbol{\sigma}(a, t), \quad \tilde{\boldsymbol{\gamma}}(t) = \boldsymbol{\sigma}(t, b),$$

respectively, where a is the constant value of u and b the constant value of v in the two cases. Thus,

$$\begin{aligned}& u(t) = a, \quad v(t) = t, \quad \tilde{u}(t) = t, \quad \tilde{v}(t) = b, \\ \therefore \quad & \dot{u} = 0, \quad \dot{v} = 1, \quad \dot{\tilde{u}} = 1, \quad \dot{\tilde{v}} = 0.\end{aligned}$$

These parameter curves intersect at the point $\boldsymbol{\sigma}(a, b)$ of the surface. By Eq. (5), their angle of intersection θ is given by

$$\cos\theta = \frac{F}{\sqrt{EG}}.$$

In particular, the parameter curves are orthogonal if and only if $F = 0$.

Corresponding to the Definition 5.1 of an isometry, we make

Definition 5.2

If $\mathcal{S}_1$ and $\mathcal{S}_2$ are surfaces, a diffeomorphism $f : \mathcal{S}_1 \to \mathcal{S}_2$ is said to be *conformal* if, whenever f takes two intersecting curves $\boldsymbol{\gamma}_1$ and $\tilde{\boldsymbol{\gamma}}_1$ on $\mathcal{S}_1$ to curves $\boldsymbol{\gamma}_2$ and $\tilde{\boldsymbol{\gamma}}_2$

on $\mathcal{S}_2$, the angle of intersection of $\boldsymbol{\gamma}_1$ and $\tilde{\boldsymbol{\gamma}}_1$ is equal to the angle of intersection of $\boldsymbol{\gamma}_2$ and $\tilde{\boldsymbol{\gamma}}_2$.

In short, *f is conformal if it preserves angles.*

As a special case, if $\boldsymbol{\sigma} : U \to \mathbf{R}^3$ is a surface, then $\boldsymbol{\sigma}$ may be viewed as a map from part of the plane (namely U), parametrised by (u, v) in the usual way, and the image $\mathcal{S}$ of $\boldsymbol{\sigma}$, and we say that $\boldsymbol{\sigma}$ is a *conformal parametrisation* or a *conformal surface patch* of $\mathcal{S}$ if this map between surfaces is conformal.

Theorem 5.2

A diffeomorphism $f : \mathcal{S}_1 \to \mathcal{S}_2$ is conformal if and only if, for any surface patch $\boldsymbol{\sigma}_1$ on $\mathcal{S}_1$, the first fundamental forms of $\boldsymbol{\sigma}_1$ and $f \circ \boldsymbol{\sigma}_1$ are proportional.

Proof 5.2

As in the proof of Theorem 5.1, we can assume that $\mathcal{S}_1$ and $\mathcal{S}_2$ are covered by the single surface patches $\boldsymbol{\sigma}_1 : U \to \mathbf{R}^3$ and $\boldsymbol{\sigma}_2 = f \circ \boldsymbol{\sigma}_1$, respectively. Suppose that their first fundamental forms

$$E_1 du^2 + 2F_1 du dv + G_1 dv^2 \quad \text{and} \quad E_2 du^2 + 2F_2 du dv + G_2 dv^2$$

are proportional, say

$$E_2 du^2 + 2F_2 du dv + G_2 dv^2 = \lambda(E_1 du^2 + 2F_1 du dv + G_1 dv^2)$$

for some smooth function $\lambda(u, v)$, where (u, v) are coordinates on U. Note that $\lambda > 0$ everywhere, since (for example) E_1 and E_2 are both > 0. If $\boldsymbol{\gamma}(t) = \boldsymbol{\sigma}_1(u(t), v(t))$ and $\tilde{\boldsymbol{\gamma}}(t) = \boldsymbol{\sigma}_1(\tilde{u}(t), \tilde{v}(t))$ are curves in $\mathcal{S}_1$, then f takes $\boldsymbol{\gamma}$ and $\tilde{\boldsymbol{\gamma}}$ to the curves $\boldsymbol{\sigma}_2(u(t), v(t))$ and $\boldsymbol{\sigma}_2(\tilde{u}(t), \tilde{v}(t))$ in $\mathcal{S}_2$, respectively. Using Eq. (5), the angle θ of intersection of the latter curves on $\mathcal{S}_2$ is given by

$$\begin{aligned}
\cos\theta &= \frac{E_2 \dot{u}\dot{\tilde{u}} + F_2(\dot{u}\dot{\tilde{v}} + \dot{\tilde{u}}\dot{v}) + G_2 \dot{v}\dot{\tilde{v}}}{(E_2\dot{u}^2 + 2F_2\dot{u}\dot{v} + G_2\dot{v}^2)^{1/2}(E_2\dot{\tilde{u}}^2 + 2F_2\dot{\tilde{u}}\dot{\tilde{v}} + G_2\dot{\tilde{v}}^2)^{1/2}} \\
&= \frac{\lambda E_1 \dot{u}\dot{\tilde{u}} + \lambda F_1(\dot{u}\dot{\tilde{v}} + \dot{\tilde{u}}\dot{v}) + \lambda G_1 \dot{v}\dot{\tilde{v}}}{(\lambda E_1\dot{u}^2 + 2\lambda F_1\dot{u}\dot{v} + \lambda G_1\dot{v}^2)^{1/2}(\lambda E_1\dot{\tilde{u}}^2 + 2\lambda F_1\dot{\tilde{u}}\dot{\tilde{v}} + \lambda G_1\dot{\tilde{v}}^2)^{1/2}} \\
&= \frac{E_1 \dot{u}\dot{\tilde{u}} + F_1(\dot{u}\dot{\tilde{v}} + \dot{\tilde{u}}\dot{v}) + G_1 \dot{v}\dot{\tilde{v}}}{(E_1\dot{u}^2 + 2F_1\dot{u}\dot{v} + G_1\dot{v}^2)^{1/2}(E_1\dot{\tilde{u}}^2 + 2F_1\dot{\tilde{u}}\dot{\tilde{v}} + G_1\dot{\tilde{v}}^2)^{1/2}},
\end{aligned}$$

since the λ's cancel. But, using Eq. (5) again, we see that the right-hand side is the cosine of the angle of intersection of the curves $\boldsymbol{\gamma}$ and $\tilde{\boldsymbol{\gamma}}$ on $\mathcal{S}_1$. Hence, f is conformal.

For the converse, we must show that if

$$\frac{E_1\dot{u}\dot{\tilde{u}} + F_1(\dot{u}\dot{\tilde{v}} + \dot{\tilde{u}}\dot{v}) + G_1\dot{v}\dot{\tilde{v}}}{(E_1\dot{u}^2 + 2F_1\dot{u}\dot{v} + G_1\dot{v}^2)^{1/2}(E_1\dot{\tilde{u}}^2 + 2F_1\dot{\tilde{u}}\dot{\tilde{v}} + G_1\dot{\tilde{v}}^2)^{1/2}} = \frac{E_2\dot{u}\dot{\tilde{u}} + F_2(\dot{u}\dot{\tilde{v}} + \dot{\tilde{u}}\dot{v}) + G_2\dot{v}\dot{\tilde{v}}}{(E_2\dot{u}^2 + 2F_2\dot{u}\dot{v} + G_2\dot{v}^2)^{1/2}(E_2\dot{\tilde{u}}^2 + 2F_2\dot{\tilde{u}}\dot{\tilde{v}} + G_2\dot{\tilde{v}}^2)^{1/2}} \tag{6}$$

for *all* pairs of intersecting curves

$$\boldsymbol{\gamma}(t) = \boldsymbol{\sigma}_1(u(t), v(t)) \quad \text{and} \quad \tilde{\boldsymbol{\gamma}}(t) = \boldsymbol{\sigma}_1(\tilde{u}(t), \tilde{v}(t))$$

in $\mathcal{S}_1$, then the first fundamental forms of $\boldsymbol{\sigma}_1$ and $\boldsymbol{\sigma}_2$ are proportional. Fix $(a, b) \in U$ and consider the curves

$$\boldsymbol{\gamma}(t) = \boldsymbol{\sigma}_1(a + t, b), \quad \tilde{\boldsymbol{\gamma}}(t) = \boldsymbol{\sigma}_1(a + t\cos\phi, b + t\sin\phi),$$

where ϕ is a constant, for which

$$\dot{u} = 1, \ \dot{v} = 0, \ \dot{\tilde{u}} = \cos\phi, \ \dot{\tilde{v}} = \sin\phi.$$

Substituting in Eq. (6) gives

$$\frac{E_1\cos\phi + F_1\sin\phi}{\sqrt{E_1(E_1\cos^2\phi + 2F_1\sin\phi\cos\phi + G_1\sin^2\phi)}} = \frac{E_2\cos\phi + F_2\sin\phi}{\sqrt{E_2(E_2\cos^2\phi + 2F_2\sin\phi\cos\phi + G_2\sin^2\phi)}}. \tag{7}$$

Squaring both sides of Eq. (7) and writing

$$(E_1\cos\phi + F_1\sin\phi)^2 = E_1(E_1\cos^2\phi + 2F_1\sin\phi\cos\phi + G_1\sin^2\phi) - (E_1G_1 - F_1^2)\sin^2\phi,$$

we get

$$(E_1G_1 - F_1^2)E_2(E_2\cos^2\phi + 2F_2\sin\phi\cos\phi + G_2\sin^2\phi) = (E_2G_2 - F_2^2)E_1(E_1\cos^2\phi + 2F_1\sin\phi\cos\phi + G_1\sin^2\phi), \tag{16}$$

or, setting $\lambda = (E_2G_2 - F_2^2)E_1/(E_1G_1 - F_1^2)E_2$,

$$(E_2 - \lambda E_1)\cos^2\phi + 2(F_2 - \lambda F_1)\sin\phi\cos\phi + (G_2 - \lambda G_1)\sin^2\phi = 0.$$

Taking $\phi = 0$ and then $\phi = \pi/2$ gives $E_2 = \lambda E_1$, $G_2 = \lambda G_1$, and then substituting in the last equation gives $F_2 = \lambda F_1$. □

Example 5.7

We consider the unit sphere $x^2 + y^2 + z^2 = 1$. If $P = (u, v, 0)$ is any point in the xy-plane, draw the straight line through P and the north pole $N = (0, 0, 1)$.

This line intersects the sphere at a point Q, say. Every point Q of the sphere arises as such a point of intersection, with the sole exception of the north pole itself.

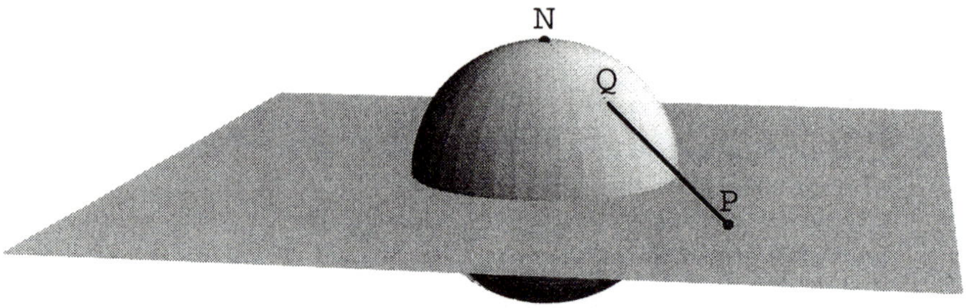

The vector $\mathbf{NQ}$ is parallel to the vector $\mathbf{NP}$, so there is a scalar, say ρ, such that the position vector $\mathbf{q}$ of Q is related to those of N and P by

$$\mathbf{q} - \mathbf{n} = \rho(\mathbf{p} - \mathbf{n}),$$

and hence

$$\mathbf{q} = (0,0,1) + \rho((u,v,0) - (0,0,1)) = (\rho u, \rho v, 1-\rho).$$

Since Q lies on the sphere,

$$\rho^2 u^2 + \rho^2 v^2 + (1-\rho)^2 = 1$$

which gives $\rho = 2/(u^2+v^2+1)$ (the other root $\rho = 0$ corresponds to the other intersection point N between the line and the sphere). Hence,

$$\mathbf{q} = \left(\frac{2u}{u^2+v^2+1}, \frac{2v}{u^2+v^2+1}, \frac{u^2+v^2-1}{u^2+v^2+1} \right).$$

If we denote the right-hand side by $\boldsymbol{\sigma}_1(u,v)$, then $\boldsymbol{\sigma}_1$ is a parametrisation of the whole sphere minus the north pole. Parametrising the plane $z = 0$ by $\boldsymbol{\sigma}_2(u,v) = (u,v,0)$, the map that takes Q to P takes $\boldsymbol{\sigma}_1(u,v)$ to $\boldsymbol{\sigma}_2(u,v)$. This map is called *stereographic projection.* We are going to prove that it is conformal.

According to Theorem 5.5, we have to show that the first fundamental forms of $\boldsymbol{\sigma}_1$ and $\boldsymbol{\sigma}_2$ are proportional. The first fundamental form of $\boldsymbol{\sigma}_2$ is $du^2 + dv^2$.

As to $\boldsymbol{\sigma}_1$, we get

$$(\boldsymbol{\sigma}_1)_u = \left(\frac{2(v^2-u^2+1)}{(u^2+v^2+1)^2}, \frac{-4uv}{(u^2+v^2+1)^2}, \frac{4u}{(u^2+v^2+1)^2}\right),$$

$$(\boldsymbol{\sigma}_1)_v = \left(\frac{-4uv}{(u^2+v^2+1)^2}, \frac{2(u^2-v^2+1)}{(u^2+v^2+1)^2}, \frac{4v}{(u^2+v^2+1)^2}\right).$$

This gives

$$E_1 = (\boldsymbol{\sigma}_1)_u.(\boldsymbol{\sigma}_1)_u = \frac{4(v^2-u^2+1)^2+16u^2v^2+16u^2}{(u^2+v^2+1)^4}$$

which simplifies to $E_1 = 4/(u^2+v^2+1)^2$. Similarly, $F_1 = 0$, $G_1 = E_1$. Thus, the first fundamental form of $\boldsymbol{\sigma}_2$ is λ times that of $\boldsymbol{\sigma}_1$, where $\lambda = \frac{1}{4}(u^2+v^2+1)^2$.

EXERCISES

5.9 Show that every isometry is a conformal map. Give an example of a conformal map that is not an isometry.

5.10 Show that the curve on the cone in Exercise 5.2 intersects all the rulings of the cone at the same angle.

5.11 Show that Mercator's parametrisation of the sphere

$$\boldsymbol{\sigma}(u,v) = (\operatorname{sech} u \cos v, \operatorname{sech} u \sin v, \tanh u)$$

is conformal.

5.12 Let $f(x)$ be a smooth function and let

$$\boldsymbol{\sigma}(u,v) = (u\cos v, u\sin v, f(u))$$

be the surface obtained by rotating the curve $z = f(x)$ in the xz-plane around the z-axis. Find all functions f for which $\boldsymbol{\sigma}$ is conformal.

5.13 Let $\boldsymbol{\sigma}$ be the ruled surface

$$\boldsymbol{\sigma}(u,v) = \boldsymbol{\gamma}(u) + v\boldsymbol{\delta}(u),$$

where $\boldsymbol{\gamma}$ is a unit-speed curve in $\mathbf{R}^3$ and $\boldsymbol{\delta}(u)$ is a unit vector for all u. Prove that $\boldsymbol{\sigma}$ is conformal if and only if $\boldsymbol{\delta}(u)$ is independent of u and $\boldsymbol{\gamma}$ lies in a plane perpendicular to $\boldsymbol{\delta}$. What kind of surface is $\boldsymbol{\sigma}$ in this case?

5.14 Show that the surface patch

$$\boldsymbol{\sigma}(u,v) = (f(u,v), g(u,v), 0),$$

where f and g are smooth functions on the uv-plane, is conformal if and only if either

$$f_u = g_v \quad \text{and} \quad f_v = -g_u,$$

or

$$f_u = -g_v \quad \text{and} \quad f_v = g_u.$$

The first pair of equations are called the *Cauchy–Riemann equations*; they are the condition for the map from the complex plane to itself given by $u + iv \mapsto f(u,v) + ig(u,v)$ to be *holomorphic*. The second pair of equations says that this map is *anti-holomorphic*, i.e. that its complex-conjugate is holomorphic. We shall say more about holomorphic functions in relation to surfaces in Section 9.4.

5.4. Surface Area

Suppose that $\boldsymbol{\sigma} : U \to \mathbf{R}^3$ is a surface patch on a surface $\mathcal{S}$. The image of $\boldsymbol{\sigma}$ is covered by the two families of parameter curves obtained by setting $u =$ constant and $v =$ constant, respectively. Fix $(u_0, v_0) \in U$, and let Δu and Δv be very small. Since the change in $\boldsymbol{\sigma}(u,v)$ corresponding to a small change Δu in u is approximately $\boldsymbol{\sigma}_u \Delta u$ and that corresponding to a small change Δv in v is approximately $\boldsymbol{\sigma}_v \Delta v$, the part of the surface contained by the parameter curves in the surface corresponding to $u = u_0$, $u = u_0 + \Delta u$, $v = v_0$ and $v = v_0 + \Delta v$ is almost a parallelogram in the plane with sides given by the vectors $\boldsymbol{\sigma}_u \Delta u$ and $\boldsymbol{\sigma}_v \Delta v$ (the derivatives being evaluated at (u_0, v_0)):

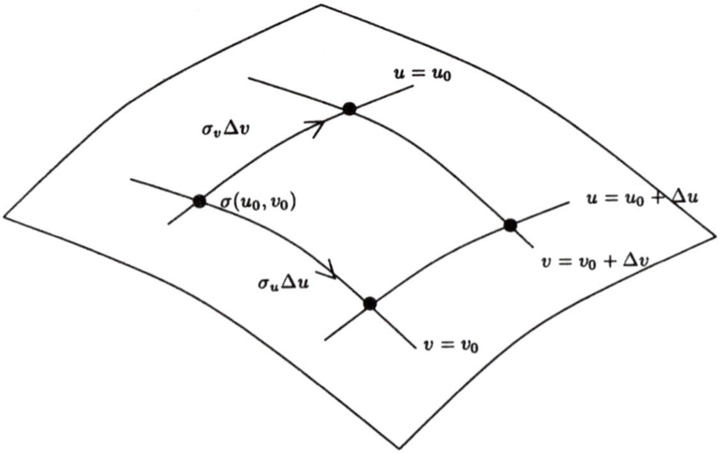

Recalling that the area of a parallelogram in the plane with sides **a** and **b** is $\| \mathbf{a} \times \mathbf{b} \|$, we see that the area of the parallelogram on the surface is approximately

$$\| \boldsymbol{\sigma}_u \Delta u \times \boldsymbol{\sigma}_v \Delta v \| = \| \boldsymbol{\sigma}_u \times \boldsymbol{\sigma}_v \| \Delta u \Delta v.$$

This suggests the following definition.

Definition 5.3

The *area* $\mathcal{A}_{\boldsymbol{\sigma}}(R)$ of the part $\boldsymbol{\sigma}(R)$ of surface patch $\boldsymbol{\sigma} : U \to \mathbf{R}^3$ corresponding to a region $R \subseteq U$ is

$$\mathcal{A}_{\boldsymbol{\sigma}}(R) = \int\int_R \| \boldsymbol{\sigma}_u \times \boldsymbol{\sigma}_v \| \, dudv.$$

Of course, this integral may be infinite – think of the area of a whole plane, for example. However, the integral will be finite if, say, R is contained in a rectangle that is entirely contained, along with its boundary, in U.

The quantity $\| \boldsymbol{\sigma}_u \times \boldsymbol{\sigma}_v \|$ that appears in the definition of area is easily computed in terms of the first fundamental form $Edu^2 + 2Fdudv + Gdv^2$ of $\boldsymbol{\sigma}$:

Proposition 5.2

$\| \boldsymbol{\sigma}_u \times \boldsymbol{\sigma}_v \| = (EG - F^2)^{1/2}$.

Proof 5.2

We use a result from vector algebra: if $\mathbf{a}, \mathbf{b}, \mathbf{c}$ and $\mathbf{d}$ are vectors in $\mathbf{R}^3$, then

$$(\mathbf{a} \times \mathbf{b}).(\mathbf{c} \times \mathbf{d}) = (\mathbf{a}.\mathbf{c})(\mathbf{b}.\mathbf{d}) - (\mathbf{a}.\mathbf{d})(\mathbf{b}.\mathbf{c}).$$

Applying this to $\| \boldsymbol{\sigma}_u \times \boldsymbol{\sigma}_v \|^2 = (\boldsymbol{\sigma}_u \times \boldsymbol{\sigma}_v).(\boldsymbol{\sigma}_u \times \boldsymbol{\sigma}_v)$, we get

$$\| \boldsymbol{\sigma}_u \times \boldsymbol{\sigma}_v \|^2 = (\boldsymbol{\sigma}_u.\boldsymbol{\sigma}_u)(\boldsymbol{\sigma}_v.\boldsymbol{\sigma}_v) - (\boldsymbol{\sigma}_u.\boldsymbol{\sigma}_v)^2 = EG - F^2. \qquad \square$$

Note that, for a regular surface, $EG - F^2 > 0$ everywhere, since for a regular surface $\boldsymbol{\sigma}_u \times \boldsymbol{\sigma}_v$ is never zero.

Thus, our definition of area is

$$\mathcal{A}_{\boldsymbol{\sigma}}(R) = \int\int_R (EG - F^2)^{1/2} dudv. \tag{8}$$

We sometimes denote $(EG - F^2)^{1/2}dudv$ by $d\mathcal{A}_{\boldsymbol{\sigma}}$. But we have still to check that this definition is sensible, i.e. that it is unchanged if $\boldsymbol{\sigma}$ is reparametrised.

This is certainly not obvious, since E, F and G change under reparametrisation (see Exercise 5.4).

Proposition 5.3

The area of a surface patch is unchanged by reparametrisation.

Proof 5.3

Let $\boldsymbol{\sigma} : U \to \mathbf{R}^3$ be a surface patch and let $\tilde{\boldsymbol{\sigma}} : \tilde{U} \to \mathbf{R}^3$ be a reparametrisation of $\boldsymbol{\sigma}$, with reparametrisation map $\Phi : \tilde{U} \to U$. Thus, if $\Phi(\tilde{u}, \tilde{v}) = (u, v)$, we have

$$\tilde{\boldsymbol{\sigma}}(\tilde{u}, \tilde{v}) = \boldsymbol{\sigma}(u, v).$$

Let $\tilde{R} \subseteq \tilde{U}$ be a region, and let $R = \Phi(\tilde{R}) \subseteq U$. We have to prove that

$$\int\int_R \| \boldsymbol{\sigma}_u \times \boldsymbol{\sigma}_v \| \, dudv = \int\int_{\tilde{R}} \| \tilde{\boldsymbol{\sigma}}_{\tilde{u}} \times \tilde{\boldsymbol{\sigma}}_{\tilde{v}} \| \, d\tilde{u}d\tilde{v}.$$

We showed in the proof of Proposition 4.2 that

$$\tilde{\boldsymbol{\sigma}}_{\tilde{u}} \times \tilde{\boldsymbol{\sigma}}_{\tilde{v}} = \det(J(\Phi)) \, \boldsymbol{\sigma}_u \times \boldsymbol{\sigma}_v,$$

where $J(\Phi)$ is the jacobian matrix of Φ. Hence,

$$\int\int_{\tilde{R}} \| \tilde{\boldsymbol{\sigma}}_{\tilde{u}} \times \tilde{\boldsymbol{\sigma}}_{\tilde{v}} \| \, d\tilde{u}d\tilde{v} = \int\int_{\tilde{R}} |\det(J(\Phi))| \, \| \boldsymbol{\sigma}_u \times \boldsymbol{\sigma}_v \| \, d\tilde{u}d\tilde{v}.$$

By the change of variables formula for double integrals, the right-hand side of this equation is exactly

$$\int\int_R \| \boldsymbol{\sigma}_u \times \boldsymbol{\sigma}_v \| \, dudv. \qquad \square$$

This proposition implies that we can calculate the area of any surface $\mathcal{S}$ by breaking $\mathcal{S}$ up into pieces, each of which are contained in a single surface patch, calculating the area of each piece using Eq. (8), and adding up the results (cf. Section 11.3, where an analogous procedure is carried out).

EXERCISES

5.15 Determine the area of the part of the paraboloid $z = x^2 + y^2$ with $z \leq 1$ and compare with the area of the hemisphere $x^2 + y^2 + z^2 = 1$, $z \leq 0$.

5.16 A surface is obtained by rotating about the z-axis a unit-speed curve γ in the xz-plane that does not intersect the z-axis. Using the standard parametrisation of this surface, calculate its first fundamental form, and deduce that its area is

$$2\pi \int \rho(u)\,du,$$

where $\rho(u)$ is the distance of $\gamma(u)$ from the z-axis. Hence find the area of

(i) the unit sphere;

(ii) the torus in Exercise 4.10.

5.17 Let $\boldsymbol{\gamma}(s)$ be a unit-speed curve in $\mathbf{R}^3$ with principal normal $\mathbf{n}$ and binormal $\mathbf{b}$. The *tube* of radius $a > 0$ around $\boldsymbol{\gamma}$ is the surface parametrised by

$$\boldsymbol{\sigma}(s,\theta) = \boldsymbol{\gamma}(s) + a(\mathbf{n}(s)\cos\theta + \mathbf{b}(s)\sin\theta).$$

Give a geometrical description of this surface. Prove that $\boldsymbol{\sigma}$ is regular if the curvature κ of $\boldsymbol{\gamma}$ is less than a^{-1} everywhere. Assuming that this condition holds, prove that the area of the part of the surface given by $s_0 < s < s_1$, $0 < \theta < 2\pi$, where s_0 and s_1 are constants, is $2\pi a(s_1 - s_0)$.

The tube around a circular helix

5.5. Equiareal Maps and a Theorem of Archimedes

We are going to use the formula (8) for the area of a surface to prove a theorem due to Archimedes which, legend has it, was inscribed onto his tombstone by the Roman general Marcellus who led the siege of Syracuse in which Archimedes perished. Naturally, since calculus was not available to him, Archimedes's proof of his theorem was quite different from ours. From his theorem, we shall deduce a beautiful formula for the area of any triangle on a sphere whose sides are arcs of great circles.

In modern language, the Theorem of Archimedes asserts that a certain map between surfaces is equiareal, in the following sense:

Definition 5.4

Let $\mathcal{S}_1$ and $\mathcal{S}_2$ be two surfaces. A diffeomorphism $f : \mathcal{S}_1 \to \mathcal{S}_2$ is said to be *equiareal* if it takes any region in $\mathcal{S}_1$ to a region of the same area in $\mathcal{S}_2$.

We have the following analogue of Theorem 5.1.

Theorem 5.3

A diffeomorphism $f : \mathcal{S}_1 \to \mathcal{S}_2$ is equiareal if and only if, for any surface patch $\boldsymbol{\sigma}(u, v)$ on $\mathcal{S}_1$, the first fundamental forms

$$E_1 du^2 + 2F_1 dudv + G_1 dv^2 \quad \text{and} \quad E_2 du^2 + 2F_2 dudv + G_2 dv^2$$

of the patches $\boldsymbol{\sigma}$ on $\mathcal{S}_1$ and $f \circ \boldsymbol{\sigma}$ on $\mathcal{S}_2$ satisfy

$$E_1 G_1 - F_1^2 = E_2 G_2 - F_2^2. \tag{9}$$

The proof is very similar to that of Theorem 5.1 and we leave it as Exercise 5.22.

For Archimedes's theorem, we consider the unit sphere $x^2 + y^2 + z^2 = 1$ and the cylinder $x^2 + y^2 = 1$. The sphere is contained inside the cylinder, and the two surfaces touch along the circle $x^2 + y^2 = 1$ in the xy-plane. For each point P on the sphere other than the poles $(0, 0, \pm 1)$, there is a unique straight line parallel to the xy-plane and passing through the point P and the z-axis. This line intersects the cylinder in two points, one of which, say Q, is closest to P. Let f be the map from the sphere (minus the two poles) to the cylinder that takes P to Q.

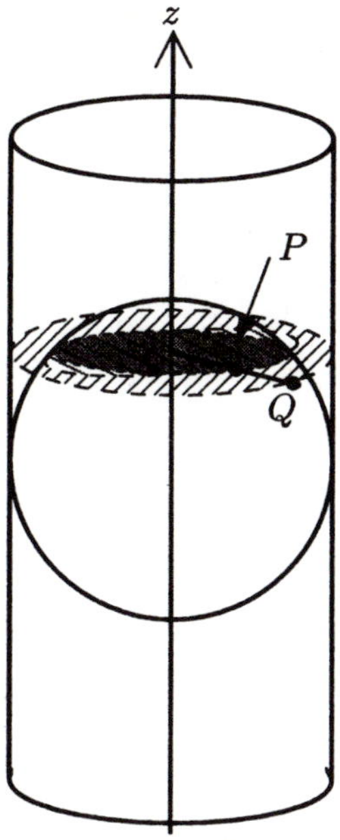

To find a formula for f, let (x, y, z) be the cartesian coordinates of P, and (X, Y, Z) those of Q. Since the line PQ is parallel to the xy-plane, we have $Z = z$ and $(X, Y) = \lambda(x, y)$ for some scalar λ. Since (X, Y, Z) is on the cylinder,

$$1 = X^2 + Y^2 = \lambda^2(x^2 + y^2),$$
$$\therefore \quad \lambda = \pm(x^2 + y^2)^{-1/2}.$$

Taking the $+$ sign gives the point Q, so we get

$$f(x, y, z) = \left(\frac{x}{(x^2 + y^2)^{1/2}}, \frac{y}{(x^2 + y^2)^{1/2}}, z\right).$$

We shall show in the proof of the next theorem that f is a diffeomorphism.

Theorem 5.4 (Archimedes's Theorem)

The map f is equiareal.

Proof 5.4

We take the atlas for the surface $\mathcal{S}_1$ consisting of the sphere minus the north and south poles with two patches, both given by the formula

$$\boldsymbol{\sigma}_1(\theta, \varphi) = (\cos\theta\cos\varphi, \cos\theta\sin\varphi, \sin\theta),$$

and defined on the open sets

$$\{-\pi/2 < \theta < \pi/2,\ 0 < \varphi < 2\pi\} \text{ and } \{-\pi/2 < \theta < \pi/2,\ -\pi < \varphi < \pi\}.$$

The image of $\boldsymbol{\sigma}_1(\theta, \varphi)$ under the map f is the point

$$\boldsymbol{\sigma}_2(\theta, \varphi) = (\cos\varphi, \sin\varphi, \sin\theta) \tag{10}$$

of the cylinder. It is easy to check that this gives an atlas for the surface $\mathcal{S}_2$, consisting of the part of the cylinder between the planes $z = 1$ and $z = -1$, with two patches, both given by Eq. (10) and defined on the same two open sets as $\boldsymbol{\sigma}_1$. We have to show that Eq. (9) holds.

We computed the coefficients E_1, F_1 and G_1 of the first fundamental form of $\boldsymbol{\sigma}_1$ in Example 5.3:

$$E_1 = 1, \quad F_1 = 0, \quad G_1 = \cos^2\theta.$$

For $\boldsymbol{\sigma}_2$, we get

$$(\boldsymbol{\sigma}_2)_\theta = (0, 0, \cos\theta), \quad (\boldsymbol{\sigma}_2)_\varphi = (-\sin\varphi, \cos\varphi, 0),$$
$$\therefore \quad E_2 = \cos^2\theta, \quad F_2 = 0, \quad G_2 = 1.$$

It is now clear that Eq. (9) holds.

Note that, since f corresponds simply to the identity map $(\theta, \varphi) \mapsto (\theta, \varphi)$ in terms of the parametrisations $\boldsymbol{\sigma}_1$ and $\boldsymbol{\sigma}_2$ of the sphere and cylinder, respectively, it follows that f is a diffeomorphism. □

Example 5.8

We use Archimedes's theorem to compute the area of a 'lune', i.e. the area enclosed between two great circles:

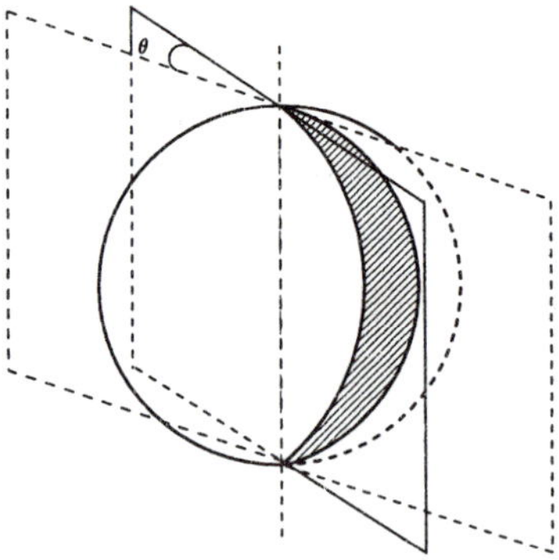

We can assume that the great circles intersect at the poles, since this can be achieved by applying a rotation of the sphere, and this does not change areas (see Exercise 5.3). If θ is the angle between them, the image of the lune under the map f is a curved rectangle on the cylinder of width θ and height 2:

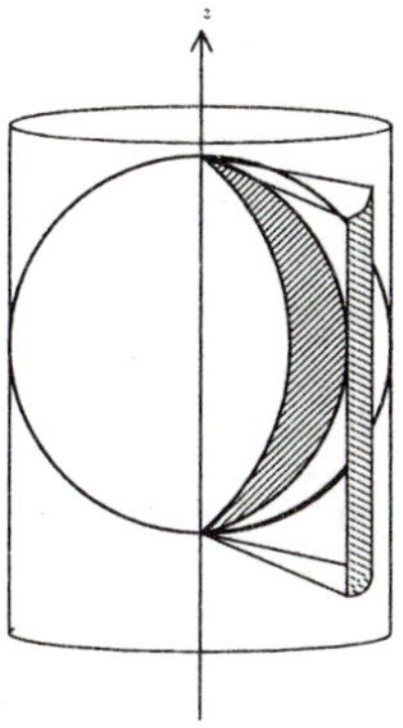

If we now apply the isometry which unwraps the cylinder on the plane, this curved rectangle on the cylinder will map to a genuine rectangle on the plane, with width θ and height 2. By Archimedes' theorem, the lune has the same area as the curved rectangle on the cylinder, and since every isometry is an equiareal map (see Exercise 5.18), this has the same area as the genuine rectangle in the plane, namely 2θ. Note that this gives the area of the whole sphere to be 4π.

Theorem 5.5

Let ABC be a triangle on a sphere of unit radius whose sides are arcs of great circles. Then, the area of the triangle is

$$\angle A + \angle B + \angle C - \pi,$$

where $\angle A$ is the angle of the triangle at A, etc.

Proof 5.5

The three great circles, of which the sides of the triangle are arcs, divide the sphere into 8 triangles, as shown in the following diagram (in which A' is the antipodal point of A, etc.). Denoting the area of triangle ABC by $\mathcal{A}(ABC)$, etc., we have, by Example 5.8,

$$\mathcal{A}(ABC) + \mathcal{A}(A'BC) = 2\angle A,$$
$$\mathcal{A}(ABC) + \mathcal{A}(AB'C) = 2\angle B,$$
$$\mathcal{A}(ABC) + \mathcal{A}(ABC') = 2\angle C.$$

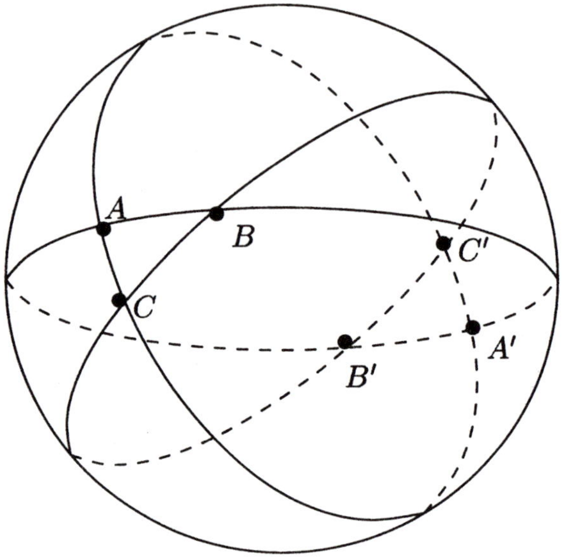

Adding these equations, we get

$$2\mathcal{A}(ABC) + \{\mathcal{A}(ABC) + \mathcal{A}(A'BC) + \mathcal{A}(AB'C) + \mathcal{A}(ABC')\} \\ = 2\angle A + 2\angle B + 2\angle C. \tag{11}$$

Now, the triangles ABC, $AB'C$, $AB'C'$ and ABC' together make a hemisphere, so

$$\mathcal{A}(ABC) + \mathcal{A}(AB'C) + \mathcal{A}(AB'C') + \mathcal{A}(ABC') = 2\pi. \tag{12}$$

Finally, since the map which takes each point of the sphere to its antipodal point is clearly an isometry, and hence equiareal (Exercise 5.18), we have

$$\mathcal{A}(A'BC) = \mathcal{A}(AB'C').$$

Inserting this into Eq. (12), we see that the term in { } on the right-hand side of Eq. (11) is equal to 2π. Rearranging now gives the result. □

In Chapter 11, we shall obtain a far-reaching generalization of this result in which the sphere is replaced by an arbitrary surface, and great circles by arbitrary curves on the surface.

EXERCISES

5.18 Show that every isometry is an equiareal map. Give an example of an equiareal map that is not an isometry.

5.19 Show that a map between surfaces that is both conformal and equiareal is an isometry.

5.20 A sailor circumnavigates Australia by a route consisting of a triangle whose sides are arcs of great circles. Prove that at least one interior angle of the triangle is $\geq \frac{\pi}{3} + \frac{10}{169}$ radians. (Take the Earth to be a sphere of radius 6500km and assume that the area of Australia is 7.5 million square km.)

5.21 The unit sphere in $\mathbf{R}^3$ is covered by triangles whose sides are arcs of great circles, and such that the intersection of any two triangles is either empty or a common edge or vertex of each triangle. Suppose that there are F triangles, E edges (a common edge of two triangles being counted only once) and V vertices (a common vertex of several triangles being counted only once). Show that $3F = 2E$. Deduce from Theorem 5.5 that $2V - F = 4$. Hence show that $V - E + F = 2$. (This result will be generalised in Chapter 11.)

5.22 Prove Theorem 5.3.

6

Curvature of Surfaces

In this chapter, we introduce several ways to measure how 'curved' a surface is. All of these rest ultimately on the *second fundamental form* of a surface patch. It turns out (see Theorem 10.4) that a surface patch is determined up to a rigid motion of $\mathbf{R}^3$ by its first and second fundamental forms, just as a unit-speed plane curve is determined up to a rigid motion by its signed curvature.

6.1. The Second Fundamental Form

To see how we might define the curvature of a surface, we start by finding a new interpretation of the curvature of a plane curve. Suppose then that $\boldsymbol{\gamma}$ is a unit-speed curve in $\mathbf{R}^2$. As the parameter t of $\boldsymbol{\gamma}$ changes to $t + \Delta t$, the curve moves away from its tangent line at $\boldsymbol{\gamma}(t)$ by a distance $(\boldsymbol{\gamma}(t + \Delta t) - \boldsymbol{\gamma}(t)).\mathbf{n}$, where $\mathbf{n}$ is the principal normal to $\boldsymbol{\gamma}$ at $\boldsymbol{\gamma}(t)$. By Taylor's theorem,

$$\boldsymbol{\gamma}(t + \Delta t) = \boldsymbol{\gamma}(t) + \dot{\boldsymbol{\gamma}}(t)\Delta t + \frac{1}{2}\ddot{\boldsymbol{\gamma}}(t)(\Delta t)^2 + \text{ remainder},$$

where (remainder)$/(\Delta t)^2$ tends to zero as Δt tends to zero. Now, $\mathbf{n}$ is perpendicular to the unit tangent vector $\mathbf{t} = \dot{\boldsymbol{\gamma}}$, and $\ddot{\boldsymbol{\gamma}} = \dot{\mathbf{t}} = \kappa\mathbf{n}$, where κ is the curvature of $\boldsymbol{\gamma}$. Hence, $\ddot{\boldsymbol{\gamma}}.\mathbf{n} = \kappa$ and the deviation of $\boldsymbol{\gamma}$ from its tangent line is

$$(\dot{\boldsymbol{\gamma}}(t)\Delta t + \frac{1}{2}\ddot{\boldsymbol{\gamma}}(t)(\Delta t)^2 + \cdots).\mathbf{n} = \frac{1}{2}\kappa(\Delta t)^2 + \text{remainder}. \quad (1)$$

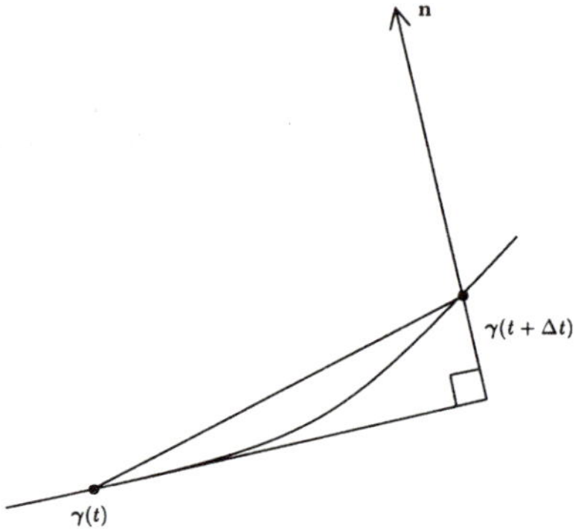

Now let $\boldsymbol{\sigma}$ be a surface patch in $\mathbf{R}^3$ with standard unit normal $\mathbf{N}$. As the parameters (u, v) of $\boldsymbol{\sigma}$ change to $(u + \Delta u, v + \Delta v)$, the surface moves away from its tangent plane at $\boldsymbol{\sigma}(u, v)$ by a distance

$$(\boldsymbol{\sigma}(u + \Delta u, v + \Delta v) - \boldsymbol{\sigma}(u, v)).\mathbf{N}.$$

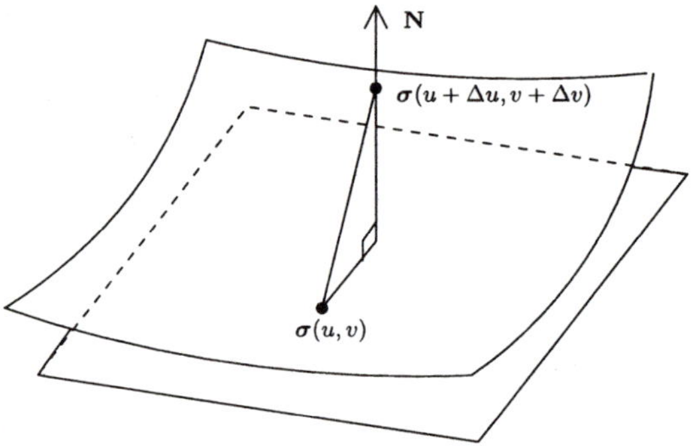

By the two variable form of Taylor's theorem, $\boldsymbol{\sigma}(u + \Delta u, v + \Delta v) - \boldsymbol{\sigma}(u, v)$ is equal to

$$\boldsymbol{\sigma}_u \Delta u + \boldsymbol{\sigma}_v \Delta v + \frac{1}{2}\left(\boldsymbol{\sigma}_{uu}(\Delta u)^2 + 2\boldsymbol{\sigma}_{uv}\Delta u \Delta v + \boldsymbol{\sigma}_{vv}(\Delta v)^2\right) + \text{ remainder},$$

where (remainder)/$((\Delta u)^2 + (\Delta v)^2)$ tends to zero as $(\Delta u)^2 + (\Delta v)^2$ tends to zero. Now $\boldsymbol{\sigma}_u$ and $\boldsymbol{\sigma}_v$ are tangent to the surface, hence perpendicular to $\mathbf{N}$, so the deviation of $\boldsymbol{\sigma}$ from its tangent plane is

$$\frac{1}{2}\left(L(\Delta u)^2 + 2M\Delta u \Delta v + N(\Delta v)^2\right) + \text{ remainder}, \tag{2}$$

where

$$L = \boldsymbol{\sigma}_{uu}.\mathbf{N}, \quad M = \boldsymbol{\sigma}_{uv}.\mathbf{N}, \quad N = \boldsymbol{\sigma}_{vv}.\mathbf{N}. \tag{3}$$

Comparing Eq. (2) with Eq. (1), we see that the expression

$$L(\Delta u)^2 + 2M\Delta u \Delta v + N(\Delta v)^2$$

is the analogue for the surface of the curvature term $\kappa(\Delta t)^2$ in the case of a curve. One calls the expression

$$Ldu^2 + 2Mdudv + Ndv^2 \tag{4}$$

the *second fundamental form* of $\boldsymbol{\sigma}$. As in the case of the first fundamental form, we regard the expression (4) simply as a convenient way of keeping track of the three functions L, M and N. We shall soon see that a knowledge of these functions (together with that of the first fundamental form) will enable us to compute the curvature of any curve on the surface $\boldsymbol{\sigma}$.

Example 6.1

Consider the plane

$$\boldsymbol{\sigma}(u, v) = \mathbf{a} + u\mathbf{p} + v\mathbf{q}$$

(see Example 4.1). Since $\boldsymbol{\sigma}_u = \mathbf{p}$ and $\boldsymbol{\sigma}_v = \mathbf{q}$ are constant vectors, we have $\boldsymbol{\sigma}_{uu} = \boldsymbol{\sigma}_{uv} = \boldsymbol{\sigma}_{vv} = \mathbf{0}$. Hence, the second fundamental form of a plane is zero.

Example 6.2

Consider a patch $\boldsymbol{\sigma}$ on a surface of revolution:

$$\boldsymbol{\sigma}(u, v) = (f(u)\cos v, f(u)\sin v, g(u)).$$

Recall from Example 4.12 that we can assume that $f(u) > 0$ for all values of u and that the profile curve $u \mapsto (f(u), 0, g(u))$ is unit-speed, i.e. $\dot{f}^2 + \dot{g}^2 = 1$ (a dot denoting d/du). Then:

$$\begin{aligned}
&\boldsymbol{\sigma}_u = (\dot{f}\cos v, \dot{f}\sin v, \dot{g}), \quad \boldsymbol{\sigma}_v = (-f\sin v, f\cos v, 0),\\
\therefore \quad & E = \| \boldsymbol{\sigma}_u \|^2 = \dot{f}^2 + \dot{g}^2 = 1, \ F = \boldsymbol{\sigma}_u . \boldsymbol{\sigma}_v = 0, \ G = \| \boldsymbol{\sigma}_v \|^2 = f^2,\\
\therefore \quad & \boldsymbol{\sigma}_u \times \boldsymbol{\sigma}_v = (-f\dot{g}\cos v, -f\dot{g}\sin v, f\dot{f}),\\
\therefore \quad & \| \boldsymbol{\sigma}_u \times \boldsymbol{\sigma}_v \| = f \quad (\text{since } \dot{f}^2 + \dot{g}^2 = 1),\\
\therefore \quad & \mathbf{N} = \frac{\boldsymbol{\sigma}_u \times \boldsymbol{\sigma}_v}{\| \boldsymbol{\sigma}_u \times \boldsymbol{\sigma}_v \|} = (-\dot{g}\cos v, -\dot{g}\sin v, \dot{f}),\\
& \boldsymbol{\sigma}_{uu} = (\ddot{f}\cos v, \ddot{f}\sin v, \ddot{g}),\\
& \boldsymbol{\sigma}_{uv} = (-\dot{f}\sin v, \dot{f}\cos v, 0),\\
& \boldsymbol{\sigma}_{vv} = (-f\cos v, -f\sin v, 0),\\
\therefore \quad & L = \boldsymbol{\sigma}_{uu}.\mathbf{N} = \dot{f}\ddot{g} - \ddot{f}\dot{g}, \ M = \boldsymbol{\sigma}_{uv}.\mathbf{N} = 0, \ N = \boldsymbol{\sigma}_{vv}.\mathbf{N} = f\dot{g},
\end{aligned}$$

so the second fundamental form is

$$(\dot{f}\ddot{g} - \ddot{f}\dot{g})du^2 + f\dot{g}dv^2.$$

If the surface is the unit sphere, we can take $f(u) = \cos u$, $g(u) = \sin u$, with $-\pi/2 < u < \pi/2$, giving

$$L = 1, \quad M = 0, \quad N = \cos^2 u.$$

(Note that the conditions $f > 0$ and $\dot{f}^2 + \dot{g}^2 = 1$ are satisfied.) Replacing u and v by the more usual θ and φ, we get the second fundamental form of the unit sphere:

$$d\theta^2 + \cos^2\theta \, d\varphi^2.$$

If the surface is a circular cylinder of unit radius, we can take $f(u) = 1$, $g(u) = u$ (again, the conditions $f > 0$ and $\dot{f}^2 + \dot{g}^2 = 1$ are satisfied). This gives

$$L = M = 0, \quad N = 1,$$

so the second fundamental form of the cylinder is dv^2.

EXERCISES

6.1 Compute the second fundamental form of the elliptic paraboloid

$$\boldsymbol{\sigma}(u, v) = (u, v, u^2 + v^2).$$

6.2 The second fundamental form of a surface patch $\boldsymbol{\sigma}$ is zero everywhere. Prove that $\boldsymbol{\sigma}$ is part of a plane. (By computing expressions such as $(\boldsymbol{\sigma}_u.\mathbf{N})_u$, prove that $\mathbf{N}_u$ and $\mathbf{N}_v$ are perpendicular to $\boldsymbol{\sigma}_u$ and $\boldsymbol{\sigma}_v$, and deduce that the unit normal $\mathbf{N}$ of $\boldsymbol{\sigma}$ is a constant vector.) This is the analogue for surfaces of the theorem that a curve with zero curvature everywhere is part of a straight line.

6.3 Let a surface patch $\tilde{\boldsymbol{\sigma}}(\tilde{u}, \tilde{v})$ be a reparametrisation of a surface patch $\boldsymbol{\sigma}(u, v)$ with reparametrisation map $(u, v) = \Phi(\tilde{u}, \tilde{v})$. Prove that

$$\begin{pmatrix} \tilde{L} & \tilde{M} \\ \tilde{M} & \tilde{N} \end{pmatrix} = \pm J^t \begin{pmatrix} L & M \\ M & N \end{pmatrix} J,$$

where J is the jacobian matrix of Φ and we take the plus sign if $\det(J) > 0$ and the minus sign if $\det(J) < 0$.

6.4 Show that the second fundamental form of a surface patch is unchanged by applying a rigid motion to the patch.

6.2. The Curvature of Curves on a Surface

Another natural way to investigate how much a surface curves is to look at the curvature of various curves on the surface. If $\boldsymbol{\gamma}(t) = \boldsymbol{\sigma}(u(t), v(t))$ is a unit-speed curve in a surface patch $\boldsymbol{\sigma}$, then $\dot{\boldsymbol{\gamma}}$ is a unit vector and is, by definition, a tangent vector to $\boldsymbol{\sigma}$. Hence, $\dot{\boldsymbol{\gamma}}$ is perpendicular to the standard unit normal $\mathbf{N}$ of $\boldsymbol{\sigma}$, so $\dot{\boldsymbol{\gamma}}$, $\mathbf{N}$ and $\mathbf{N} \times \dot{\boldsymbol{\gamma}}$ are mutually perpendicular unit vectors. Again since $\boldsymbol{\gamma}$ is unit-speed, $\ddot{\boldsymbol{\gamma}}$ is perpendicular to $\dot{\boldsymbol{\gamma}}$, and hence is a linear combination of $\mathbf{N}$ and $\mathbf{N} \times \dot{\boldsymbol{\gamma}}$:

$$\ddot{\boldsymbol{\gamma}} = \kappa_n \mathbf{N} + \kappa_g \mathbf{N} \times \dot{\boldsymbol{\gamma}}. \qquad (5)$$

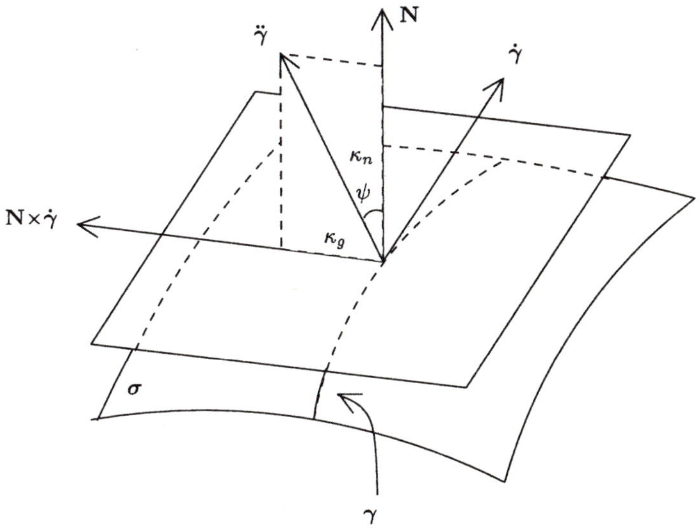

The scalars κ_n and κ_g are called the *normal curvature* and the *geodesic curvature* of $\boldsymbol{\gamma}$, respectively. Since $\mathbf{N}$ and $\mathbf{N} \times \dot{\boldsymbol{\gamma}}$ are perpendicular unit vectors, Eq. (5) implies that

$$\kappa_n = \ddot{\boldsymbol{\gamma}}.\mathbf{N}, \quad \kappa_g = \ddot{\boldsymbol{\gamma}}.(\mathbf{N} \times \dot{\boldsymbol{\gamma}})$$

and

$$\| \ddot{\boldsymbol{\gamma}} \|^2 = \kappa_n^2 + \kappa_g^2.$$

Hence, the curvature $\kappa = \| \ddot{\boldsymbol{\gamma}} \|$ of $\boldsymbol{\gamma}$ is given by

$$\kappa^2 = \kappa_n^2 + \kappa_g^2. \qquad (6)$$

Moreover, if $\mathbf{n}$ is the principal normal of $\boldsymbol{\gamma}$, so that $\ddot{\boldsymbol{\gamma}} = \kappa\mathbf{n}$, we have

$$\kappa_n = \kappa\mathbf{n}.\mathbf{N} = \kappa\cos\psi, \qquad (7)$$

where ψ is the angle between $\mathbf{n}$ and $\mathbf{N}$. Then, from Eq. (6),

$$\kappa_g = \pm\kappa\sin\psi. \qquad (8)$$

It is clear from their definition that κ_n and κ_g either stay the same or both change sign when $\boldsymbol{\sigma}$ is reparametrised (since this is the case for $\mathbf{N}$).

If $\boldsymbol{\gamma}$ is regular, but not necessarily unit-speed, we define the geodesic and normal curvatures of $\boldsymbol{\gamma}$ to be those of a unit-speed reparametrisation of $\boldsymbol{\gamma}$. When a unit-speed parameter t is changed to another such parameter $\pm t + c$, where c is a constant, it is clear that $\kappa_n \mapsto \kappa_n$ and $\kappa_g \mapsto \pm\kappa_g$, so κ_n is well defined for any regular curve, while κ_g is well defined up to sign. Equations (7) and (8) continue to hold in this more general situation.

An important special case is that where $\boldsymbol{\gamma}$ is a *normal section* of the surface, i.e. $\boldsymbol{\gamma}$ is the intersection of the surface with a plane Π that is perpendicular to the tangent plane of the surface at every point of $\boldsymbol{\gamma}$.

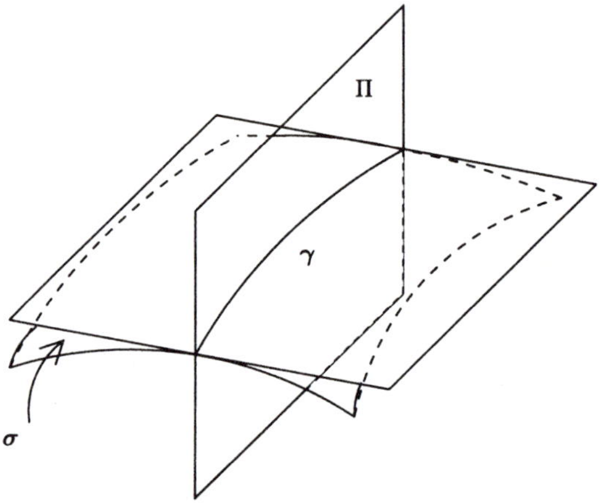

Since $\boldsymbol{\gamma}$ lies in Π, the principal normal $\mathbf{n}$ is parallel to Π, and since Π is perpendicular to the tangent plane, $\mathbf{N}$ is also parallel to Π. Since $\mathbf{n}$ and $\mathbf{N}$ are both perpendicular to $\dot{\boldsymbol{\gamma}}$, and since $\dot{\boldsymbol{\gamma}}$ is parallel to Π, $\mathbf{n}$ and $\mathbf{N}$ must be parallel to each other, i.e. $\psi = 0$ or π. From Eqs. (7) and (8), we deduce that

$$\kappa_n = \pm\kappa, \quad \kappa_g = 0$$

for a normal section.

We shall now study the normal curvature κ_n in more detail. The study of κ_g will be taken up in Chapter 8.

EXERCISES

6.5 Compute the normal curvature of the circle $\boldsymbol{\gamma}(t) = (\cos t, \sin t, 1)$ on the elliptic paraboloid $\boldsymbol{\sigma}(u, v) = (u, v, u^2 + v^2)$ (see Exercise 6.1).

6.6 Show that if a curve on a surface has zero normal and geodesic curvature everywhere, it is part of a straight line.

6.7 Show that the normal curvature of any curve on a sphere of radius r is $\pm 1/r$.

6.8 Compute the geodesic curvature of any circle on a sphere (not necessarily a great circle).

6.9 Consider the surface of revolution

$$\boldsymbol{\sigma}(u, v) = (f(u) \cos v, f(u) \sin v, g(u)),$$

where $u \mapsto (f(u), 0, g(u))$ is a unit-speed curve in $\mathbf{R}^3$. Compute the geodesic curvature of

(i) a meridian $v =$ constant;
(ii) a parallel $u =$ constant.

6.10 A unit-speed curve $\boldsymbol{\gamma}$ with curvature $\kappa > 0$ and principal normal $\mathbf{n}$ forms the intersection of two surfaces $\mathcal{S}_1$ and $\mathcal{S}_2$ with unit normals $\mathbf{N}_1$ and $\mathbf{N}_2$. Show that, if κ_1 and κ_2 are the normal curvatures of $\boldsymbol{\gamma}$ when viewed as a curve in $\mathcal{S}_1$ and $\mathcal{S}_2$, respectively, then

$$\kappa_1 \mathbf{N}_2 - \kappa_2 \mathbf{N}_1 = \kappa(\mathbf{N}_1 \times \mathbf{N}_2) \times \mathbf{n}.$$

Deduce that, if α is the angle between the two surfaces,

$$\kappa^2 \sin^2 \alpha = \kappa_1^2 + \kappa_2^2 - 2\kappa_1 \kappa_2 \cos \alpha.$$

6.11 Let $\boldsymbol{\gamma}$ be a unit-speed curve on a surface patch $\boldsymbol{\sigma}$ with curvature $\kappa > 0$. Let ψ be the angle between $\ddot{\boldsymbol{\gamma}}$ and $\mathbf{N}$, and let $\mathbf{B} = \mathbf{t} \times \mathbf{N}$ (in the usual notation). Show that

$$\mathbf{N} = \mathbf{n} \cos \psi + \mathbf{b} \sin \psi, \quad \mathbf{B} = \mathbf{b} \cos \psi - \mathbf{n} \sin \psi.$$

Deduce that

$$\dot{\mathbf{t}} = \kappa_n \mathbf{N} - \kappa_g \mathbf{B}, \quad \dot{\mathbf{N}} = -\kappa_n \mathbf{t} + \tau_g \mathbf{B}, \quad \dot{\mathbf{B}} = \kappa_g \mathbf{t} - \tau_g \mathbf{N},$$

where $\tau_g = \tau + \dot{\psi}$. ($\tau_g$ is called the *geodesic torsion* of $\boldsymbol{\gamma}$; cf. Exercise 8.4.)

6.12 A curve $\boldsymbol{\gamma}$ on a surface $\mathcal{S}$ is called *asymptotic* if its normal curvature is everywhere zero. Show that any straight line on a surface is an asymptotic curve. Show also that a curve $\boldsymbol{\gamma}$ with positive curvature is asymptotic if and only if its binormal $\mathbf{b}$ is parallel to the unit normal of $\mathcal{S}$ at all points of $\boldsymbol{\gamma}$.

6.13 Prove that the asymptotic curves on the surface

$$\boldsymbol{\sigma}(u, v) = (u\cos v, u\sin v, \ln u)$$

are given by

$$\ln u = \pm(v + c),$$

where c is an arbitrary constant.

6.14 Show that an asymptotic curve with positive curvature has torsion equal to its geodesic torsion (see Exercise 6.11). (Show that $\mathbf{B}$ is parallel to $\mathbf{n}$.)

6.3. The Normal and Principal Curvatures

The most important single fact about the normal curvature κ_n of a curve $\boldsymbol{\gamma}$ on a surface $\boldsymbol{\sigma}$ is contained in

Proposition 6.1

If $\gamma(t) = \boldsymbol{\sigma}(u(t), v(t))$ is a unit-speed curve on a surface patch $\boldsymbol{\sigma}$, its normal curvature is given by

$$\kappa_n = L\dot{u}^2 + 2M\dot{u}\dot{v} + N\dot{v}^2,$$

where $Ldu^2 + 2Mdudv + Ndv^2$ is the second fundamental form of $\boldsymbol{\sigma}$.

This result means that two unit-speed curves passing through a point P on a surface and with the same tangent vector at P have the same normal curvature at P, since both κ_n and the tangent vector $\dot{\boldsymbol{\gamma}} = \boldsymbol{\sigma}_u\dot{u} + \boldsymbol{\sigma}_v\dot{v}$ depend only on $u, v, \dot{u}$ and $\dot{v}$ (and not on any higher derivatives of u and v).

Proof 6.1

We have, with $\mathbf{N}$ denoting the standard unit normal of $\boldsymbol{\sigma}$,

$$\begin{aligned}\kappa_n &= \mathbf{N}.\ddot{\gamma} = \mathbf{N}.\frac{d}{dt}(\dot{\gamma}) = \mathbf{N}.\frac{d}{dt}(\boldsymbol{\sigma}_u\dot{u} + \boldsymbol{\sigma}_v\dot{v}) \\ &= \mathbf{N}.(\boldsymbol{\sigma}_u\ddot{u} + \boldsymbol{\sigma}_v\ddot{v} + (\boldsymbol{\sigma}_{uu}\dot{u} + \boldsymbol{\sigma}_{uv}\dot{v})\dot{u} + (\boldsymbol{\sigma}_{uv}\dot{u} + \boldsymbol{\sigma}_{vv}\dot{v})\dot{v}) \\ &= L\dot{u}^2 + 2M\dot{u}\dot{v} + N\dot{v}^2,\end{aligned}$$

using the definition (3) of L, M and N, and the fact that $\mathbf{N}$ is perpendicular to $\boldsymbol{\sigma}_u$ and $\boldsymbol{\sigma}_v$. □

This proposition implies the following classical result, which takes longer to state than to prove.

Proposition 6.2 (Meusnier's Theorem)

Let P be a point on a surface $\mathcal{S}$ and let $\mathbf{v}$ be a unit tangent vector to $\mathcal{S}$ at P. Let Π_θ be the plane containing the line through P parallel to $\mathbf{v}$ and making an angle θ with the tangent plane to $\mathcal{S}$ at P. Suppose that Π_θ intersects $\mathcal{S}$ in a curve with curvature κ_θ. Then, $\kappa_\theta \sin\theta$ is independent of θ.

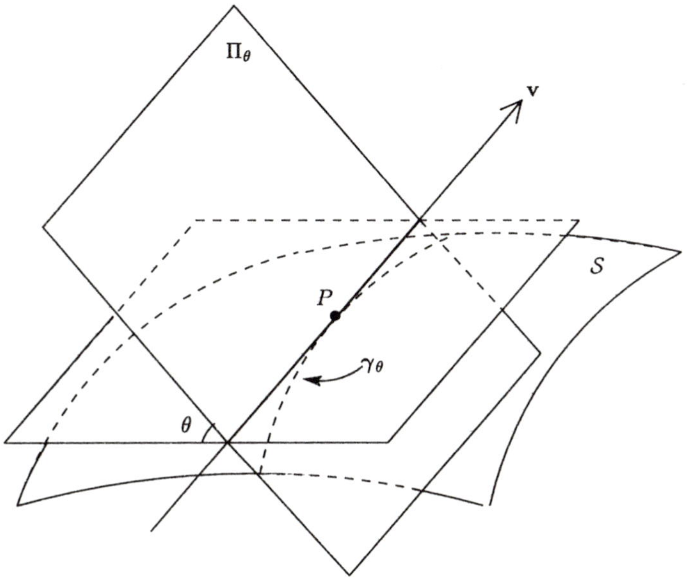

Proof 6.2

Assume that $\boldsymbol{\gamma}_\theta$ is a unit-speed parametrisation of the curve of intersection of Π_θ and $\mathcal{S}$. Then, at P, $\dot{\boldsymbol{\gamma}}_\theta = \pm\mathbf{v}$, so $\ddot{\boldsymbol{\gamma}}_\theta$ is perpendicular to $\mathbf{v}$ and is parallel to Π_θ. Thus, in the notation of Section 6.1, $\psi = \pi/2 - \theta$ and so Eq. (7) gives

$$\kappa_\theta \sin\theta = \kappa_n.$$

But κ_n depends only on P and $\mathbf{v}$, and not on θ. □

To analyse κ_n further, it is useful to use matrix notation. If

$$Edu^2 + 2Fdudv + Gdv^2 \quad \text{and} \quad Ldu^2 + 2Mdudv + Ndv^2$$

are the first and second fundamental forms of a surface $\boldsymbol{\sigma}$, we introduce the following symmetric 2×2 matrices:

$$\mathcal{F}_I = \begin{pmatrix} E & F \\ F & G \end{pmatrix}, \quad \mathcal{F}_{II} = \begin{pmatrix} L & M \\ M & N \end{pmatrix}.$$

Let

$$\mathbf{t}_1 = \xi_1 \boldsymbol{\sigma}_u + \eta_1 \boldsymbol{\sigma}_v, \quad \mathbf{t}_2 = \xi_2 \boldsymbol{\sigma}_u + \eta_2 \boldsymbol{\sigma}_v$$

be two tangent vectors at some point of $\boldsymbol{\sigma}$. Then,

$$\begin{aligned} \mathbf{t}_1 . \mathbf{t}_2 &= (\xi_1 \boldsymbol{\sigma}_u + \eta_1 \boldsymbol{\sigma}_v).(\xi_2 \boldsymbol{\sigma}_u + \eta_2 \boldsymbol{\sigma}_v) \\ &= E\xi_1\xi_2 + F(\xi_1\eta_2 + \xi_2\eta_1) + G\eta_1\eta_2 \\ &= (\,\xi_1 \quad \eta_1\,) \begin{pmatrix} E & F \\ F & G \end{pmatrix} \begin{pmatrix} \xi_2 \\ \eta_2 \end{pmatrix}. \end{aligned}$$

Thus, writing

$$T_1 = \begin{pmatrix} \xi_1 \\ \eta_1 \end{pmatrix}, \quad T_2 = \begin{pmatrix} \xi_2 \\ \eta_2 \end{pmatrix},$$

we get

$$\mathbf{t}_1 . \mathbf{t}_2 = T_1^t \mathcal{F}_I T_2. \tag{9}$$

On the other hand, the tangent vector $\dot{\boldsymbol{\gamma}} = \dot{u}\boldsymbol{\sigma}_u + \dot{v}\boldsymbol{\sigma}_v$, and if $T = \begin{pmatrix} \dot{u} \\ \dot{v} \end{pmatrix}$, then by using Proposition 6.1 we see by a similar matrix calculation that

$$\kappa_n = T^t \mathcal{F}_{II} T. \tag{10}$$

The justification for the next two definitions will appear in Proposition 6.3 and Corollary 6.1.

Definition 6.1

The *principal curvatures* of a surface patch are the roots of the equation

$$\det(\mathcal{F}_{II} - \kappa \mathcal{F}_I) = 0, \tag{11}$$

i.e.

$$\begin{vmatrix} L - \kappa E & M - \kappa F \\ M - \kappa F & N - \kappa G \end{vmatrix} = 0. \tag{12}$$

Since (12) is a quadratic equation for κ, there are two roots. *A priori*, these may be complex numbers. However, we shall prove in the next proposition that the principal curvatures are always real. To motivate this result, note that if $\mathcal{F}_I$ happens to be the identity matrix (as it is for the standard parametrisation of the plane, for example), Eq. (11) would become the equation for the eigenvalues of $\mathcal{F}_{II}$. But a standard result from linear algebra states that the eigenvalues of any real symmetric matrix, such as $\mathcal{F}_{II}$, are real numbers. In general, $\mathcal{F}_I$ is not the identity matrix, but it is always invertible (see the remark following Proposition 5.2), so Eq. (11) is equivalent to

$$\det(\mathcal{F}_I(\mathcal{F}_I^{-1}\mathcal{F}_{II} - \kappa)) = 0,$$
$$\therefore \quad \det(\mathcal{F}_I)\det(\mathcal{F}_I^{-1}\mathcal{F}_{II} - \kappa) = 0,$$
$$\therefore \quad \det(\mathcal{F}_I^{-1}\mathcal{F}_{II} - \kappa) = 0,$$

and hence *the principal curvatures are the eigenvalues of* $\mathcal{F}_I^{-1}\mathcal{F}_{II}$. However, $\mathcal{F}_I^{-1}\mathcal{F}_{II}$ is *not usually symmetric*, so the above result from linear algebra does not immediately imply the reality of the principal curvatures in the general case.

If κ is one of the principal curvatures, Eq. (11) says that $\mathcal{F}_{II} - \kappa\mathcal{F}_I$ is not invertible, so, assuming that κ is real, there is a non-zero 2×1 column matrix T with real number entries such that

$$(\mathcal{F}_{II} - \kappa\mathcal{F}_I)T = 0. \tag{13}$$

Definition 6.2

If $T = \begin{pmatrix} \xi \\ \eta \end{pmatrix}$ satisfies Eq. (13), the corresponding tangent vector $\mathbf{t} = \xi\boldsymbol{\sigma}_u + \eta\boldsymbol{\sigma}_v$ to the surface $\boldsymbol{\sigma}(u, v)$ is called a *principal vector* corresponding to the principal curvature κ.

Proposition 6.3

Let κ_1 and κ_2 be the principal curvatures at a point P of a surface patch $\boldsymbol{\sigma}$. Then,

(i) κ_1 and κ_2 are real numbers;

(ii) if $\kappa_1 = \kappa_2 = \kappa$, say, then $\mathcal{F}_{II} = \kappa\mathcal{F}_I$ and (hence) every tangent vector to $\boldsymbol{\sigma}$ at P is a principal vector;

(iii) if $\kappa_1 \neq \kappa_2$, then any two (non-zero) principal vectors $\mathbf{t}_1$ and $\mathbf{t}_2$ corresponding to κ_1 and κ_2, respectively, are perpendicular.

In case (ii), P is called an *umbilic*.

Proof 6.3

For (i), let $\mathbf{t}_1$ and $\mathbf{t}_2$ be *any* two perpendicular unit tangent vectors to the surface at P (not yet known to be principal vectors). Define ξ_i, η_i and T_i, for $i = 1, 2$ as in the discussion preceding Definition 6.1, and let

$$A = \begin{pmatrix} \xi_1 & \xi_2 \\ \eta_1 & \eta_2 \end{pmatrix}.$$

By multiplying out the matrices, it is easy to check that

$$\begin{aligned} A^t \mathcal{F}_I A &= \begin{pmatrix} T_1^t \mathcal{F}_I T_1 & T_1^t \mathcal{F}_I T_2 \\ T_2^t \mathcal{F}_I T_1 & T_2^t \mathcal{F}_I T_2 \end{pmatrix} \\ &= \begin{pmatrix} \mathbf{t}_1.\mathbf{t}_1 & \mathbf{t}_1.\mathbf{t}_2 \\ \mathbf{t}_2.\mathbf{t}_1 & \mathbf{t}_2.\mathbf{t}_2 \end{pmatrix} \quad \text{(by Eq. (9))} \\ &= \begin{pmatrix} 1 & 0 \\ 0 & 1 \end{pmatrix}, \end{aligned}$$

since $\mathbf{t}_1$ and $\mathbf{t}_2$ are perpendicular unit vectors. Let

$$\mathcal{G}_{II} = A^t \mathcal{F}_{II} A.$$

Then, $\mathcal{G}_{II}$ is still (real and) symmetric because

$$\mathcal{G}_{II}^t = A^t \mathcal{F}_{II}^t (A^t)^t = A^t \mathcal{F}_{II} A = \mathcal{G}_{II}.$$

By the theorem from linear algebra referred to above, there is an orthogonal matrix B (so that $B^t B = I$), say, such that

$$B^t \mathcal{G}_{II} B = \begin{pmatrix} \lambda_1 & 0 \\ 0 & \lambda_2 \end{pmatrix},$$

for some *real* numbers λ_1 and λ_2. Let $C = AB$. Then,

$$C^t \mathcal{F}_I C = B^t (A^t \mathcal{F}_I A) B = B^t B = \begin{pmatrix} 1 & 0 \\ 0 & 1 \end{pmatrix}, \tag{14}$$

because B is orthogonal, and

$$C^t \mathcal{F}_{II} C = B^t (A^t \mathcal{F}_{II} A) B = B^t \mathcal{G}_{II} B = \begin{pmatrix} \lambda_1 & 0 \\ 0 & \lambda_2 \end{pmatrix}. \tag{15}$$

Now C is invertible (being the product of two invertible matrices), so

$$\det(\mathcal{F}_{II} - \kappa \mathcal{F}_I) = 0 \ \text{ if and only if } \ \det(C^t(\mathcal{F}_{II} - \kappa \mathcal{F}_I)C) = 0,$$

$$\therefore \det(\mathcal{F}_{II} - \kappa \mathcal{F}_I) = 0 \ \text{ if and only if } \ \det\left(\begin{pmatrix} \lambda_1 & 0 \\ 0 & \lambda_2 \end{pmatrix} - \kappa \begin{pmatrix} 1 & 0 \\ 0 & 1 \end{pmatrix} \right) = 0.$$

Hence, the principal curvatures are the roots of

$$\begin{vmatrix} \lambda_1 - \kappa & 0 \\ 0 & \lambda_2 - \kappa \end{vmatrix} = 0,$$

i.e. λ_1 and λ_2.

For (ii), suppose that the principal curvatures are equal, to κ, say. Then $\lambda_1 = \lambda_2 = \kappa$ and Eqs. (14) and (15) give

$$C^t\mathcal{F}_I C = I, \quad C^t\mathcal{F}_{II} C = \kappa I,$$
$$\therefore \quad C^t(\mathcal{F}_{II} - \kappa\mathcal{F}_I)C = 0,$$
$$\therefore \quad \mathcal{F}_{II} - \kappa\mathcal{F}_I = 0,$$

because C and C^t are invertible. Obviously then, if T is *any* 2×1 column matrix,

$$(\mathcal{F}_{II} - \kappa\mathcal{F}_I)T = 0.$$

It follows that *every* tangent vector to $\boldsymbol{\sigma}$ at P is a principal vector.

Finally, for (iii) let

$$\mathbf{t}_i = \xi_i\boldsymbol{\sigma}_u + \eta_i\boldsymbol{\sigma}_v, \quad T_i = \begin{pmatrix} \xi_i \\ \eta_i \end{pmatrix},$$

for $i = 1, 2$. Then, by Eq. (9),

$$\mathbf{t}_1.\mathbf{t}_2 = T_1^t\mathcal{F}_I T_2,$$

and by Eq. (13),

$$\mathcal{F}_{II}T_1 = \kappa_1\mathcal{F}_I T_1, \quad \mathcal{F}_{II}T_2 = \kappa_2\mathcal{F}_I T_2. \tag{16}$$

Hence,

$$T_2^t\mathcal{F}_{II}T_1 = \kappa_1(\mathbf{t}_1.\mathbf{t}_2), \quad T_1^t\mathcal{F}_{II}T_2 = \kappa_2(\mathbf{t}_1.\mathbf{t}_2). \tag{17}$$

But since $T_1^t\mathcal{F}_{II}T_2$ is a 1×1 matrix, it is equal to its transpose:

$$T_1^t\mathcal{F}_{II}T_2 = (T_1^t\mathcal{F}_{II}T_2)^t = T_2^t\mathcal{F}_{II}^t T_1 = T_2^t\mathcal{F}_{II}T_1,$$

where the last equality uses the fact that $\mathcal{F}_{II}$ is symmetric. Hence, Eq. (17) gives

$$\kappa_1(\mathbf{t}_1.\mathbf{t}_2) = \kappa_2(\mathbf{t}_1.\mathbf{t}_2),$$

so if $\kappa_1 \neq \kappa_2$, then $\mathbf{t}_1.\mathbf{t}_2 = 0$, i.e. $\mathbf{t}_1$ and $\mathbf{t}_2$ are perpendicular. □

Example 6.3

It is intuitively clear that a sphere curves the same amount in every direction, and at every point of the sphere. Thus, we expect that the principal curvatures of a sphere are equal to each other at every point, and are constant over the sphere. To confirm this by calculation, we use the latitude longitude parametrisation as usual. We found in Example 5.2 that

$$E = 1, \quad F = 0, \quad G = \cos^2\theta,$$

and in Example 6.2 that

$$L = 1, \quad M = 0, \quad N = \cos^2\theta.$$

So the principal curvatures are the roots of

$$\begin{vmatrix} 1-\kappa & 0 \\ 0 & \cos^2\theta - \kappa\cos^2\theta \end{vmatrix} = 0,$$

i.e. $\kappa = 1$ (repeated root), as we expected. Any tangent vector is a principal vector.

Example 6.4

We consider the circular cylinder of radius one and axis the z-axis, parametrised in the usual way:

$$\boldsymbol{\sigma}(u, v) = (\cos v, \sin v, u).$$

We found in Example 5.3 that

$$E = 1, \quad F = 0, \quad G = 1,$$

and in Example 6.2 that

$$L = 0, \quad M = 0, \quad N = 1.$$

So the principal curvatures are the roots of

$$\begin{vmatrix} 0-\kappa & 0 \\ 0 & 1-\kappa \end{vmatrix} = 0,$$

$$\therefore \quad \kappa(\kappa - 1) = 0,$$

$$\therefore \quad \kappa = 0 \text{ or } 1.$$

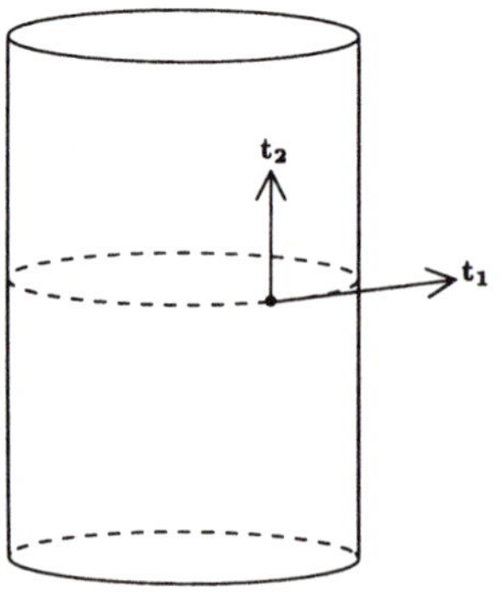

To find the principal vectors $\mathbf{t}_1$ and $\mathbf{t}_2$, recall that $\mathbf{t}_i = \xi_i \boldsymbol{\sigma}_u + \eta_i \boldsymbol{\sigma}_v$, where $T_i = \begin{pmatrix} \xi_i \\ \eta_i \end{pmatrix}$ satisfies

$$(\mathcal{F}_{II} - \kappa_i \mathcal{F}_I)T_i = 0,$$

$$\text{i.e.} \quad \begin{pmatrix} L - \kappa_i E & M - \kappa_i F \\ M - \kappa_i F & N - \kappa_i G \end{pmatrix} \begin{pmatrix} \xi_i \\ \eta_i \end{pmatrix} = 0.$$

For κ_1 $(= 1)$, we get

$$\begin{pmatrix} -1 & 0 \\ 0 & 0 \end{pmatrix} \begin{pmatrix} \xi_1 \\ \eta_1 \end{pmatrix} = 0,$$

$$\therefore \quad \xi_1 = 0.$$

So T_1 is a multiple of $\begin{pmatrix} 0 \\ 1 \end{pmatrix}$, and hence $\mathbf{t}_1$ is a multiple of $0\boldsymbol{\sigma}_u + 1\boldsymbol{\sigma}_v = \boldsymbol{\sigma}_v = (-\sin v, \cos v, 0)$. Similarly, for κ_2 $(= 0)$, one finds that T_2 is a multiple of $\begin{pmatrix} 1 \\ 0 \end{pmatrix}$, and hence $\mathbf{t}_2$ is a multiple of $\boldsymbol{\sigma}_u = (0, 0, 1)$.

As we mentioned above, one reason for introducing the principal curvatures and principal vectors is contained in the following result, which shows that, if we know the principal curvatures and principal vectors of a surface, it is easy to calculate the normal curvature of any curve on the surface:

Corollary 6.1 (Euler's Theorem)

Let $\boldsymbol{\gamma}$ be a curve on a surface patch $\boldsymbol{\sigma}$, and let κ_1 and κ_2 be the principal curvatures of $\boldsymbol{\sigma}$, with non-zero principal vectors $\mathbf{t}_1$ and $\mathbf{t}_2$. Then, the normal curvature of $\boldsymbol{\gamma}$ is

$$\kappa_n = \kappa_1 \cos^2 \theta + \kappa_2 \sin^2 \theta,$$

where θ is the angle between $\dot{\boldsymbol{\gamma}}$ and $\mathbf{t}_1$.

Proof 6.1

We can assume that $\boldsymbol{\gamma}$ is unit-speed. Let $\mathbf{t}$ be the tangent vector of $\boldsymbol{\gamma}$, and let $\mathbf{t} = \xi \boldsymbol{\sigma}_u + \eta \boldsymbol{\sigma}_v$, $T = \begin{pmatrix} \xi \\ \eta \end{pmatrix}$. Suppose first that $\kappa_1 = \kappa_2 = \kappa$, say. By Proposition 6.3(ii), the normal curvature of $\boldsymbol{\gamma}$ is

$$\kappa_n = T^t \mathcal{F}_{II} T = \kappa T^t \mathcal{F}_I T = \kappa\, \mathbf{t}.\mathbf{t} = \kappa.$$

This agrees with the formula in the statement of the corollary, since

$$\kappa_1 \cos^2 \theta + \kappa_2 \sin^2 \theta = \kappa(\cos^2 \theta + \sin^2 \theta) = \kappa.$$

Assume now that $\kappa_1 \neq \kappa_2$, so that $\mathbf{t}_1$ and $\mathbf{t}_2$ are perpendicular by Proposition 6.3(iii). We might as well assume that $\mathbf{t}_1$ and $\mathbf{t}_2$ are unit vectors. Let

$$\mathbf{t}_i = \xi_i \boldsymbol{\sigma}_u + \eta_i \boldsymbol{\sigma}_v, \quad T_i = \begin{pmatrix} \xi_i \\ \eta_i \end{pmatrix},$$

for $i = 1, 2$. Now

$$\dot{\boldsymbol{\gamma}} = \cos\theta \mathbf{t}_1 + \sin\theta \mathbf{t}_2,$$

so

$$\cos\theta(\xi_1 \boldsymbol{\sigma}_u + \eta_1 \boldsymbol{\sigma}_v) + \sin\theta(\xi_2 \boldsymbol{\sigma}_u + \eta_2 \boldsymbol{\sigma}_v) = \xi \boldsymbol{\sigma}_u + \eta \boldsymbol{\sigma}_v,$$
$$\therefore \quad \xi_1 \cos\theta + \xi_2 \sin\theta = \xi, \quad \eta_1 \cos\theta + \eta_2 \sin\theta = \eta,$$
$$\therefore \quad \begin{pmatrix} \xi \\ \eta \end{pmatrix} = \cos\theta \begin{pmatrix} \xi_1 \\ \eta_1 \end{pmatrix} + \sin\theta \begin{pmatrix} \xi_2 \\ \eta_2 \end{pmatrix},$$
$$\therefore \quad T = \cos\theta T_1 + \sin\theta T_2.$$

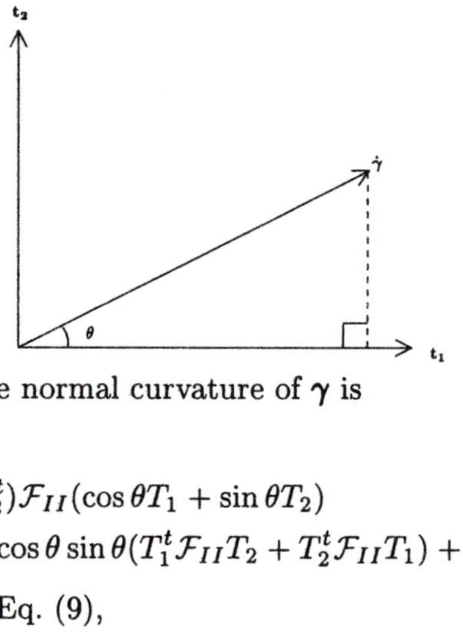

Hence, by Eq. (10), the normal curvature of $\boldsymbol{\gamma}$ is

$$\begin{aligned} \kappa_n &= T^t \mathcal{F}_{II} T \\ &= (\cos\theta T_1^t + \sin\theta T_2^t)\mathcal{F}_{II}(\cos\theta T_1 + \sin\theta T_2) \\ &= \cos^2\theta T_1^t \mathcal{F}_{II} T_1 + \cos\theta\sin\theta(T_1^t \mathcal{F}_{II} T_2 + T_2^t \mathcal{F}_{II} T_1) + \sin^2\theta T_2^t \mathcal{F}_{II} T_2. \end{aligned} \tag{18}$$

By Definition 6.2 and Eq. (9),

$$T_i^t \mathcal{F}_{II} T_j = \kappa_i T_i^t \mathcal{F}_I T_j = \begin{cases} \kappa_i & \text{if } i = j, \\ 0 & \text{otherwise.} \end{cases} \tag{19}$$

Substituting this into Eq. (18) gives the result. □

Corollary 6.2

The principal curvatures at a point of a surface are the maximum and minimum values of the normal curvature of all curves on the surface that pass through the point. Moreover, the principal vectors are the tangent vectors of the curves giving these maximum and minimum values.

Proof 6.2

If the principal curvatures κ_1 and κ_2 are different, we might as well suppose that $\kappa_1 > \kappa_2$. Let κ_n be the normal curvature of a curve $\boldsymbol{\gamma}$ on the surface. Then, since

$$\kappa_n = \kappa_1 \cos^2\theta + \kappa_2 \sin^2\theta = \kappa_1 - (\kappa_1 - \kappa_2)\sin^2\theta,$$

it is clear that $\kappa_n \leq \kappa_1$ with equality if and only if $\theta = 0$ or π, i.e. if and only if the tangent vector $\dot{\boldsymbol{\gamma}}$ of $\boldsymbol{\gamma}$ is parallel to the principal vector $\mathbf{t}_1$. Similarly, one shows that $\kappa_n \geq \kappa_2$ with equality if and only if $\dot{\boldsymbol{\gamma}}$ is parallel to $\mathbf{t}_2$.

If $\kappa_1 = \kappa_2$, the normal curvature of every curve is equal to κ_1 by Euler's Theorem and every tangent vector to the surface is a principal vector by Proposition 6.3(ii). □

We conclude this section with the following computation which will be very useful on several occasions later in the book.

Proposition 6.4

Let $\mathbf{N}$ *be the standard unit normal of a surface patch* $\boldsymbol{\sigma}(u, v)$. *Then,*

$$\mathbf{N}_u = a\boldsymbol{\sigma}_u + b\boldsymbol{\sigma}_v, \quad \mathbf{N}_v = c\boldsymbol{\sigma}_u + d\boldsymbol{\sigma}_v, \tag{20}$$

where

$$\begin{pmatrix} a & c \\ b & d \end{pmatrix} = -\mathcal{F}_I^{-1}\mathcal{F}_{II}.$$

The matrix $\mathcal{F}_I^{-1}\mathcal{F}_{II}$ is called the *Weingarten matrix* of the surface patch $\boldsymbol{\sigma}$, and is denoted by $\mathcal{W}$.

Proof 6.4

Since $\mathbf{N}$ is a unit vector, we know that $\mathbf{N}_u$ and $\mathbf{N}_v$ are perpendicular to $\mathbf{N}$, hence are in the tangent plane to $\boldsymbol{\sigma}$, and hence are linear combinations of $\boldsymbol{\sigma}_u$ and $\boldsymbol{\sigma}_v$. So scalars a, b, c and d satisfying Eq. (20) exist.

To calculate them, note that $\mathbf{N}.\boldsymbol{\sigma}_u = 0$ implies, on differentiating with respect to u, that

$$\mathbf{N}_u.\boldsymbol{\sigma}_u + \mathbf{N}.\boldsymbol{\sigma}_{uu} = 0,$$
$$\therefore \quad \mathbf{N}_u.\boldsymbol{\sigma}_u = -L.$$

Similarly,

$$\mathbf{N}_u.\boldsymbol{\sigma}_v = \mathbf{N}_v.\boldsymbol{\sigma}_u = -M, \quad \mathbf{N}_v.\boldsymbol{\sigma}_v = -N.$$

Taking the dot product of each of the equations in (20) with $\boldsymbol{\sigma}_u$ and $\boldsymbol{\sigma}_v$ thus gives

$$-L = aE + bF, \quad -M = cE + dF,$$
$$-M = aF + bG, \quad -N = cF + dG.$$

These four scalar equations are equivalent to the single matrix equation

$$-\begin{pmatrix} L & M \\ M & N \end{pmatrix} = \begin{pmatrix} E & F \\ F & G \end{pmatrix} \begin{pmatrix} a & c \\ b & d \end{pmatrix}$$
$$\therefore \quad -\mathcal{F}_{II} = \mathcal{F}_I \begin{pmatrix} a & c \\ b & d \end{pmatrix}$$
$$\therefore \quad \begin{pmatrix} a & c \\ b & d \end{pmatrix} = -\mathcal{F}_I^{-1}\mathcal{F}_{II}.$$

□

EXERCISES

6.15 Calculate the principal curvatures of the helicoid and the catenoid, defined in Exercises 4.14 and 4.18, respectively.

6.16 Let $\boldsymbol{\gamma}(t) = \boldsymbol{\sigma}(u(t), v(t))$ be a regular, but not necessarily unit-speed, curve on a surface $\boldsymbol{\sigma}$, and denote d/dt by a dot. Prove that the normal curvature of $\boldsymbol{\gamma}$ is

$$\kappa_n = \frac{L\dot{u}^2 + 2M\dot{u}\dot{v} + N\dot{v}^2}{E\dot{u}^2 + 2F\dot{u}\dot{v} + G\dot{v}^2}.$$

6.17 By using the results of Exercises 5.4 and 6.3, show that the principal curvatures of a surface either stay the same or both change sign when the surface is reparametrised, according to whether the unit normal stays the same or changes sign. Show also that the principal vectors are unchanged by reparametrisation.

6.18 A curve $\mathcal{C}$ on a surface $\mathcal{S}$ is called a *line of curvature* if the tangent vector of $\mathcal{C}$ is a principal vector of $\mathcal{S}$ at all points of $\mathcal{C}$. If $\boldsymbol{\gamma}$ is a parametrisation of the part of a curve $\mathcal{C}$ lying in a surface patch $\boldsymbol{\sigma}$ of $\mathcal{S}$, and if $\mathbf{N}$ is the standard unit normal of $\boldsymbol{\sigma}$, show that $\mathcal{C}$ is a line of curvature if and only if

$$\dot{\mathbf{N}} = -\lambda\dot{\boldsymbol{\gamma}},$$

for some scalar λ, and that in this case the corresponding principal curvature is λ. (This is called *Rodrigues' Formula.*)

Show that the meridians and parallels of a surface of revolution are lines of curvature.

6.19 Show that a curve on a surface is a line of curvature if and only if its geodesic torsion vanishes everywhere (see Exercise 6.11).

6.20 Two surfaces $\mathcal{S}_1$ and $\mathcal{S}_2$ intersect in a curve $\mathcal{C}$ that is a line of curvature of $\mathcal{S}_1$. Show that $\mathcal{C}$ is a line of curvature of $\mathcal{S}_2$ if and only if the angle between the tangent planes of $\mathcal{S}_1$ and $\mathcal{S}_2$ is constant along $\mathcal{C}$.

6.21 Let $\boldsymbol{\sigma} : W \to \mathbf{R}^3$ be a smooth function defined on an open subset W of $\mathbf{R}^3$ such that, for each fixed value of u (resp. v, w), $\boldsymbol{\sigma}(u, v, w)$ is a (regular) surface patch. Assume also that

$$\boldsymbol{\sigma}_u.\boldsymbol{\sigma}_v = \boldsymbol{\sigma}_v.\boldsymbol{\sigma}_w = \boldsymbol{\sigma}_w.\boldsymbol{\sigma}_u = 0. \tag{21}$$

This means that the three families of surfaces formed by fixing the values of u, v or w is a triply orthogonal system (see Section 4.6).

(i) Show that $\boldsymbol{\sigma}_u.\boldsymbol{\sigma}_{vw} = \boldsymbol{\sigma}_v.\boldsymbol{\sigma}_{uw} = \boldsymbol{\sigma}_w.\boldsymbol{\sigma}_{uv} = 0$. (Differentiate the Eqs. (21).)

(ii) Show that, for each of the surfaces in the triply orthogonal system, the matrices $\mathcal{F}_I$ and $\mathcal{F}_{II}$ are diagonal. (Note that the standard unit normal of the surface obtained by fixing u is parallel to $\boldsymbol{\sigma}_u$, etc.)

(iii) Deduce that the intersection of any surface from one family of the triply orthogonal system with any surface from another family is a line of curvature on both surfaces. (This is called *Dupin's Theorem.*)

6.22 The *third fundamental form* of a surface patch $\boldsymbol{\sigma}(u, v)$ is

$$\| \mathbf{N}_u \|^2 \, du^2 + 2\mathbf{N}_u.\mathbf{N}_v du dv + \| \mathbf{N}_v \|^2 \, dv^2,$$

where $\mathbf{N}(u, v)$ is the standard unit normal at $\boldsymbol{\sigma}(u, v)$. Let $\mathcal{F}_{III}$ be the symmetric 2×2 matrix associated to the third fundamental form in the same way as $\mathcal{F}_I$ and $\mathcal{F}_{II}$ are associated to the first and second fundamental forms (see Section 6.3). Show that $\mathcal{F}_{III} = \mathcal{F}_{II}\mathcal{F}_I^{-1}\mathcal{F}_{II}$. (Use Proposition 6.4.)

6.4. Geometric Interpretation of Principal Curvatures

The relative values of the principal curvatures at a point P of a surface patch tell us much about the shape of the surface near P. To see this, note first that, by applying a rigid motion of $\mathbf{R}^3$ and a reparametrisation of $\boldsymbol{\sigma}$ (which does not change the shape of the surface), we can assume that

(i) P is the origin and $\boldsymbol{\sigma}(0,0) = P$;
(ii) the tangent plane to $\boldsymbol{\sigma}$ at P is the xy-plane;
(iii) the vectors parallel to the x and y-axes are principal vectors at P, corresponding to principal curvatures κ_1 and κ_2, say.

The unit principal vectors can be expressed in terms of $\boldsymbol{\sigma}_u$ and $\boldsymbol{\sigma}_v$ by

$$(1,0) = \xi_1\boldsymbol{\sigma}_u + \eta_1\boldsymbol{\sigma}_v, \quad (0,1) = \xi_2\boldsymbol{\sigma}_u + \eta_2\boldsymbol{\sigma}_v,$$

say. Then, if

$$T_1 = \begin{pmatrix} \xi_1 \\ \eta_1 \end{pmatrix}, \quad T_2 = \begin{pmatrix} \xi_2 \\ \eta_2 \end{pmatrix},$$

the point $(x, y, 0)$ in the tangent plane at P is equal to

$$x(\xi_1\boldsymbol{\sigma}_u + \eta_1\boldsymbol{\sigma}_v) + y(\xi_2\boldsymbol{\sigma}_u + \eta_2\boldsymbol{\sigma}_v) = (x\xi_1 + y\xi_2)\boldsymbol{\sigma}_u + (x\eta_1 + y\eta_2)\boldsymbol{\sigma}_v = s\boldsymbol{\sigma}_u + t\boldsymbol{\sigma}_v,$$

say. Thus, neglecting higher order terms,

$$\begin{aligned}\boldsymbol{\sigma}(s,t) &= \boldsymbol{\sigma}(0,0) + s\boldsymbol{\sigma}_u + t\boldsymbol{\sigma}_v + \frac{1}{2}(s^2\boldsymbol{\sigma}_{uu} + 2st\boldsymbol{\sigma}_{uv} + t^2\boldsymbol{\sigma}_{vv}) \\ &= (x, y, 0) + \frac{1}{2}(s^2\boldsymbol{\sigma}_{uu} + 2st\boldsymbol{\sigma}_{uv} + t^2\boldsymbol{\sigma}_{vv}),\end{aligned}$$

all derivatives being evaluated at the origin. Thus, neglecting higher order terms, the coordinates of $\boldsymbol{\sigma}(s,t)$ are (x, y, z), where

$$\begin{aligned} z &= \boldsymbol{\sigma}(s,t).\mathbf{N} \\ &= \frac{1}{2}(Ls^2 + 2Mst + Nt^2) \\ &= \frac{1}{2}\begin{pmatrix} s & t \end{pmatrix}\begin{pmatrix} L & M \\ M & N \end{pmatrix}\begin{pmatrix} s \\ t \end{pmatrix}.\end{aligned}$$

Now,

$$\begin{pmatrix} s \\ t \end{pmatrix} = \begin{pmatrix} x\xi_1 + y\xi_2 \\ x\eta_1 + y\eta_2 \end{pmatrix} = x\begin{pmatrix} \xi_1 \\ \eta_1 \end{pmatrix} + y\begin{pmatrix} \xi_2 \\ \eta_2 \end{pmatrix} = xT_1 + yT_2,$$

so

$$\begin{aligned} z &= \frac{1}{2}(xT_1 + yT_2)^t\mathcal{F}_{II}(xT_1 + yT_2) \\ &= \frac{1}{2}(x^2T_1^t\mathcal{F}_{II}T_1 + xy(T_1^t\mathcal{F}_{II}T_2 + T_2^t\mathcal{F}_{II}T_1) + y^2T_2^t\mathcal{F}_{II}T_2) \\ &= \frac{1}{2}(\kappa_1x^2 + \kappa_2y^2),\end{aligned}$$

using Eq. (19). We conclude that, *near a point P of a surface at which the principal curvatures are κ_1 and κ_2, the surface coincides with the quadric surface*

$$z = \frac{1}{2}(\kappa_1x^2 + \kappa_2y^2) \tag{22}$$

if we neglect terms of order greater than two.

We distinguish four cases:

(i) *κ_1 and κ_2 are both > 0 or both < 0.* Then, (22) is the equation of an elliptic paraboloid (see Proposition 4.5) and one says that P is an *elliptic point* of the surface.

(ii) *κ_1 and κ_2 are of opposite sign (both non-zero).* Then, (22) is the equation of a hyperbolic paraboloid and one says that P is a *hyperbolic point* of the surface.

(iii) *One of κ_1 and κ_2 is zero, the other is non-zero.* Then, (22) is the equation of a parabolic cylinder and one says that P is a *parabolic point* of the surface.

(iv) *Both principal curvatures are zero at P.* Then, (22) is the equation of a plane, and one says that P is a *planar point* of the surface. In this case, one cannot determine the shape of the surface near P without examining derivatives of order higher than the second (in the non-planar case, these terms are small compared to $\kappa_1 x^2 + \kappa_2 y^2$ when x and y are small). For example, the surfaces below (the one on the right is called the *monkey saddle*) both have the origin as a planar point, but they have quite different shapes.

$z = y^4$

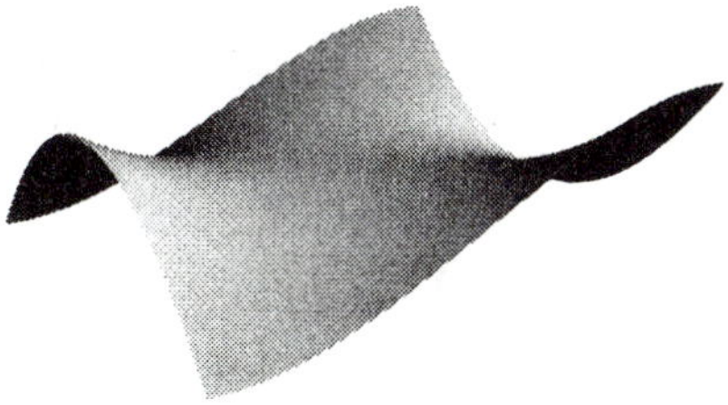

$z = x^3 - 3xy^2$

Note that this classification is independent of the surface patch $\boldsymbol{\sigma}$, since reparametrising either leaves the principal curvatures unchanged or changes the sign of both of them.

Example 6.5

On the unit sphere, $\kappa_1 = \kappa_2 = \pm 1$ (the sign depending on the parametrisation) so all points are elliptic (and umbilics). On a circular cylinder, $\kappa_1 = \pm 1, \kappa_2 = 0$, so every point is parabolic (and there are no umbilics). On a plane, $\kappa_1 = \kappa_2 = 0$ so all points are planar (!) (and umbilics).

Example 6.6

For the torus

$$\boldsymbol{\sigma}(\theta, \varphi) = ((a + b\cos\theta)\cos\varphi, (a + b\cos\theta)\sin\varphi, b\sin\theta)$$

(see Exercise 4.10), we find that the first and second fundamental forms are

$$b^2 d\theta^2 + (a + b\cos\theta)^2 d\varphi^2 \quad \text{and} \quad b\, d\theta^2 + (a + b\cos\theta)\cos\theta\, d\varphi^2,$$

respectively, so the principal curvatures are

$$\kappa_1 = \frac{1}{b}, \quad \kappa_2 = \frac{\cos\theta}{a + b\cos\theta}.$$

Since $\kappa_1 > 0$ (everywhere), the point $\boldsymbol{\sigma}(\theta, \varphi)$ of the torus is elliptic, parabolic or hyperbolic according as κ_2 is > 0, $= 0$ or < 0, respectively; from the formula for κ_2, these are the regions of the torus given by $-\pi/2 < \theta < \pi/2$, $\theta = \pm\pi/2$ and $\pi/2 < \theta < 3\pi/2$, respectively. Pictures of the elliptic and hyperbolic regions can be found in the solution to Exercise 7.18 (where they are labelled $\mathcal{S}^+$ and $\mathcal{S}^-$, respectively); the parabolic region consists of two circles of radius a centred on the z-axis.

We conclude this chapter with the analogue for surfaces of Example 2.2, which tells us that a plane curve with constant curvature is part of a circle.

Proposition 6.5

Let $\mathcal{S}$ be a (connected) surface of which every point is an umbilic. Then, $\mathcal{S}$ is either part of a plane or part of a sphere.

Proof 6.5

Let $\boldsymbol{\sigma} : U \to \mathbf{R}^3$ be a surface patch in the atlas of $\mathcal{S}$ with U a (connected) open subset of $\mathbf{R}^2$. Let κ be the commmon value of the principal curvatures of $\boldsymbol{\sigma}$. By Proposition 6.3(ii),

$$\mathcal{F}_{II} = \kappa\mathcal{F}_I,$$

so the Weingarten matrix

$$\mathcal{W} = \mathcal{F}_I^{-1}\mathcal{F}_{II} = \begin{pmatrix} \kappa & 0 \\ 0 & \kappa \end{pmatrix}.$$

By Proposition 6.4,

$$\mathbf{N}_u = -\kappa \boldsymbol{\sigma}_u, \quad \mathbf{N}_v = -\kappa \boldsymbol{\sigma}_v. \tag{23}$$

Hence,

$$(\kappa \boldsymbol{\sigma}_u)_v = -(\mathbf{N}_u)_v = -(\mathbf{N}_v)_u = (\kappa \boldsymbol{\sigma}_v)_u,$$

so

$$\kappa_v \boldsymbol{\sigma}_u = \kappa_u \boldsymbol{\sigma}_v.$$

Since $\boldsymbol{\sigma}$ is regular, $\boldsymbol{\sigma}_u$ and $\boldsymbol{\sigma}_v$ are linearly independent, so the last equation implies that $\kappa_u = \kappa_v = 0$. Thus, κ is constant.

There are now two cases to consider. If $\kappa = 0$, Eq. (23) shows that $\mathbf{N}$ is constant. Then,

$$(\mathbf{N}.\boldsymbol{\sigma})_u = \mathbf{N}.\boldsymbol{\sigma}_u = 0, \quad (\mathbf{N}.\boldsymbol{\sigma})_v = \mathbf{N}.\boldsymbol{\sigma}_v = 0,$$

so $\mathbf{N}.\boldsymbol{\sigma}$ is a constant, say c. Then $\boldsymbol{\sigma}(U)$ is part of the plane $\mathbf{r}.\mathbf{N} = c$.

If $\kappa \neq 0$, Eq. (23) shows that

$$\mathbf{N} = -\kappa \boldsymbol{\sigma} + \mathbf{a},$$

where $\mathbf{a}$ is a constant vector. Hence,

$$\| \boldsymbol{\sigma} - \frac{1}{\kappa}\mathbf{a} \|^2 = \| -\frac{1}{\kappa}\mathbf{N} \|^2 = \frac{1}{\kappa^2},$$

so $\boldsymbol{\sigma}(U)$ is part of the sphere with centre $\kappa^{-1}\mathbf{a}$ and radius κ^{-1}.

We have now proved the proposition when $\mathcal{S}$ is covered by a single surface patch. For an arbitrary surface $\mathcal{S}$, the preceding argument shows that each patch in the atlas of $\mathcal{S}$ is contained in a plane or a sphere. But clearly two overlapping patches must then be part of the same plane or the same sphere. It follows that the whole of $\mathcal{S}$ is contained in a plane or a sphere. □

EXERCISES

6.23 Show that every point on the surface of revolution

$$\boldsymbol{\sigma}(u, v) = (f(u)\cos v, f(u)\sin v, g(u))$$

is parabolic if and only if $\boldsymbol{\sigma}$ is part of a circular cylinder or a circular cone.

6.24 Show that, if p, q and r are distinct positive numbers, there are exactly four umbilics on the ellipsoid

$$\frac{x^2}{p^2} + \frac{y^2}{q^2} + \frac{z^2}{r^2} = 1.$$

(Use Proposition 6.3(ii).)

7

Gaussian Curvature and the Gauss Map

We shall now introduce two new measures of the curvature of a surface, called its gaussian and mean curvatures. Although these together contain the same information as the two principal curvatures, they turn out to have greater geometrical significance. The gaussian curvature, in particular, has the remarkable property, established in Chapter 10, that it is unchanged when the surface is bent without stretching, a property that is not shared by the principal curvatures. In the present chapter, we discuss some more elementary properties of the gaussian and mean curvatures, and what a knowledge of them implies about the geometry of the surface.

7.1. The Gaussian and Mean Curvatures

We start by defining two new measures of the curvature of a surface.

Definition 7.1

Let κ_1 and κ_2 be the principal curvatures of a surface patch. Then, the *gaussian curvature* of the surface patch is

$$K = \kappa_1\kappa_2,$$

and its *mean curvature* is

$$H = \frac{1}{2}(\kappa_1 + \kappa_2).$$

Note Some authors omit the 1/2 in the definition of H, even though this conflicts with the usual meaning of 'mean'.

Note from Exercise 6.17 that the gaussian curvature stays the same when a surface patch is reparametrised, while the mean curvature either stays the same or changes sign. It follows that *the gaussian curvature is well defined for any surface* $\mathcal{S}$.

It is easy to get explicit formulas for H and K:

Proposition 7.1

Let $\boldsymbol{\sigma}(u, v)$ *be a surface patch with first and second fundamental forms*

$$Edu^2 + 2Fdudv + Gdv^2 \quad \text{and} \quad Ldu^2 + 2Mdudv + Ndv^2,$$

respectively. Then,

(i) $K = \frac{LN-M^2}{EG-F^2}$;

(ii) $H = \frac{LG-2MF+NE}{2(EG-F^2)}$;

(iii) the principal curvatures are $H \pm \sqrt{H^2 - K}$.

Proof 7.1

By Definition 6.1, the principal curvatures are the roots of

$$\begin{vmatrix} L - \kappa E & M - \kappa F \\ M - \kappa F & N - \kappa G \end{vmatrix} = 0,$$

$$\therefore \quad (L - \kappa E)(N - \kappa G) - (M - \kappa F)^2 = 0,$$

$$\therefore \quad (EG - F^2)\kappa^2 - (LG - 2MF + NE)\kappa + LN - M^2 = 0.$$

Now, recall that in a quadratic equation $a\kappa^2 + b\kappa + c = 0$, the sum of the roots is $-b/a$ and the product of the roots is c/a. So,

$$K = \kappa_1\kappa_2 = \text{ product of roots } = \frac{LN - M^2}{EG - F^2},$$

$$H = \frac{1}{2}(\kappa_1 + \kappa_2) = \frac{1}{2}(\text{sum of roots}) = \frac{1}{2}\frac{LG - 2MF + NE}{EG - F^2}.$$

By the definition of H and K, κ_1 and κ_2 are the roots of

$$\kappa^2 - 2H\kappa + K = 0,$$

i.e. $H \pm \sqrt{H^2 - K}$. □

Example 7.1

For the unit sphere, we found in Example 6.3 that $\kappa_1 = \kappa_2 = 1$, so $K = H = 1$. For a circular cylinder of radius one, we found in Example 6.4 that $\kappa_1 = 1$, $\kappa_2 = 0$, so $H = \frac{1}{2}$, $K = 0$.

Example 7.2

In Example 6.2 we considered the surface of revolution

$$\boldsymbol{\sigma}(u,v) = (f(u)\cos v, f(u)\sin v, g(u)),$$

where we can assume that $f > 0$ and $\dot{f}^2 + \dot{g}^2 = 1$ everywhere (a dot denoting d/du). We found that

$$E = 1, \quad F = 0, \quad G = f^2,$$
$$L = \dot{f}\ddot{g} - \ddot{f}\dot{g}, \quad M = 0, \quad N = f\dot{g}.$$

By Proposition 7.1(i), the gaussian curvature is

$$K = \frac{LN - M^2}{EG - F^2} = \frac{(\dot{f}\ddot{g} - \ddot{f}\dot{g})f\dot{g}}{f^2}. \tag{1}$$

We can simplify this formula by noting that $\dot{f}^2 + \dot{g}^2 = 1$ implies (by differentiating with respect to u) that

$$\dot{f}\ddot{f} + \dot{g}\ddot{g} = 0,$$
$$\therefore \quad (\dot{f}\ddot{g} - \ddot{f}\dot{g})\dot{g} = -\dot{f}^2\ddot{f} - \ddot{f}\dot{g}^2 = -\ddot{f}(\dot{f}^2 + \dot{g}^2) = -\ddot{f},$$
$$\therefore \quad K = -\frac{\ddot{f}f}{f^2} = -\frac{\ddot{f}}{f}.$$

Example 7.3

For a ruled surface, take a patch

$$\boldsymbol{\sigma}(u,v) = \boldsymbol{\gamma}(u) + v\boldsymbol{\delta}(u),$$

(see Example 4.12). Denoting d/du by a dot, we have

$$\boldsymbol{\sigma}_u = \dot{\boldsymbol{\gamma}} + v\dot{\boldsymbol{\delta}}, \quad \boldsymbol{\sigma}_v = \boldsymbol{\delta},$$
$$\therefore \quad \boldsymbol{\sigma}_{uv} = \dot{\boldsymbol{\delta}}, \quad \boldsymbol{\sigma}_{vv} = \mathbf{0}.$$

Hence, if $\mathbf{N} = (\boldsymbol{\sigma}_u \times \boldsymbol{\sigma}_v)/ \parallel \boldsymbol{\sigma}_u \times \boldsymbol{\sigma}_v \parallel$ is the standard unit normal of $\boldsymbol{\sigma}$, then $M = \boldsymbol{\sigma}_{uv}.\mathbf{N} = \dot{\boldsymbol{\delta}}.\mathbf{N}$ and $N = 0$. So

$$K = \frac{LN - M^2}{EG - f^2} = \frac{-(\dot{\boldsymbol{\delta}}.\mathbf{N})^2}{EG - F^2} \leq 0,$$

i.e. *the gaussian curvature of a ruled surface is negative or zero*, a result that is proved by a different method in Exercise 7.8. We shall return to this example in Section 7.3.

EXERCISES

7.1 Calculate the gaussian and mean curvatures of the surface

$$\boldsymbol{\sigma}(u,v) = (u+v, u-v, uv)$$

at the point $(2,0,1)$.

7.2 Calculate the gaussian curvature of the helicoid and catenoid (see Exercises 4.14 and 4.18).

7.3 Calculate the gaussian curvature of the surface $z = f(x,y)$, where f is a smooth function.

7.4 In the notation of Example 7.3, show that
(i) if $\boldsymbol{\delta}$ is the principal normal $\mathbf{n}$ of $\boldsymbol{\gamma}$ or its binormal $\mathbf{b}$, then $K=0$ if and only if $\boldsymbol{\gamma}$ is planar (use Proposition 2.4);
(ii) if $\boldsymbol{\gamma}$ is a curve on a surface $\mathcal{S}$ and $\boldsymbol{\delta}$ is the unit normal of $\mathcal{S}$, then $K=0$ if and only if $\boldsymbol{\gamma}$ is a line of curvature of $\mathcal{S}$ (use Exercise 6.18.)

7.5 Let $\boldsymbol{\sigma}$ be the parametrisation of the torus in Exercise 4.10, and let K be its gaussian curvature. Show that

$$\iint K dA_{\boldsymbol{\sigma}} = 0.$$

(The 'explanation' of this result will appear in Section 11.3.)

7.6 Show that the gaussian and mean curvatures are unchanged by applying a rigid motion, and that the dilation $(x,y,z) \mapsto a(x,y,z)$, where a is a non-zero constant, multiplies them by a^{-2} and a^{-1}, respectively.

7.7 Show that the gaussian and mean curvatures of a surface patch $\boldsymbol{\sigma} : U \to \mathbf{R}^3$ are smooth functions on U. Show that the principal curvatures are smooth functions on any open subset of U on which $\boldsymbol{\sigma}$ has no umbilics. (Use Proposition 7.1.)

7.8 Show that $K \leq 0$ at every point of an asymptotic curve on a surface (use Corollary 6.2). Hence give another proof that the gaussian curvature of a ruled surface is ≤ 0 everywhere (use Exercise 6.12).

7.9 Show that, if $\mathcal{F}_{III}$ is the third fundamental form of a surface patch $\boldsymbol{\sigma}$ (see Exercise 6.22), then

$$\mathcal{F}_{III} - 2H\mathcal{F}_{II} + K\mathcal{F}_I = 0,$$

where K and H are the gaussian and mean curvatures of $\boldsymbol{\sigma}$. (Use the fact that any 2×2 matrix $A = \begin{pmatrix} a & c \\ b & d \end{pmatrix}$ satisfies the equation $A^2 - (a+d)A + (ad-bc)I = 0$.)

7.10 Use Exercise 7.9 to show that, if $\boldsymbol{\gamma}(t)$ is a curve in a surface patch $\boldsymbol{\sigma}$, then along $\boldsymbol{\gamma}$,

$$\dot{\mathbf{N}}.\dot{\mathbf{N}} + 2H\dot{\mathbf{N}}.\dot{\boldsymbol{\gamma}} + K\dot{\boldsymbol{\gamma}}.\dot{\boldsymbol{\gamma}} = 0.$$

(Note that, if $\boldsymbol{\gamma}(t) = \boldsymbol{\sigma}(u(t), v(t))$, $T = \begin{pmatrix} \dot{u} \\ \dot{v} \end{pmatrix}$, then $\dot{\mathbf{N}}.\dot{\mathbf{N}} = T^t \mathcal{F}_{III} T$, etc.) Deduce that the torsion τ of an asymptotic curve on a surface is related to the gaussian curvature K of the surface by $\tau^2 = -K$. (Use Exercise 6.12.)

7.2. The Pseudosphere

We have seen in the examples in Section 7.1 some surfaces of zero and constant positive curvature. For an example of a surface with *constant negative* gaussian curvature, however, we have to construct a new surface. To this end, we examine again the surface of revolution

$$\boldsymbol{\sigma}(u, v) = (f(u)\cos v, f(u)\sin v, g(u)).$$

We found in Example 7.2 that its gaussian curvature is

$$K = -\frac{\ddot{f}}{f}. \tag{2}$$

Suppose first that $K = 0$ everywhere. Then, Eq. (2) gives $\ddot{f} = 0$, so $f(u) = au + b$ for some contants a and b. Since $\dot{f}^2 + \dot{g}^2 = 1$, we get $\dot{g} = \pm\sqrt{1 - a^2}$ (so we must have $|a| \leq 1$) and hence $g(u) = \pm\sqrt{1 - a^2}u + c$, where c is another constant. By applying a translation along the z-axis we can assume that $c = 0$, and by applying a rotation about the x-axis (say) we can assume that the sign is $+$. This gives the ruled surface

$$\boldsymbol{\sigma}(u, v) = (b\cos v, b\sin v, 0) + u(a\cos v, a\sin v, \sqrt{1 - a^2}).$$

If $a = 0$ this is a circular cylinder; if $|a| = 1$ it is the xy-plane; and if $0 < |a| < 1$ it is part of a cone (to see this, put $\tilde{u} = au + b$).

Now suppose that $K = 1$ everywhere. (Any surface with constant positive gaussian curvature can be reduced to this case by applying a dilation of $\mathbf{R}^3$ – see Exercise 7.6.) Then, Eq. (2) becomes

$$\ddot{f} + f = 0,$$

which has the general solution

$$f(u) = a\cos(u + b),$$

where a and b are constants. We can assume that $b = 0$ by performing a reparametrisation $\tilde{u} = u + b, \tilde{v} = v$. Then, up to a change of sign and adding a constant,

$$g(u) = \int \sqrt{1 - a^2 \sin^2 u}\, du.$$

This integral cannot be evaluated in terms of 'elementary' functions unless $a = 0$ or ± 1. The case $a = 0$ does not give a surface, so we consider the case $a = 1$ (the case $a = -1$ can be reduced to this by rotating the surface by π around the z-axis). Then, $f(u) = \cos u$, $g(u) = \sin u$, and we have the unit sphere.

Suppose finally that $K = -1$. The general solution of Eq. (2) is then

$$f(u) = ae^u + be^{-u},$$

where a and b are arbitrary constants. For most values of a and b we cannot express g in terms of elementary functions, so we consider only the case $a = 1$ and $b = 0$. Then, $f(u) = e^u$ and we can take

$$g(u) = \int \sqrt{1 - e^{2u}}\, du. \tag{3}$$

Note that we must have $u \leq 0$ for the integral in Eq. (3) to make sense, since otherwise $1 - e^{2u}$ would be negative.

The integral in (3) can be evaluated by putting $v = e^u$. Then,

$$\begin{aligned}\int \sqrt{1 - e^{2u}}\, du &= \int \frac{\sqrt{1 - v^2}}{v}\, dv \\ &= \int \left(\frac{1}{v} - v\right) \frac{dv}{\sqrt{1 - v^2}} \\ &= \sqrt{1 - v^2} + \int \frac{dv}{v\sqrt{1 - v^2}}.\end{aligned}$$

Put $w = v^{-1}$ in the last integral. This gives

$$\begin{aligned}\int \sqrt{1 - e^{2u}}\, du &= \sqrt{1 - v^2} - \int \frac{dw}{\sqrt{w^2 - 1}} \\ &= \sqrt{1 - v^2} - \cosh^{-1} w \\ &= \sqrt{1 - v^2} - \cosh^{-1}\left(\frac{1}{v}\right) \\ &= \sqrt{1 - e^{2u}} - \cosh^{-1}(e^{-u}).\end{aligned}$$

We have omitted the arbitrary constant, but we can take it to be zero by suitably translating the surface in the direction of the z-axis. So

$$f(u) = e^u, \quad g(u) = \sqrt{1 - e^{2u}} - \cosh^{-1}(e^{-u}).$$

Putting $x = f(u)$, $z = g(u)$, we see that the profile curve in the xz-plane has equation

$$z = \sqrt{1 - x^2} - \cosh^{-1}\left(\frac{1}{x}\right). \tag{4}$$

Rotating this curve around the z-axis thus gives a surface called the *pseudosphere*, which has gaussian curvature -1 everywhere. Note that, since $u \le 0$, $x = e^u$ is restricted to the range $0 < x \le 1$.

The curve defined by Eq. (4) is called the *tractrix*, and it has an interesting geometrical property. Consider the tangent line at a point P of its graph, and suppose that it intersects the z-axis at the point Q. Let us compute the distance PQ.

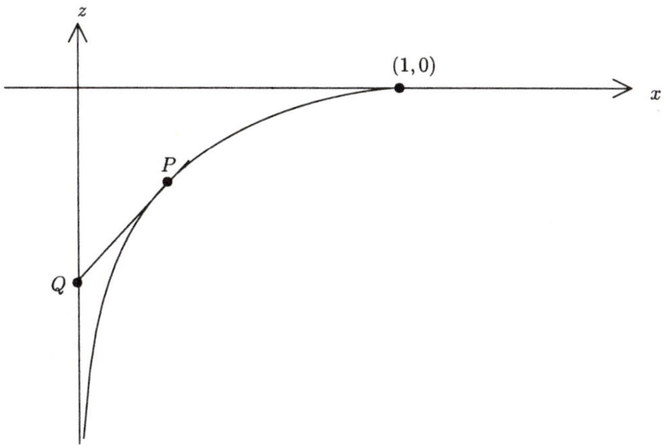

Suppose that P is the point (x_0, z_0). Either by a direct calculation or by inspecting the calculation of the integral (3), one finds that

$$\frac{dz}{dx} = \frac{\sqrt{1-x^2}}{x}.$$

Hence, the tangent line at P has equation

$$z - z_0 = \frac{\sqrt{1-x_0^2}}{x_0}(x - x_0).$$

This meets the z-axis at the point $(0, z_1)$, where

$$z_1 - z_0 = \frac{\sqrt{1-x_0^2}}{x_0}(0 - x_0) = -\sqrt{1-x_0^2}.$$

Hence, the distance PQ is given by

$$(PQ)^2 = x_0^2 + (z_1 - z_0)^2 = x_0^2 + 1 - x_0^2 = 1,$$

i.e. the distance PQ is *constant* and equal to one.

This means that the tractrix has the following description. Let a donkey pull a box of stones by a rope of length one. Suppose that the donkey is initially at $(0,0)$, the box is initially at $(1,0)$, and let the donkey walk slowly along the negative z-axis. Then, the box of stones moves along the tractrix.

The study of geometry on a pseudosphere is a subject in its own right called *non-euclidean geometry.* Many of the results of plane euclidean geometry have analogues for the pseudosphere, but there are several differences. For example, the sum of the interior angles of a triangle on the pseudosphere with 'straight' sides is always *less* than π. (We shall explain the meaning of 'straight' here in Chapter 8 – see especially Example 8.8.) This should be compared with Theorem 5.5 which shows that the sum of the interior angles of a triangle on the unit sphere whose sides are arcs of great circles is always *greater* than π.

EXERCISES

7.11 For the pseudosphere:

(i) calculate the length of a parallel;

(ii) calculate its total area;

(iii) calculate the principal curvatures;

(iv) show that all points are hyperbolic.

7.12 Show that

(i) setting $w = e^{-u}$ gives a reparametrisation $\boldsymbol{\sigma}_1(v, w)$ of the pseudosphere with first fundamental form

$$\frac{dv^2 + dw^2}{w^2}$$

(called the *upper half-plane model*);

(ii) setting

$$U = \frac{v^2 + w^2 - 1}{v^2 + (w+1)^2}, \quad V = \frac{-2v}{v^2 + (w+1)^2}$$

defines a reparametrisation $\boldsymbol{\sigma}_2(U, V)$ of the pseudosphere with first fundamental form

$$\frac{dU^2 + dV^2}{(1 - U^2 - V^2)^2}.$$

This is called the *disc model*, because the region $w > 0$ of the vw-plane corresponds to the disc $U^2 + V^2 < 1$ in the UV-plane.

Which regions in the half-plane model and the disc model correspond to the region $u < 0$, $-\pi < v < \pi$ where the parametrisation $\boldsymbol{\sigma}$ of the pseudosphere given in this section is defined?

7.13 Let $\mathcal{S}$ be a surface of revolution with axis the z-axis, and let its profile curve be a unit-speed curve $\boldsymbol{\gamma}(u)$ in the xz-plane. Suppose that $\boldsymbol{\gamma}$ intersects the z-axis at right angles when $u = \pm\pi/2$, but does not intersect the z-axis when $-\pi/2 < u < \pi/2$. Prove that, if the gaussian curvature K of $\mathcal{S}$ is constant, that constant is equal to one and $\mathcal{S}$ is the unit sphere.

7.3. Flat Surfaces

In Section 7.2, we gave some examples of surfaces of constant gaussian curvature K, but this certainly falls well short of a complete classification of such surfaces. It is possible, however, to give a fairly complete description of *flat surfaces*, i.e. surfaces for which $K = 0$ everywhere. To do so, we shall make use of a special parametrisation, valid for any surface, described in the following proposition.

Proposition 7.2

Let P be a point of a surface $\mathcal{S}$, and suppose that P is not an umbilic. Then, there is a surface patch $\boldsymbol{\sigma}(u, v)$ of $\mathcal{S}$ containing P whose first and second fundamental forms are

$$Edu^2 + Gdv^2 \quad \text{and} \quad Ldu^2 + Ndv^2,$$

respectively, for some smooth functions E, G, L and N.

We recall from Section 6.3 that a point P of a surface $\mathcal{S}$ is an umbilic if the two principal curvatures of $\mathcal{S}$ at P are equal. From Section 6.3, we see that for the patch $\boldsymbol{\sigma}$ in the statement of the proposition, $\boldsymbol{\sigma}_u$ and $\boldsymbol{\sigma}_v$ are principal vectors with corresponding principal curvatures L/E and N/G. We call $\boldsymbol{\sigma}$ a *principal patch.*

We assume Proposition 7.2 for the moment, and use it to give the proof of

Proposition 7.3

Let P be a point of a flat surface S, and assume that P is not an umbilic. Then, there is a patch of S containing P that is a ruled *surface.*

Proof 7.3

We take a patch $\boldsymbol{\sigma} : U \to \mathbf{R}^3$ containing P as in Proposition 7.2, say $P = \boldsymbol{\sigma}(u_0, v_0)$. By Proposition 7.1(ii), the gaussian curvature $K = LN/EG$. Since the gaussian curvature is zero everywhere, either $L = 0$ or $N = 0$ at each point of U, and since P is not an umbilic L and N are not both zero. Suppose that $L(u_0, v_0) \neq 0$. Then, $L(u, v) \neq 0$ for (u, v) in some open subset of U containing (u_0, v_0). Hence, by shrinking U if necessary, we can assume that $L \neq 0$ at every point of U. Then, $N = 0$ everywhere, and the second fundamental form of $\boldsymbol{\sigma}$ is Ldu^2.

We shall prove that the parameter curves $u =$ constant are straight lines. Such a curve can be parametrised by $v \mapsto \boldsymbol{\sigma}(u_0, v)$, where u_0 is the constant value of u. A unit tangent vector to this curve is $\mathbf{t} = \boldsymbol{\sigma}_v / G^{1/2}$, so by Proposition 1.1 what we have to prove is that $\mathbf{t}_v = \mathbf{0}$.

By Proposition 6.4, the derivatives of the unit normal are

$$\mathbf{N}_u = -E^{-1} L \boldsymbol{\sigma}_u, \quad \mathbf{N}_v = \mathbf{0}. \tag{5}$$

Hence, $\mathbf{t}_v.\boldsymbol{\sigma}_u = -EL^{-1}\mathbf{t}_v.\mathbf{N}_u$. Now, $\mathbf{t}.\mathbf{N}_u = 0$ and $\mathbf{N}_{uv} = \mathbf{0}$ by Eq. (5), so $\mathbf{t}_v.\mathbf{N}_u = -\mathbf{t}.\mathbf{N}_{uv} = 0$. Hence, $\mathbf{t}_v.\boldsymbol{\sigma}_u = 0$. Next, $\mathbf{t}_v.\mathbf{t} = 0$ since $\mathbf{t}$ is a unit vector by construction, so $\mathbf{t}_v.\boldsymbol{\sigma}_v = 0$. Finally, $\mathbf{t}_v.\mathbf{N} = -\mathbf{t}.\mathbf{N}_v = 0$ by Eq. (5) again. Since the vectors $\boldsymbol{\sigma}_u, \boldsymbol{\sigma}_v$ and $\mathbf{N}$ form a basis of $\mathbf{R}^3$, we have proved that $\mathbf{t}_v = \mathbf{0}$.

□

Our task, then, is to describe the structure of flat ruled surfaces. We parametrise the ruled surface as in Example 7.3:

$$\boldsymbol{\sigma}(u, v) = \boldsymbol{\gamma}(u) + v\boldsymbol{\delta}(u).$$

We found there that $\boldsymbol{\sigma}_u = \dot{\boldsymbol{\gamma}} + v\dot{\boldsymbol{\delta}}$, $\boldsymbol{\sigma}_v = \boldsymbol{\delta}$, the dot denoting d/du, and that the gaussian curvature of $\boldsymbol{\sigma}$ is zero if and only if

$$\dot{\boldsymbol{\delta}}.(\boldsymbol{\sigma}_u \times \boldsymbol{\sigma}_v) = 0.$$

Since

$$\boldsymbol{\sigma}_u \times \boldsymbol{\sigma}_v = \dot{\boldsymbol{\gamma}} \times \boldsymbol{\delta} + v\dot{\boldsymbol{\delta}} \times \boldsymbol{\delta},$$

and $\dot{\boldsymbol{\delta}}.(\dot{\boldsymbol{\delta}} \times \boldsymbol{\delta}) = 0$,

$$K = 0 \quad \text{if and only if} \quad \dot{\boldsymbol{\delta}}.(\dot{\boldsymbol{\gamma}} \times \boldsymbol{\delta}) = 0. \tag{6}$$

Thus, $K = 0$ *if and only if* $\dot{\boldsymbol{\gamma}}$, $\boldsymbol{\delta}$ *and* $\dot{\boldsymbol{\delta}}$ *are everywhere linearly dependent.*

To proceed further, let us assume, as we may, that $\boldsymbol{\delta}(u)$ is a unit vector for all values of u. Then, $\boldsymbol{\delta}.\dot{\boldsymbol{\delta}} = 0$. Suppose first that $\dot{\boldsymbol{\delta}}(u) = \mathbf{0}$ for all values of u. Then, $\boldsymbol{\delta}$ is a constant vector and $\boldsymbol{\sigma}$ is a *generalised cylinder*.

Suppose now that $\dot{\boldsymbol{\delta}}$ is never zero. Then, $\boldsymbol{\delta}$ and $\dot{\boldsymbol{\delta}}$ are linearly independent as they are non-zero and perpendicular, so if $\dot{\boldsymbol{\gamma}}, \boldsymbol{\delta}$ and $\dot{\boldsymbol{\delta}}$ are linearly dependent, then

$$\dot{\boldsymbol{\gamma}}(u) = f(u)\boldsymbol{\delta}(u) + g(u)\dot{\boldsymbol{\delta}}(u)$$

for some smooth functions f and g. Assume first that $f = \dot{g}$ everywhere. Then, $\dot{\boldsymbol{\gamma}} = (g\boldsymbol{\delta})^{\cdot}$ and so $\boldsymbol{\gamma} = g\boldsymbol{\delta} + \mathbf{a}$, where $\mathbf{a}$ is a constant vector; hence,

$$\boldsymbol{\sigma}(u, v) = \mathbf{a} + (v + g(u))\boldsymbol{\delta}(u).$$

Putting $\tilde{u} = u$, $\tilde{v} = v + g(u)$, we see that this is a reparametrisation of a *generalised cone.*

Suppose finally that $\dot{\boldsymbol{\delta}}$ and $f - \dot{g}$ are both nowhere zero. If we define

$$\tilde{\boldsymbol{\gamma}}(u) = \boldsymbol{\gamma}(u) - g(u)\boldsymbol{\delta}(u), \quad \tilde{v} = \frac{v + g(u)}{f(u) - \dot{g}(u)},$$

a short calculation gives

$$\boldsymbol{\sigma}(u, v) = \tilde{\boldsymbol{\gamma}}(u) + \tilde{v}\dot{\tilde{\boldsymbol{\gamma}}}(u),$$

so $\boldsymbol{\sigma}$ is a reparametrisation of part of the *tangent developable* of $\tilde{\boldsymbol{\gamma}}$.

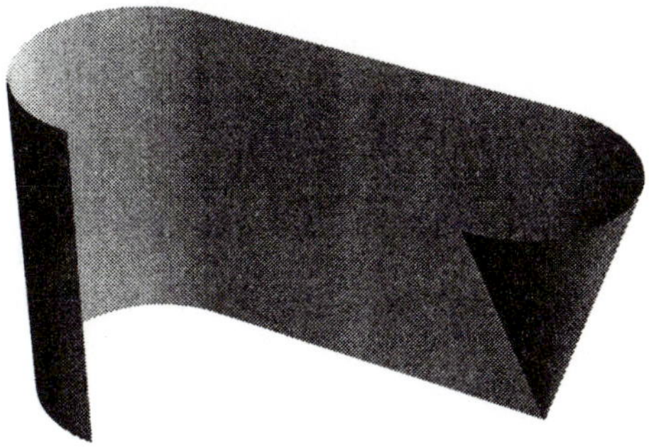

Of course, it could be that none of the conditions on $\boldsymbol{\delta}$, f and g considered above are satisfied. In fact, we have only shown that the parts of the surface corresponding to certain open subsets of U are parts of generalised cylinders, generalised cones or tangent developables. It is not true that the whole surface must be one of these three types, since flat surfaces of different types can be joined together to make a smooth surface, as shown in the diagram above. In general, a flat surface is a patchwork consisting of pieces of generalised cylinders, generalised cones and tangent developables, joined together along segments of straight lines.

The remainder of this section is devoted to the proof of Proposition 7.2 and can safely be omitted by readers uncomfortable with the use of the inverse function theorem. In fact, we can prove a more general result with no additional effort:

Proposition 7.4

Let $\tilde{\boldsymbol{\sigma}} : \tilde{U} \to \mathbf{R}^3$ be a surface patch, and suppose that for all $(\tilde{u}, \tilde{v}) \in \tilde{U}$ we are given tangent vectors

$$\mathbf{e}_1(\tilde{u}, \tilde{v}) = a(\tilde{u}, \tilde{v})\tilde{\boldsymbol{\sigma}}_{\tilde{u}} + b(\tilde{u}, \tilde{v})\tilde{\boldsymbol{\sigma}}_{\tilde{v}}, \quad \mathbf{e}_2(\tilde{u}, \tilde{v}) = c(\tilde{u}, \tilde{v})\tilde{\boldsymbol{\sigma}}_{\tilde{u}} + d(\tilde{u}, \tilde{v})\tilde{\boldsymbol{\sigma}}_{\tilde{v}},$$

whose components a, b, c, d are smooth functions of $(\tilde{u}, \tilde{v})$. Assume that, at some point $(\tilde{u}_0, \tilde{v}_0) \in \tilde{U}$, the vectors $\mathbf{e}_1(\tilde{u}_0, \tilde{v}_0)$ and $\mathbf{e}_2(\tilde{u}_0, \tilde{v}_0)$ are linearly independent. Then, there is an open subset $\tilde{V}$ of $\tilde{U}$ containing $(\tilde{u}_0, \tilde{v}_0)$ and a reparametrisation $\boldsymbol{\sigma}(u, v)$ of $\tilde{\boldsymbol{\sigma}}(\tilde{u}, \tilde{v})$, for $(\tilde{u}, \tilde{v}) \in \tilde{V}$, such that $\boldsymbol{\sigma}_u$ and $\boldsymbol{\sigma}_v$ are parallel to $\mathbf{e}_1$ and $\mathbf{e}_2$, respectively.

Proposition 7.2 is a special case of Proposition 7.4. In fact, let $\tilde{\boldsymbol{\sigma}}$ be any surface patch of $\mathcal{S}$ containing P, and let $P = \tilde{\boldsymbol{\sigma}}(\tilde{u}_0, \tilde{v}_0)$. Since the principal curvatures κ_1 and κ_2 of $\tilde{\boldsymbol{\sigma}}$ are distinct at P, and are continuous functions by Exercise 7.7, they remain distinct for $(\tilde{u}, \tilde{v})$ in some open set $\tilde{U}$ containing $(\tilde{u}_0, \tilde{v}_0)$ on which $\tilde{\boldsymbol{\sigma}}$ is defined. With the notation of Section 6.3, let $\begin{pmatrix} \xi_i \\ \eta_i \end{pmatrix}$ be non-zero eigenvectors of $\tilde{\mathcal{F}}_{II} - \kappa_i \tilde{\mathcal{F}}_I$ for $i = 1, 2$, where $\tilde{\mathcal{F}}_I$ and $\tilde{\mathcal{F}}_{II}$ are the matrices associated to the first and second fundamental forms of $\tilde{\boldsymbol{\sigma}}$. Then,

$$\mathbf{e}_1 = \xi_1 \tilde{\boldsymbol{\sigma}}_{\tilde{u}} + \eta_1 \tilde{\boldsymbol{\sigma}}_{\tilde{v}}, \quad \mathbf{e}_2 = \xi_2 \tilde{\boldsymbol{\sigma}}_{\tilde{u}} + \eta_2 \tilde{\boldsymbol{\sigma}}_{\tilde{v}}$$

are principal vectors corresponding to κ_1 and κ_2, and they are perpendicular by Proposition 6.3(iii). We can assume that $\mathbf{e}_1$ and $\mathbf{e}_2$ are unit vectors (by using $\mathbf{e}_1 / \parallel \mathbf{e}_1 \parallel$ and $\mathbf{e}_2 / \parallel \mathbf{e}_2 \parallel$ instead). Let $\boldsymbol{\sigma}(u, v)$ be a reparametrisation of $\tilde{\boldsymbol{\sigma}}$ as in Proposition 7.4. Then, $\boldsymbol{\sigma}_u . \boldsymbol{\sigma}_v = 0$ because $\mathbf{e}_1$ and $\mathbf{e}_2$ are perpendicular, so the first fundamental form of $\boldsymbol{\sigma}$ is of the form $Edu^2 + Gdv^2$. Also, $\boldsymbol{\sigma}_u$ and $\boldsymbol{\sigma}_v$ are principal vectors corresponding to κ_1 and κ_2, so we have

$$(\mathcal{F}_{II} - \kappa_1 \mathcal{F}_I) \begin{pmatrix} 1 \\ 0 \end{pmatrix} = (\mathcal{F}_{II} - \kappa_2 \mathcal{F}_I) \begin{pmatrix} 0 \\ 1 \end{pmatrix} = \begin{pmatrix} 0 \\ 0 \end{pmatrix},$$

where $\mathcal{F}_I$ and $\mathcal{F}_{II}$ are the matrices associated to the first and second fundamental forms of $\boldsymbol{\sigma}$. Since $\mathcal{F}_I = \begin{pmatrix} E & 0 \\ 0 & G \end{pmatrix}$, these equations imply that $\mathcal{F}_{II} = \begin{pmatrix} \kappa_1 E & 0 \\ 0 & \kappa_2 G \end{pmatrix}$, so the second fundamental form of $\boldsymbol{\sigma}$ is $Ldu^2 + Ndv^2$, where $L = \kappa_1 E$ and $N = \kappa_2 G$.

We are thus left with the proof of Proposition 7.4. To begin, we observe that, if

$$\mathbf{e} = A\tilde{\boldsymbol{\sigma}}_{\tilde{u}} + B\tilde{\boldsymbol{\sigma}}_{\tilde{v}},$$

where A and B are any given smooth functions of $(\tilde{u}, \tilde{v}) \in \tilde{U}$, we can find a curve $\boldsymbol{\gamma}$ in $\tilde{\boldsymbol{\sigma}}$ with $\dot{\boldsymbol{\gamma}} = \mathbf{e}$ and with any given point $Q = \tilde{\boldsymbol{\sigma}}(\alpha, \beta)$ as starting point $\boldsymbol{\gamma}(0)$. For, finding such a curve $\boldsymbol{\gamma}(t) = \tilde{\boldsymbol{\sigma}}(\tilde{u}(t), \tilde{v}(t))$ is equivalent to solving the pair of ordinary differential equations

$$\dot{\tilde{u}} = A(\tilde{u}, \tilde{v}), \quad \dot{\tilde{v}} = B(\tilde{u}, \tilde{v})$$

with initial conditions $\tilde{u}(0) = \alpha$, $\tilde{v}(0) = \beta$. It is proved in the theory of ordinary differential equations that this problem has a unique solution $\tilde{u}(t), \tilde{v}(t)$ defined on some open interval containing $t = 0$. Moreover, $\tilde{u}$ and $\tilde{v}$ are smooth functions of the three variables t, α and β.

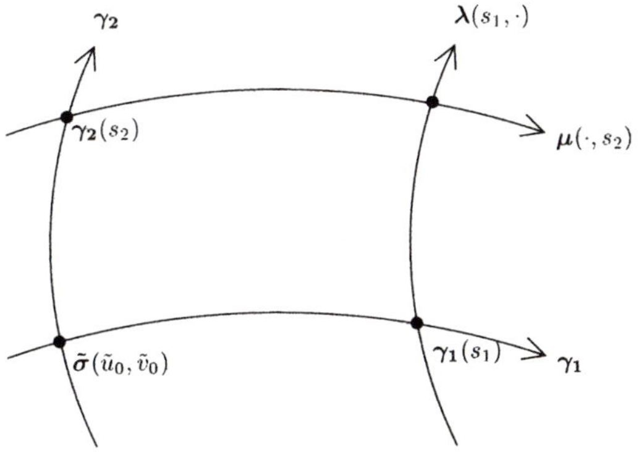

Applying this observation to $\mathbf{e} = \mathbf{e}_1$, we can find a curve $\boldsymbol{\gamma}_1(s_1)$ in $\tilde{\boldsymbol{\sigma}}$ with $\boldsymbol{\gamma}_1(0) = \tilde{\boldsymbol{\sigma}}(\tilde{u}_0, \tilde{v}_0)$ and $d\boldsymbol{\gamma}_1/ds_1 = \mathbf{e}_1$. Now applying the same observation to $\mathbf{e} = \mathbf{e}_2$, we can find, for each value of s_1 close to 0, a curve $s_2 \mapsto \boldsymbol{\lambda}(s_1, s_2)$ in $\tilde{\boldsymbol{\sigma}}$ with $\partial\boldsymbol{\lambda}/\partial s_2 = \mathbf{e}_2$ and $\boldsymbol{\lambda}(s_1, 0) = \boldsymbol{\gamma}_1(s_1)$. Define $(\tilde{u}, \tilde{v})$ as functions of (s_1, s_2) by

$$\tilde{\boldsymbol{\sigma}}(\tilde{u}, \tilde{v}) = \boldsymbol{\lambda}(s_1, s_2). \tag{7}$$

Differentiating with respect to s_1 and s_2 gives

$$\tilde{\boldsymbol{\sigma}}_{\tilde{u}}\frac{\partial \tilde{u}}{\partial s_1} + \tilde{\boldsymbol{\sigma}}_{\tilde{v}}\frac{\partial \tilde{v}}{\partial s_1} = \boldsymbol{\lambda}_{s_1}, \quad \tilde{\boldsymbol{\sigma}}_{\tilde{u}}\frac{\partial \tilde{u}}{\partial s_2} + \tilde{\boldsymbol{\sigma}}_{\tilde{v}}\frac{\partial \tilde{v}}{\partial s_2} = \boldsymbol{\lambda}_{s_2}.$$

We have

$$\boldsymbol{\lambda}_{s_1}|_{s_2=0} = \frac{d}{ds_1}\boldsymbol{\lambda}(s_1, 0) = \frac{d\boldsymbol{\gamma}_1}{ds_1} = \mathbf{e}_1, \quad \boldsymbol{\lambda}_{s_2} = \frac{\partial \lambda}{\partial s_2} = \mathbf{e}_2. \tag{8}$$

Equating coefficients of $\tilde{\boldsymbol{\sigma}}_{\tilde{u}}$ and $\tilde{\boldsymbol{\sigma}}_{\tilde{v}}$, we see from the last two sets of equations that, at the point $\tilde{\boldsymbol{\sigma}}(\tilde{u}_0, \tilde{v}_0)$, where $s_1 = s_2 = 0$, the jacobian matrix

$$\begin{pmatrix} \frac{\partial \tilde{u}}{\partial s_1} & \frac{\partial \tilde{u}}{\partial s_2} \\ \frac{\partial \tilde{v}}{\partial s_1} & \frac{\partial \tilde{v}}{\partial s_2} \end{pmatrix} = \begin{pmatrix} a & c \\ b & d \end{pmatrix}. \tag{9}$$

Since $\mathbf{e}_1$ and $\mathbf{e}_2$ are linearly independent at $(\tilde{u}_0, \tilde{v}_0)$, this matrix is invertible. By the inverse function theorem 4.2, Eq. (7) can be solved for (s_1, s_2) as smooth functions of $(\tilde{u}, \tilde{v})$ when $(\tilde{u}, \tilde{v})$ is in some open set $\tilde{W}$ of $\tilde{U}$ containing $(\tilde{u}_0, \tilde{v}_0)$. Thus, $\boldsymbol{\lambda}$ is an allowable surface patch; by Eq. (8), it has the property that $\boldsymbol{\lambda}_{s_1} = \mathbf{e}_1$ when $s_2 = 0$, and $\boldsymbol{\lambda}_{s_2} = \mathbf{e}_2$ everywhere.

We now repeat the procedure, this time starting with a curve $\boldsymbol{\gamma}_2(t_2)$ with $d\boldsymbol{\gamma}_2/dt_2 = \mathbf{e}_2$ and $\boldsymbol{\gamma}_2(0) = \tilde{\boldsymbol{\sigma}}(\tilde{u}_0, \tilde{v}_0)$, and then taking a curve $t_1 \mapsto \boldsymbol{\mu}(t_1, t_2)$ with $\partial \boldsymbol{\mu}/\partial t_1 = \mathbf{e}_1$ and $\boldsymbol{\mu}(0, t_2) = \boldsymbol{\gamma}_2(t_2)$. This gives an allowable patch $\boldsymbol{\mu}(t_1, t_2)$ such that

$$\boldsymbol{\mu}(t_1, t_2) = \tilde{\boldsymbol{\sigma}}(\tilde{u}, \tilde{v})$$

for $(\tilde{u}, \tilde{v})$ in some open subset $\tilde{Z}$ of $\tilde{U}$ containing $(\tilde{u}_0, \tilde{v}_0)$. This patch has the property that $\boldsymbol{\mu}_{t_1} = \mathbf{e}_1$ everywhere and $\boldsymbol{\mu}_{t_2} = \mathbf{e}_2$ when $t_1 = 0$.

The parametrisation we want is $\boldsymbol{\sigma}(u, v)$, where $\boldsymbol{\sigma}(u, v)$ is the intersection of the curve $s_2 \mapsto \boldsymbol{\lambda}(u, s_2)$ with the curve $t_1 \mapsto \boldsymbol{\mu}(t_1, v)$. Thus, we consider the equations

$$\tilde{\boldsymbol{\sigma}}(\tilde{u}, \tilde{v}) = \boldsymbol{\lambda}(u, s_2) = \boldsymbol{\mu}(t_1, v).$$

From Eq. (9),

$$\frac{\partial \tilde{u}}{\partial u} = a, \quad \frac{\partial \tilde{v}}{\partial u} = b,$$

and similarly

$$\frac{\partial \tilde{u}}{\partial v} = c, \quad \frac{\partial \tilde{v}}{\partial v} = d.$$

Hence, the jacobian matrix

$$\begin{pmatrix} \frac{\partial \tilde{u}}{\partial u} & \frac{\partial \tilde{u}}{\partial v} \\ \frac{\partial \tilde{v}}{\partial u} & \frac{\partial \tilde{v}}{\partial v} \end{pmatrix} = \begin{pmatrix} a & c \\ b & d \end{pmatrix}.$$

As usual, the fact that this matrix is invertible means that (u, v) can be expressed as smooth functions of $(\tilde{u}, \tilde{v})$, for $(\tilde{u}, \tilde{v})$ in some open subset $\tilde{V}$ of $\tilde{W} \cap \tilde{Z}$ containing $(\tilde{u}_0, \tilde{v}_0)$, and we get a reparametrisation $\boldsymbol{\sigma}(u, v)$ of $\tilde{\boldsymbol{\sigma}}(\tilde{u}, \tilde{v})$. Finally, the equation $\boldsymbol{\sigma}(u, v) = \boldsymbol{\mu}(t_1, v)$ implies that

$$\boldsymbol{\sigma}_u = \frac{\partial t_1}{\partial u}\boldsymbol{\mu}_{t_1} = \frac{\partial t_1}{\partial u}\mathbf{e}_1,$$

and similarly

$$\boldsymbol{\sigma}_v = \frac{\partial s_2}{\partial v}\mathbf{e}_2,$$

so $\boldsymbol{\sigma}_u$ and $\boldsymbol{\sigma}_v$ are parallel to $\mathbf{e}_1$ and $\mathbf{e}_2$ everywhere.

EXERCISES

7.14 Let P be a hyperbolic point of a surface $\mathcal{S}$ (see Section 6.4). Show that there is a patch of $\mathcal{S}$ containing P whose parameter curves are asymptotic curves (see Exercise 6.12).

7.4. Surfaces of Constant Mean Curvature

We now consider surfaces whose mean curvature H is constant. As we shall see in Chapter 9, such surfaces are encountered in real life as the shapes taken up by soap films.

We shall discuss the surfaces for which H is everywhere zero in detail in Chapter 9. In this section, we are going to describe a construction which gives a correspondence between surfaces of constant non-zero mean curvature and surfaces of constant positive gaussian curvature.

Definition 7.2

Let $\boldsymbol{\sigma}$ be a surface patch with standard unit normal $\mathbf{N}$, and let λ be a constant scalar. The *parallel surface* $\boldsymbol{\sigma}^\lambda$ of $\boldsymbol{\sigma}$ is

$$\boldsymbol{\sigma}^\lambda = \boldsymbol{\sigma} + \lambda\mathbf{N}.$$

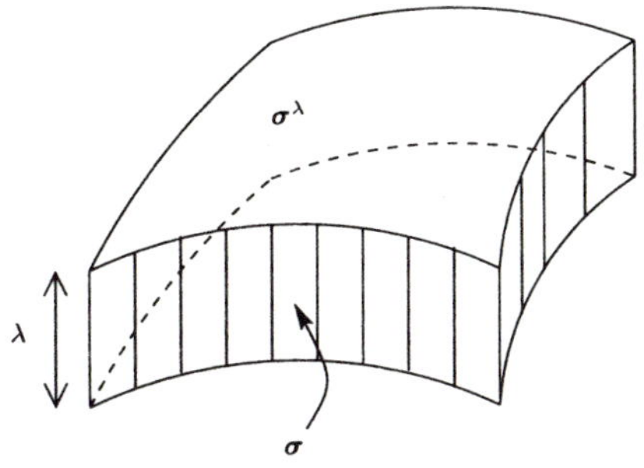

Roughly speaking, $\boldsymbol{\sigma}^\lambda$ is obtained by translating the surface $\boldsymbol{\sigma}$ a distance λ perpendicular to itself (but this is not a genuine translation since $\mathbf{N}$ will in general vary over the surface).

Proposition 7.5

Let κ_1 and κ_2 be the principal curvatures of a surface patch $\boldsymbol{\sigma} : U \to \mathbf{R}^3$, and suppose that there is a constant C such that $|\kappa_1|$ and $|\kappa_2|$ are both $\leq C$ everywhere. Let λ be a constant with $|\lambda| < 1/C$, and let $\boldsymbol{\sigma}^\lambda$ be the corresponding parallel surface of $\boldsymbol{\sigma}$. Then,

(i) $\boldsymbol{\sigma}^\lambda$ is a (regular) surface patch;

(ii) the standard unit normal of $\boldsymbol{\sigma}^\lambda$ at $\boldsymbol{\sigma}^\lambda(u,v)$ is the same as that of $\boldsymbol{\sigma}$ at $\boldsymbol{\sigma}(u,v)$, for all $(u,v) \in U$;

(iii) the principal curvatures of $\boldsymbol{\sigma}^\lambda$ are $\kappa_1/(1-\lambda\kappa_1)$ and $\kappa_2/(1-\lambda\kappa_2)$, and the corresponding principal vectors are the same as those of $\boldsymbol{\sigma}$ for the principal curvatures κ_1 and κ_2, respectively;

(iv) the gaussian and mean curvatures of $\boldsymbol{\sigma}^\lambda$ are

$$\frac{K}{1-2\lambda H+\lambda^2 K} \quad \text{and} \quad \frac{H-\lambda K}{1-2\lambda H+\lambda^2 K},$$

respectively.

Proof 7.5

By Proposition 6.4,

$$\begin{aligned} \boldsymbol{\sigma}_u^\lambda &= \boldsymbol{\sigma}_u + \lambda \mathbf{N}_u = (1+\lambda a)\,\boldsymbol{\sigma}_u + \lambda b\,\boldsymbol{\sigma}_v, \\ \boldsymbol{\sigma}_v^\lambda &= \boldsymbol{\sigma}_v + \lambda \mathbf{N}_v = \lambda c\,\boldsymbol{\sigma}_u + (1+\lambda d)\,\boldsymbol{\sigma}_v, \end{aligned} \tag{10}$$

where the Weingarten matrix

$$\mathcal{W} = -\begin{pmatrix} a & c \\ b & d \end{pmatrix}.$$

Hence,

$$\boldsymbol{\sigma}_u^\lambda \times \boldsymbol{\sigma}_v^\lambda = (1+\lambda(a+d)+\lambda^2(ad-bc))\,\boldsymbol{\sigma}_u \times \boldsymbol{\sigma}_v.$$

Since κ_1 and κ_2 are the eigenvalues of $\mathcal{W}$ (see Section 6.3), and since the sum and product of the eigenvalues of a matrix are equal to the sum of the diagonal entries and the determinant of the matrix, respectively,

$$\kappa_1 + \kappa_2 = -(a+d), \quad \kappa_1\kappa_2 = ad - bc.$$

Hence,

$$\boldsymbol{\sigma}_u^\lambda \times \boldsymbol{\sigma}_v^\lambda = (1-\lambda\kappa_1)(1-\lambda\kappa_2)\,\boldsymbol{\sigma}_u \times \boldsymbol{\sigma}_v. \tag{11}$$

Since $|\lambda| < 1/C$ and $|\kappa_1|$ and $|\kappa_2|$ are $\leq C$, it follows that $|\lambda\kappa_1|$ and $|\lambda\kappa_2|$ are < 1, so $(1-\lambda\kappa_1)(1-\lambda\kappa_2) > 0$, and Eq. (11) shows that $\boldsymbol{\sigma}^\lambda$ is regular and that its standard unit normal is

$$\mathbf{N}^\lambda = \frac{\boldsymbol{\sigma}_u^\lambda \times \boldsymbol{\sigma}_v^\lambda}{\| \boldsymbol{\sigma}_u^\lambda \times \boldsymbol{\sigma}_v^\lambda \|} = \frac{\boldsymbol{\sigma}_u \times \boldsymbol{\sigma}_v}{\| \boldsymbol{\sigma}_u \times \boldsymbol{\sigma}_v \|} = \mathbf{N}.$$

The principal curvatures of $\boldsymbol{\sigma}^\lambda$ are the eigenvalues of the Weingarten matrix $\mathcal{W}^\lambda$ of $\boldsymbol{\sigma}^\lambda$. By Proposition 6.4, this is the negative of the matrix expressing $\mathbf{N}_u^\lambda$ and $\mathbf{N}_v^\lambda$ in terms of $\boldsymbol{\sigma}_u^\lambda$ and $\boldsymbol{\sigma}_v^\lambda$. Equation (10) says that the matrix expressing $\boldsymbol{\sigma}_u^\lambda$ and $\boldsymbol{\sigma}_v^\lambda$ in terms of $\boldsymbol{\sigma}_u$ and $\boldsymbol{\sigma}_v$ is $I - \lambda\mathcal{W}$, and the fact that $\mathbf{N}^\lambda = \mathbf{N}$ implies that $-\mathcal{W}$ is the matrix expressing $\mathbf{N}_u^\lambda$ and $\mathbf{N}_v^\lambda$ in terms of $\boldsymbol{\sigma}_u$ and $\boldsymbol{\sigma}_v$. Combining these two observations we get

$$\mathcal{W}^\lambda = (I - \lambda\mathcal{W})^{-1}\mathcal{W}.$$

If T is an eigenvector of $\mathcal{W}$ with eigenvalue κ, then T is also an eigenvector of $\mathcal{W}^\lambda$ with eigenvalue $\kappa/(1-\lambda\kappa)$. The assertions in part (iii) follows from this.

Part (iv) follows from part (iii) by straightforward algebra. □

Corollary 7.1

If $\boldsymbol{\sigma}$ is a surface patch with constant non-zero mean curvature H, then for $\lambda = 1/2H$, $\boldsymbol{\sigma}^\lambda$ has constant gaussian curvature $4H^2$. Conversely, if $\boldsymbol{\sigma}$ has constant positive gaussian curvature K, then for $\lambda = \pm 1/\sqrt{K}$, $\boldsymbol{\sigma}^\lambda$ has constant mean curvature $\mp\frac{1}{2}\sqrt{K}$.

Proof 7.1

This follows from part (iv) of the proposition by straightforward algebra. □

EXERCISES

7.15 The first fundamental form of a surface patch $\boldsymbol{\sigma}(u,v)$ is of the form $E(du^2 + dv^2)$. Prove that $\boldsymbol{\sigma}_{uu} + \boldsymbol{\sigma}_{vv}$ is perpendicular to $\boldsymbol{\sigma}_u$ and $\boldsymbol{\sigma}_v$. Deduce that the mean curvature $H = 0$ everywhere if and only if the laplacian

$$\boldsymbol{\sigma}_{uu} + \boldsymbol{\sigma}_{vv} = \mathbf{0}.$$

Show that the surface patch

$$\boldsymbol{\sigma}(u,v) = \left(u - \frac{u^3}{3} + uv^2, v - \frac{v^3}{3} + u^2 v, u^2 - v^2\right)$$

has $H = 0$ everywhere. (A picture of this surface can be found in Section 9.3.)

7.16 Compute the mean curvature of the surface with cartesian equation

$$z = f(x, y)$$

where f is a smooth function of x and y. Prove that $H = 0$ for the surface

$$z = \ln\left(\frac{\cos y}{\cos x}\right).$$

(A picture of this surface can also be found in Section 9.3.)

7.17 Let $\boldsymbol{\sigma}(u, v)$ be a surface patch with first and second fundamental forms $Edu^2 + Gdv^2$ and $Ldu^2 + Ndv^2$, respectively (cf. Proposition 7.2). Define

$$\boldsymbol{\Sigma}(u, v, w) = \boldsymbol{\sigma}(u, v) + w\mathbf{N}(u, v),$$

where $\mathbf{N}$ is the standard unit normal of $\boldsymbol{\sigma}$. Show that the three families of surfaces obtained by fixing the values of u, v or w in $\boldsymbol{\Sigma}$ form a triply orthogonal system (see Section 4.6 and Exercise 6.21). The surfaces $w =$ constant are parallel surfaces of $\boldsymbol{\sigma}$. Show that the surfaces $u =$ constant and $v =$ constant are flat ruled surfaces.

7.5. Gaussian Curvature of Compact Surfaces

We have seen in Section 6.4 how the relative signs of the principal curvatures at a point P of a surface $\mathcal{S}$ determine the shape of $\mathcal{S}$ near P. In fact, since the gaussian curvature K of $\mathcal{S}$ is the product of its principal curvatures, the discussion there shows that

(i) if $K > 0$ at P, then P is an elliptic point;
(ii) if $K < 0$ at P, then P is a hyperbolic point;
(iii) if $K = 0$ at P, then P is either a parabolic point or a planar point (and in the last case we cannot say much about the shape of the surface near P).

In this section, we give a result which shows how the gaussian curvature influences the *overall* shape of a surface. We shall give another result of a similar nature in Section 10.4.

Proposition 7.6

If $\mathcal{S}$ is a compact surface, there is a point P of $\mathcal{S}$ at which its gaussian curvature K is > 0.

We recall that a subset X of $\mathbf{R}^3$ is called *compact* if it is closed (i.e. the set of points in $\mathbf{R}^3$ that are *not* in X is open) and bounded (i.e. X is contained in some open ball). In the proof, we shall make use of the following fact about compact sets: if $f : \mathbf{R}^3 \to \mathbf{R}$ is a continuous function, then there are points P and Q in X such that $f(Q) \leq f(R) \leq f(P)$ for all points R in X, so that f attains its maximum value on X at P and its minimum at Q.

Proof 7.6

Define $f : \mathbf{R}^3 \to \mathbf{R}$ by $f(\mathbf{v}) = \parallel \mathbf{v} \parallel^2$. Then, f is continuous so the fact that $\mathcal{S}$ is compact implies that there is a point P in $\mathcal{S}$ where f attains its maximum value. Let P have position vector $\mathbf{p}$; then $\mathcal{S}$ is contained inside the closed ball of radius $\parallel \mathbf{p} \parallel$ and centre the origin, and $\mathcal{S}$ intersects its boundary sphere at P. The idea is that $\mathcal{S}$ is at least as curved as the sphere at P, so its gaussian curvature should be at least that of the sphere at P, i.e. at least $1/\parallel \mathbf{p} \parallel^2$.

To make this argument precise, let $\boldsymbol{\gamma}(t)$ be any unit-speed curve in $\mathcal{S}$ passing through P when $t = 0$. Then, $f(\boldsymbol{\gamma}(t))$ has a local maximum at $t = 0$, so

$$\frac{d}{dt} f(\boldsymbol{\gamma}(t)) = 0, \quad \frac{d^2}{dt^2} f(\boldsymbol{\gamma}(t)) \leq 0$$

at $t = 0$, i.e.

$$\boldsymbol{\gamma}(0).\dot{\boldsymbol{\gamma}}(0) = 0, \quad \boldsymbol{\gamma}(0).\ddot{\boldsymbol{\gamma}}(0) + 1 \leq 0. \tag{12}$$

The equation in (12) shows that $\mathbf{p} = \boldsymbol{\gamma}(0)$ is perpendicular to every unit tangent vector to $\mathcal{S}$ at P, and hence is perpendicular to the tangent plane of $\mathcal{S}$ at P.

Choose a surface patch $\boldsymbol{\sigma}$ in the atlas of $\mathcal{S}$ containing P, and let $\mathbf{N}$ be its standard unit normal. By the preceding remark,

$$\mathbf{N} = \pm \frac{\mathbf{p}}{\parallel \mathbf{p} \parallel}. \tag{13}$$

The inequality in (12) implies that the normal curvature $\kappa_n = \ddot{\boldsymbol{\gamma}}(0).\mathbf{N}$ of $\boldsymbol{\gamma}$ at P (computed in the patch $\boldsymbol{\sigma}$) is $\leq -1/\parallel \mathbf{p} \parallel$ or $\geq 1/\parallel \mathbf{p} \parallel$, according to whether the sign in Eq. (13) is $+$ or $-$, respectively. By Corollary 6.2, the principal curvatures of $\boldsymbol{\sigma}$ at P are either both $\leq -1/\parallel \mathbf{p} \parallel$ or both $\geq 1/\parallel \mathbf{p} \parallel$. In each case, $K \geq 1/\parallel \mathbf{p} \parallel^2 > 0$ at P. □

7.6. The Gauss Map

Proposition 2.2 shows that, if $\boldsymbol{\gamma}(s)$ is a unit-speed plane curve, its signed curvature $\kappa_s = d\varphi/ds$, where φ is the angle between its tangent vector $\dot{\boldsymbol{\gamma}}$ and a fixed direction, i.e. the (signed) *curvature is the rate of change of direction of the tangent vector of* $\boldsymbol{\gamma}$ *per unit length.* We are going to look for an analogue of this for surfaces.

The 'direction' of the tangent plane to a surface patch $\boldsymbol{\sigma} : U \to \mathbf{R}^3$ is measured by its standard unit normal $\mathbf{N}$, so we might expect that the curvature of $\boldsymbol{\sigma}$ is measured by the 'rate of change of $\mathbf{N}$ per unit area'. To make sense of this, note that $\mathbf{N}$ is a point of the unit sphere

$$S^2 = \{\mathbf{v} \in \mathbf{R}^3 \mid \parallel \mathbf{v} \parallel = 1\}.$$

The *Gauss map* is the map $\mathcal{S} \to S^2$, where $\mathcal{S}$ is the image of $\boldsymbol{\sigma}$, which sends the point $\boldsymbol{\sigma}(u, v)$ of $\mathcal{S}$ to the point $\mathbf{N}(u, v)$ of S^2. We denote the Gauss map by $\mathcal{G}$. More generally, the Gauss map can be defined for any *orientable* surface $\mathcal{S}$ (see Section 4.3), since such a surface has a well defined unit normal $\mathbf{N}$ at every point.

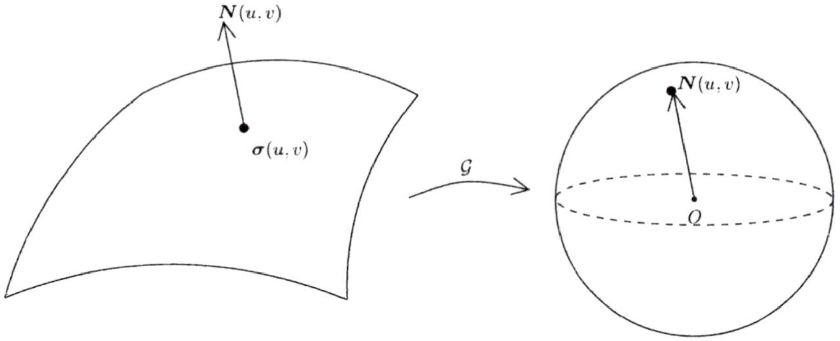

If $R \subseteq U$ is a region, the amount by which $\mathbf{N}$ varies over the part $\boldsymbol{\sigma}(R)$ of $\mathcal{S}$ is measured by the area of the part $\mathbf{N}(R)$ of the sphere. Thus, the rate of change of $\mathbf{N}$ per unit area is approximately

$$\frac{\text{area of } \mathbf{N}(R)}{\text{area of } \boldsymbol{\sigma}(R)} = \frac{\mathcal{A}_{\mathbf{N}}(R)}{\mathcal{A}_{\boldsymbol{\sigma}}(R)},$$

in the notation of Section 5.4. The following theorem shows that, as the region R shrinks to a point, this ratio becomes the absolute value of the gaussian curvature of $\boldsymbol{\sigma}$ at the point.

Theorem 7.1

Let $\boldsymbol{\sigma} : U \to \mathbf{R}^3$ be a surface, let $(u_0, v_0) \in U$, and let $\delta > 0$ be such that the closed disc

$$R_\delta = \{(u,v) \in \mathbf{R}^2 \mid (u-u_0)^2 + (v-v_0)^2 \leq \delta^2\}$$

with centre (u_0, v_0) and radius δ is contained in U (such a δ exists because U is open). Then,

$$\lim_{\delta \to 0} \frac{\mathcal{A}_{\mathbf{N}}(R_\delta)}{\mathcal{A}_{\boldsymbol{\sigma}}(R_\delta)} = |K|,$$

where K is the gaussian curvature of $\boldsymbol{\sigma}$ at $\boldsymbol{\sigma}(u_0, v_0)$.

Proof 7.1

By Definition 5.3,

$$\frac{\mathcal{A}_{\mathbf{N}}(R_\delta)}{\mathcal{A}_{\boldsymbol{\sigma}}(R_\delta)} = \frac{\iint_{R_\delta} \| \mathbf{N}_u \times \mathbf{N}_v \| \, dudv}{\iint_{R_\delta} \| \boldsymbol{\sigma}_u \times \boldsymbol{\sigma}_v \| \, dudv}. \tag{14}$$

By Proposition 6.4,

$$\begin{aligned}
\mathbf{N}_u \times \mathbf{N}_v &= (a\boldsymbol{\sigma}_u \times b\boldsymbol{\sigma}_v) \times (c\boldsymbol{\sigma}_u \times d\boldsymbol{\sigma}_v) \\
&= (ad - bc)\boldsymbol{\sigma}_u \times \boldsymbol{\sigma}_v \\
&= \det(-\mathcal{F}_I^{-1}\mathcal{F}_{II})\boldsymbol{\sigma}_u \times \boldsymbol{\sigma}_v \\
&= \frac{\det(\mathcal{F}_{II})}{\det(\mathcal{F}_I)}\boldsymbol{\sigma}_u \times \boldsymbol{\sigma}_v \\
&= \frac{\begin{vmatrix} L & M \\ M & N \end{vmatrix}}{\begin{vmatrix} E & F \\ F & G \end{vmatrix}}\boldsymbol{\sigma}_u \times \boldsymbol{\sigma}_v \quad \text{(by the definition of } \mathcal{F}_I \text{ and } \mathcal{F}_{II}) \\
&= \frac{LN - M^2}{EG - F^2}\boldsymbol{\sigma}_u \times \boldsymbol{\sigma}_v \\
&= K\boldsymbol{\sigma}_u \times \boldsymbol{\sigma}_v \quad \text{(by Proposition 7.1(i))}.
\end{aligned} \tag{15}$$

Sustituting in Eq. (14), we get

$$\frac{\mathcal{A}_{\mathbf{N}}(R_\delta)}{\mathcal{A}_{\boldsymbol{\sigma}}(R_\delta)} = \frac{\iint_{R_\delta} |K| \, \| \boldsymbol{\sigma}_u \times \boldsymbol{\sigma}_v \| \, dudv}{\iint_{R_\delta} \| \boldsymbol{\sigma}_u \times \boldsymbol{\sigma}_v \| \, dudv}.$$

Let ϵ be any positive number. Since $K(u,v)$ is a continuous function of (u,v) (see Exercise 7.7), we can choose $\delta > 0$ so small that

$$|K(u,v) - K(u_0, v_0)| < \epsilon$$

if $(u, v) \in R_\delta$. Since, for any real numbers a, b, $|a - b| \geq ||a| - |b||$, it follows that $||K(u, v)| - |K(u_0, v_0)|| < \epsilon$ if $(u, v) \in R_\delta$, i.e.

$$|K(u_0, v_0)| - \epsilon < |K(u, v)| < |K(u_0, v_0)| + \epsilon$$

if $(u, v) \in R_\delta$. Multiplying through by $\| \boldsymbol{\sigma}_u \times \boldsymbol{\sigma}_v \|$ and integrating over R_δ, we get

$$(|K(u_0, v_0)| - \epsilon)\int\int \| \boldsymbol{\sigma}_u \times \boldsymbol{\sigma}_v \| dudv < \int\int |K(u, v)| \| \boldsymbol{\sigma}_u \times \boldsymbol{\sigma}_v \| dudv$$

$$< (|K(u_0, v_0)| + \epsilon)\int\int \| \boldsymbol{\sigma}_u \times \boldsymbol{\sigma}_v \| dudv,$$

$$\therefore \quad |K(u_0, v_0)| - \epsilon < \frac{\mathcal{A}_{\mathbf{N}}(R_\delta)}{\mathcal{A}_{\boldsymbol{\sigma}}(R_\delta)} < |K(u_0, v_0)| + \epsilon \quad \text{(using Eq. (14))}$$

$$\therefore \quad \left| \frac{\mathcal{A}_{\mathbf{N}}(R_\delta)}{\mathcal{A}_{\boldsymbol{\sigma}}(R_\delta)} - |K(u_0, v_0)| \right| < \epsilon.$$

This proves the theorem. □

Although this proposition only gives the absolute value of the gaussian curvature K, the sign can be recovered from the Gauss map if we define the *signed area* of $\mathbf{N}(R)$ to be $\pm\mathcal{A}_{\mathbf{N}}(R)$, where the sign is + or − according to whether $\mathbf{N}_u \times \mathbf{N}_v$ points in the same or the opposite direction as $\mathbf{N}$. By Eq. (15), this sign is that of K, so K is the limit of the ratio

$$\frac{\text{signed area of } \mathbf{N}(R)}{\text{area of } \boldsymbol{\sigma}(R)}$$

as the region R shrinks to a point.

As the following examples show, Theorem 7.1 sometimes allows one to find the gaussian curvature of a surface with no calculation.

Example 7.4

For a plane, the unit normal is constant. Thus, for any R, $\mathbf{N}(R)$ is a single point, and thus has zero area. By the theorem, a plane has gaussian curvature zero everywhere.

For a (generalised) cylinder, the unit normal is clearly always perpendicular to the rulings of the cylinder, so the image of the Gauss map is contained in the great circle on the unit sphere formed by intersecting the sphere with the plane passing through the centre of the sphere and perpendicular to the rulings of the cylinder. Any great circle obviously has zero area, so the cylinder has zero gaussian curvature too.

Finally, for the unit sphere S^2 itself, the unit normal at a point P is clearly parallel to the radius vector from the centre of the sphere to P. In other words, the Gauss map is the identity map or the antipodal map (depending on the choice of parametrisation). Both of these maps are obviously equiareal, so the (absolute value of the) gaussian curvature of S^2 is one. In fact, this discussion shows that, for any parametrisation $\boldsymbol{\sigma}$ of S^2, we have $\mathbf{N} = \pm\boldsymbol{\sigma}$. The sign depends of the choice of parametrisation, but in any case $\mathbf{N}_u \times \mathbf{N}_v = \boldsymbol{\sigma}_u \times \boldsymbol{\sigma}_v$ so the gaussian curvature is $+1$.

EXERCISES

7.18 Let $\boldsymbol{\sigma} : U \to \mathbf{R}^3$ be a patch on a surface $\mathcal{S}$. Show that the image under the Gauss map of the part $\boldsymbol{\sigma}(R)$ of $\mathcal{S}$ corresponding to a region $R \subseteq U$ has area

$$\iint_R |K| d\mathcal{A}_{\boldsymbol{\sigma}},$$

where K is the gaussian curvature of $\mathcal{S}$. (Inspect the proof of Theorem 7.1.)

7.19 Let $\mathcal{S}$ be the torus in Exercise 4.10. Sketch the parts $\mathcal{S}^+$ and $\mathcal{S}^-$ of $\mathcal{S}$ where the gaussian curvature K of $\mathcal{S}$ is positive and negative, respectively. Show, without calculation, that

$$\iint_{\mathcal{S}^+} K\, d\mathcal{A} = -\iint_{\mathcal{S}^-} K\, d\mathcal{A} = 4\pi.$$

(The meaning of these integrals should be clear: if $\mathcal{S}^+$, say, is contained in a single surface patch $\boldsymbol{\sigma}$ of $\mathcal{S}$, so that $\mathcal{S}^+ = \boldsymbol{\sigma}(R^+)$, say, then the left-hand side means $\iint_{R^+} K\, d\mathcal{A}_{\boldsymbol{\sigma}}$; if $\mathcal{S}^+$ is *not* contained in a single patch, it can be broken into pieces, each of which *is* contained in a single patch, and then the integral over $\mathcal{S}^+$ is the sum of the integrals over each piece of $\mathcal{S}^+$. Further details can be found in Section 11.3.)
Of course, it follows that $\iint_{\mathcal{S}} K\, d\mathcal{A} = 0$, a result that will be 'explained' in Section 11.3.

8
Geodesics

Geodesics are the curves in a surface that a bug living in the surface would perceive to be straight. For example, the shortest path between two points in a surface is always a geodesic. We shall actually begin by giving a quite different definition of geodesics, since this definition is easier to work with. We give various methods of finding the geodesics on surfaces, before finally making contact with the idea of shortest paths towards the end of the chapter.

8.1. Definition and Basic Properties

Recall from Proposition 1.1 that a curve $\boldsymbol{\gamma}$ is a straight line if it has zero acceleration $\ddot{\boldsymbol{\gamma}}$ everywhere. A bug living in a surface and travelling along a curve $\boldsymbol{\gamma}$, however, would perceive only the component of $\ddot{\boldsymbol{\gamma}}$ parallel to the tangent plane, so would argue that $\boldsymbol{\gamma}$ is straight if this component of the acceleration is everywhere zero. This suggests

Definition 8.1

A curve $\boldsymbol{\gamma}$ on a surface $\mathcal{S}$ is called a *geodesic* if $\ddot{\boldsymbol{\gamma}}(t)$ is zero or perpendicular to the surface at the point $\boldsymbol{\gamma}(t)$, i.e. parallel to its unit normal, for all values of the parameter t.

Note that the unit normal is well defined up to sign for any surface $\mathcal{S}$, so this definition makes sense.

There is an interesting mechanical interpretation of geodesics: a particle moving on the surface, and subject to no forces except a force acting perpendicular to the surface that keeps the particle on the surface, would move along a geodesic. This is because Newton's second law of motion says that the force on the particle is parallel to its acceleration $\ddot{\boldsymbol{\gamma}}$, which would therefore be perpendicular to the surface.

We begin our study of geodesics by noting that there is essentially no choice in their parametrisation.

Proposition 8.1

Any geodesic has constant speed.

Proof 8.1

Let $\boldsymbol{\gamma}(t)$ be a geodesic on a surface $\mathcal{S}$. Then, denoting d/dt by a dot,

$$\frac{d}{dt} \parallel \dot{\boldsymbol{\gamma}} \parallel^2 = \frac{d}{dt}(\dot{\boldsymbol{\gamma}}.\dot{\boldsymbol{\gamma}}) = 2\ddot{\boldsymbol{\gamma}}.\dot{\boldsymbol{\gamma}}.$$

Since $\boldsymbol{\gamma}$ is a geodesic, $\ddot{\boldsymbol{\gamma}}$ is perpendicular to the tangent plane and is therefore perpendicular to the tangent vector $\dot{\boldsymbol{\gamma}}$. So $\ddot{\boldsymbol{\gamma}}.\dot{\boldsymbol{\gamma}} = 0$ and the last equation shows that $\parallel \dot{\boldsymbol{\gamma}} \parallel$ is constant. □

It is clear that a unit-speed reparametrisation of a geodesic $\boldsymbol{\gamma}$ is still a geodesic, since Proposition 8.1 shows that the effect of the reparametrisation is to multiply $\ddot{\boldsymbol{\gamma}}$ by a non-zero scalar. Thus, we can always restrict to unit-speed geodesics if we wish.

We observe next that there is an equivalent definition of a geodesic expressed in terms of the geodesic curvature κ_g (see Section 6.2). Of course, this is why κ_g is called the geodesic curvature!

Proposition 8.2

A curve on a surface is a geodesic if and only if its geodesic curvature is zero everywhere.

Proof 8.2

It is sufficient to consider a unit-speed curve $\boldsymbol{\gamma}$ contained in a patch $\boldsymbol{\sigma}$ of the surface. Let $\mathbf{N}$ be the standard unit normal of $\boldsymbol{\sigma}$, so that

$$\kappa_g = \ddot{\boldsymbol{\gamma}}.(\mathbf{N} \times \dot{\boldsymbol{\gamma}}). \tag{1}$$

If $\ddot{\boldsymbol{\gamma}}$ is parallel to $\mathbf{N}$, it is obviously perpendicular to $\mathbf{N} \times \dot{\boldsymbol{\gamma}}$, so by Eq. (1), $\kappa_g = 0$.

Conversely, suppose that $\kappa_g = 0$. Then, $\ddot{\boldsymbol{\gamma}}$ is perpendicular to $\mathbf{N} \times \dot{\boldsymbol{\gamma}}$. But then, since $\dot{\boldsymbol{\gamma}}$, $\mathbf{N}$ and $\mathbf{N} \times \dot{\boldsymbol{\gamma}}$ are perpendicular unit vectors in $\mathbf{R}^3$, and since $\ddot{\boldsymbol{\gamma}}$ is perpendicular to $\dot{\boldsymbol{\gamma}}$, it follows that $\ddot{\boldsymbol{\gamma}}$ is parallel to $\mathbf{N}$. □

The next result gives the simplest examples of geodesics.

Proposition 8.3

Any (part of a) straight line on a surface is a geodesic.

By this, we mean that every straight line can be parametrised so that it is a geodesic. A similar remark applies to other geodesics we shall consider whose parametrisation is not specified.

Proof 8.3

This is obvious, for a straight line may be parametrised by

$$\boldsymbol{\gamma}(t) = \mathbf{a} + \mathbf{b}t,$$

where $\mathbf{a}$ and $\mathbf{b}$ are constant vectors, and clearly $\ddot{\boldsymbol{\gamma}} = \mathbf{0}$. □

Example 8.1

All straight lines in the plane are geodesics, as are the rulings of any ruled surface, such as the generators of a (generalised) cylinder, or those of a (generalised) cone, or the straight lines on a hyperboloid of one sheet.

The next result is almost as simple:

Proposition 8.4

Any normal section of a surface is a geodesic.

Proof 8.4

Recall from Section 6.2 that a normal section of a surface $\mathcal{S}$ is the intersection $\mathcal{C}$ of $\mathcal{S}$ with a plane Π, such that Π is perpendicular to the surface at each point of $\mathcal{C}$. We showed in Section 6.2 that $\kappa_g = 0$ for a normal section, so the result follows from Proposition 8.2. □

Example 8.2

All great circles on a sphere are geodesics.

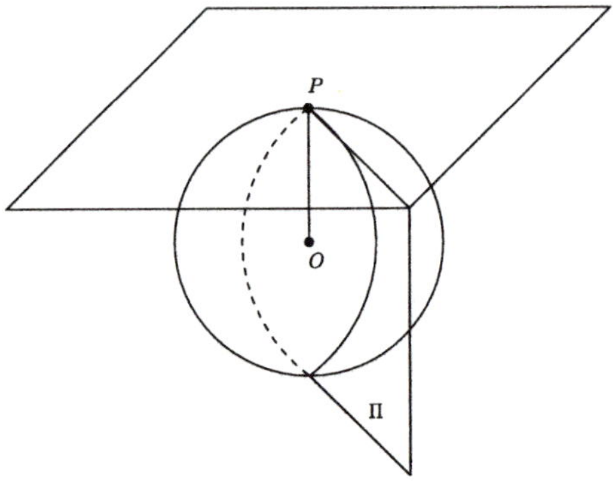

For a great circle is the intersection of the sphere with a plane Π passing through the centre O of the sphere, so if P is a point of the great circle, the vector $\mathbf{OP}$ lies in Π and is perpendicular to the tangent plane of the unit sphere at P. Hence, Π is perpendicular to the tangent plane at P.

Example 8.3

The intersection of a (generalised) cylinder with a plane Π perpendicular to the rulings of the cylinder is a geodesic. For it is clear that the unit normal $\mathbf{N}$ is perpendicular to the rulings. It follows that $\mathbf{N}$ is parallel to Π, and hence that Π is perpendicular to the tangent plane.

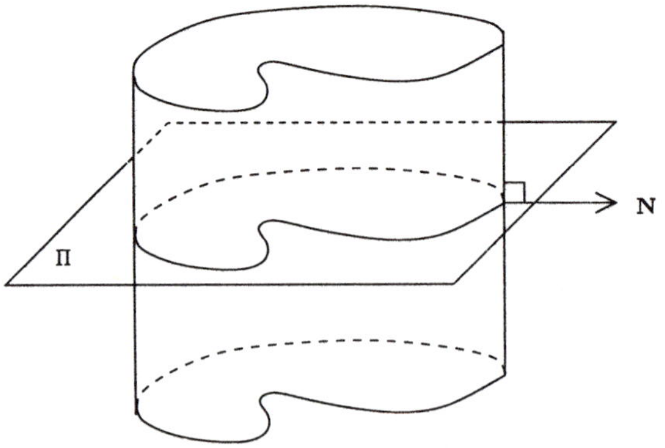

EXERCISES

8.1 Describe four different geodesics on the hyperboloid of one sheet

$$x^2 + y^2 - z^2 = 1$$

passing through the point $(1, 0, 0)$. (Use Propositions 8.3 and 8.4.)

8.2 Consider the tube of radius $a > 0$ around a unit-speed curve $\boldsymbol{\gamma}$ in $\mathbf{R}^3$ defined in Exercise 5.17:

$$\boldsymbol{\sigma}(s, \theta) = \boldsymbol{\gamma}(s) + a(\cos\theta\, \mathbf{n}(s) + \sin\theta\, \mathbf{b}(s)).$$

Show that the parameter curves on the tube obtained by fixing the value of s are circular geodesics on $\mathcal{S}$.

8.3 Let $\boldsymbol{\gamma}(t)$ be a geodesic on an ellipsoid $\mathcal{S}$ (see Proposition 4.6). Let $2R(t)$ be the length of the diameter of the ellipsoid parallel to $\dot{\boldsymbol{\gamma}}(t)$, and let $S(t)$ be the distance from the centre of the ellipsoid to the tangent plane of $\mathcal{S}$ at $\boldsymbol{\gamma}(t)$. Show that the curvature of $\boldsymbol{\gamma}$ is $S(t)/R(t)^2$, and that the product $R(t)S(t)$ is independent of t. (Take the ellipsoid to be $\frac{x^2}{p^2} + \frac{y^2}{q^2} + \frac{z^2}{r^2} = 1$ and note that $\boldsymbol{\gamma}(t) = (f(t), g(t), h(t))$ is a geodesic if and only if $(\ddot{f}, \ddot{g}, \ddot{h}) = \lambda(\frac{f}{p^2}, \frac{g}{q^2}, \frac{h}{r^2})$ for some scalar $\lambda(t)$.)

8.4 Show that the torsion of a geodesic with nowhere vanishing curvature is equal to its geodesic torsion (see Exercise 6.11). (Use the fact that the principal normal of the geodesic is parallel to the unit normal of the surface.)

8.5 A geodesic $\boldsymbol{\gamma}$ on a surface $\mathcal{S}$ lies in a plane and has nowhere vanishing curvature. Show that $\boldsymbol{\gamma}$ is a line of curvature of $\mathcal{S}$. (If Π is the plane, show that the unit normal $\mathbf{N}$ of $\mathcal{S}$ is parallel to Π and deduce that $\dot{\mathbf{N}}$ is parallel to $\dot{\boldsymbol{\gamma}}$; then use Exercise 6.18.)

8.2. Geodesic Equations

Unfortunately, Propositions 8.3 and 8.4 are not usually sufficient to determine all the geodesics on a given surface. For that, we need the following result:

Theorem 8.1

A curve $\boldsymbol{\gamma}$ on a surface $\mathcal{S}$ is a geodesic if and only if, for any part $\boldsymbol{\gamma}(t) = \boldsymbol{\sigma}(u(t), v(t))$ of $\boldsymbol{\gamma}$ contained in a surface patch $\boldsymbol{\sigma}$ of $\mathcal{S}$, the following two equations *are satisfied:*

$$\begin{aligned} \frac{d}{dt}(E\dot{u} + F\dot{v}) &= \frac{1}{2}(E_u\dot{u}^2 + 2F_u\dot{u}\dot{v} + G_u\dot{v}^2), \\ \frac{d}{dt}(F\dot{u} + G\dot{v}) &= \frac{1}{2}(E_v\dot{u}^2 + 2F_v\dot{u}\dot{v} + G_v\dot{v}^2), \end{aligned} \tag{2}$$

where $Edu^2 + 2Fdudv + Gdv^2$ *is the first fundamental form of* $\boldsymbol{\sigma}$.

The differential equations (2) are called the *geodesic equations.*

Proof 8.1

Since $\{\boldsymbol{\sigma}_u, \boldsymbol{\sigma}_v\}$ is a basis of the tangent plane of $\boldsymbol{\sigma}$, $\boldsymbol{\gamma}$ is a geodesic if and only if $\ddot{\boldsymbol{\gamma}}$ is perpendicular to $\boldsymbol{\sigma}_u$ and $\boldsymbol{\sigma}_v$. Since $\dot{\boldsymbol{\gamma}} = \dot{u}\boldsymbol{\sigma}_u + \dot{v}\boldsymbol{\sigma}_v$, this is equivalent to

$$\left(\frac{d}{dt}(\dot{u}\boldsymbol{\sigma}_u + \dot{v}\boldsymbol{\sigma}_v)\right).\boldsymbol{\sigma}_u = 0 \text{ and } \left(\frac{d}{dt}(\dot{u}\boldsymbol{\sigma}_u + \dot{v}\boldsymbol{\sigma}_v)\right).\boldsymbol{\sigma}_v = 0. \tag{3}$$

We show that these two equations are equivalent to the two geodesic equations.

The left-hand side of the first equation in (3) is equal to

$$\begin{aligned} &\frac{d}{dt}\left((\dot{u}\boldsymbol{\sigma}_u + \dot{v}\boldsymbol{\sigma}_v).\boldsymbol{\sigma}_u\right) - (\dot{u}\boldsymbol{\sigma}_u + \dot{v}\boldsymbol{\sigma}_v).\frac{d\boldsymbol{\sigma}_u}{dt} \\ &= \frac{d}{dt}(E\dot{u} + F\dot{v}) - (\dot{u}\boldsymbol{\sigma}_u + \dot{v}\boldsymbol{\sigma}_v).(\dot{u}\boldsymbol{\sigma}_{uu} + \dot{v}\boldsymbol{\sigma}_{uv}) \\ &= \frac{d}{dt}(E\dot{u} + F\dot{v}) - (\dot{u}^2(\boldsymbol{\sigma}_u.\boldsymbol{\sigma}_{uu}) + \dot{u}\dot{v}(\boldsymbol{\sigma}_u.\boldsymbol{\sigma}_{uv} + \boldsymbol{\sigma}_v.\boldsymbol{\sigma}_{uu}) + \dot{v}^2(\boldsymbol{\sigma}_v.\boldsymbol{\sigma}_{uv})). \end{aligned} \tag{4}$$

Now,

$$E_u = (\boldsymbol{\sigma}_u.\boldsymbol{\sigma}_u)_u = \boldsymbol{\sigma}_{uu}.\boldsymbol{\sigma}_u + \boldsymbol{\sigma}_u.\boldsymbol{\sigma}_{uu} = 2\boldsymbol{\sigma}_u.\boldsymbol{\sigma}_{uu},$$

so $\boldsymbol{\sigma}_u.\boldsymbol{\sigma}_{uu} = \frac{1}{2}E_u$. Similarly, $\boldsymbol{\sigma}_v.\boldsymbol{\sigma}_{uv} = \frac{1}{2}G_u$. Finally,

$$\boldsymbol{\sigma}_u.\boldsymbol{\sigma}_{uv} + \boldsymbol{\sigma}_v.\boldsymbol{\sigma}_{uu} = (\boldsymbol{\sigma}_u.\boldsymbol{\sigma}_v)_u = F_u.$$

Substituting these values into (4) gives

$$\left(\frac{d}{dt}(\dot{u}\boldsymbol{\sigma}_u + \dot{v}\boldsymbol{\sigma}_v)\right).\boldsymbol{\sigma}_u = \frac{d}{dt}(E\dot{u} + F\dot{v}) - \frac{1}{2}(E_u\dot{u}^2 + 2F_u\dot{u}\dot{v} + G_u\dot{v}^2).$$

This shows that the first equation in (3) is equivalent to the first geodesic equation in (2). Similarly for the other equations. □

The geodesic equations are non-linear differential equations, and are usually difficult or impossible to solve explicitly. The following example is one case in which this can be done. Another is given in Exercise 8.9.

Example 8.4

We determine the geodesics on the unit sphere S^2 by solving the geodesic equations. For the usual parametrisation by latitude θ and longitude φ,

$$\boldsymbol{\sigma}(\theta,\varphi) = (\cos\theta\cos\varphi, \cos\theta\sin\varphi, \sin\theta),$$

we found in Example 5.2 that the first fundamental form is

$$d\theta^2 + \cos^2\theta\, d\varphi^2.$$

We might as well restrict ourselves to unit-speed curves $\boldsymbol{\gamma}(t) = \boldsymbol{\sigma}(\theta(t), \varphi(t))$, so that

$$\dot\theta^2 + \dot\varphi^2\cos^2\theta = 1,$$

and if $\boldsymbol{\gamma}$ is a geodesic the second equation in (2) gives

$$\frac{d}{dt}(\dot\varphi\cos^2\theta) = 0,$$

so that

$$\dot\varphi\cos^2\theta = \Omega,$$

where Ω is a constant. If $\Omega = 0$, then $\dot\varphi = 0$ so φ is constant and $\boldsymbol{\gamma}$ is part of a meridian. We assume that $\Omega \neq 0$ from now on.

The unit-speed condition gives

$$\dot\theta^2 = 1 - \frac{\Omega^2}{\cos^2\theta},$$

so along the geodesic we have

$$\left(\frac{d\varphi}{d\theta}\right)^2 = \frac{\dot\varphi^2}{\dot\theta^2} = \frac{1}{\cos^2\theta(\Omega^{-2}\cos^2\theta - 1)},$$

and hence

$$\pm(\varphi - \varphi_0) = \int \frac{d\theta}{\cos\theta\sqrt{\Omega^{-2}\cos^2\theta - 1}},$$

where φ_0 is a constant. The integral can be evaluated by making the substitution $u = \tan\theta$. This gives

$$\pm(\varphi - \varphi_0) = \int \frac{du}{\sqrt{\Omega^{-2} - 1 - u^2}} = \sin^{-1}\left(\frac{u}{\sqrt{\Omega^{-2} - 1}}\right),$$

and hence

$$\tan\theta = \pm\sqrt{\Omega^{-2} - 1}\,\sin(\varphi - \varphi_0).$$

This implies that the coordinates $x = \cos\theta\cos\varphi$, $y = \cos\theta\sin\varphi$ and $z = \sin\theta$ of $\boldsymbol{\gamma}(t)$ satisfy the equation

$$z = ax + by,$$

where $a = \mp\sqrt{\Omega^{-2}-1}\sin\varphi_0$, $b = \pm\sqrt{\Omega^{-2}-1}\cos\varphi_0$. This shows that $\boldsymbol{\gamma}$ is contained in the intersection of S^2 with a plane passing through the centre of the sphere.

Hence, in all cases, $\boldsymbol{\gamma}$ is part of a great circle.

The geodesics on a sphere can actually be determined much more simply by using the following consequence of Theorem 8.1.

Corollary 8.1

Let P be a point of a surface $\mathcal{S}$, and let $\mathbf{t}$ be a unit tangent vector to $\mathcal{S}$ at P. Then, there exists a unique unit-speed geodesic $\boldsymbol{\gamma}$ on $\mathcal{S}$ which passes through P and has tangent vector $\mathbf{t}$ there.

In short, *there is a unique geodesic through any given point of a surface in any given direction.*

Proof 8.1

The geodesic equations are of the form

$$\ddot{u} = f(u, v, \dot{u}, \dot{v}), \quad \ddot{v} = g(u, v, \dot{u}, \dot{v}), \tag{5}$$

where f and g are smooth functions of the four variables $u, v, \dot{u}$ and $\dot{v}$. It is proved in the theory of ordinary differential equations that, for any given constants a, b, c, d, and any value t_0 of t, there is a solution of Eq. (5) such that

$$u(t_0) = a,\ v(t_0) = b,\ \dot{u}(t_0) = c,\ \dot{v}(t_0) = d, \tag{6}$$

and such that $u(t)$ and $v(t)$ are defined and smooth for all t satisfying $|t - t_0| < \epsilon$, where ϵ is some positive number. Moreover, any two solutions of Eq. (5) satisfying (6) agree for all values of t such that $|t - t_0| < \epsilon'$, where ϵ' is some positive number $\leq \epsilon$.

We now apply these facts to the geodesic equations. Suppose that P lies in a patch $\boldsymbol{\sigma}(u, v)$ of $\mathcal{S}$, say that P is the point $\boldsymbol{\sigma}(a, b)$, and that $\mathbf{t} = c\boldsymbol{\sigma}_u + d\boldsymbol{\sigma}_v$, where a, b, c, d are scalars and the derivatives are evaluated at $u = a$, $v = b$. A unit-speed curve $\boldsymbol{\gamma}(t) = \boldsymbol{\sigma}(u(t), v(t))$ passes through P at $t = t_0$ if and only if $u(t_0) = a$, $v(t_0) = b$, and has tangent vector $\mathbf{t}$ there if and only if

$$c\boldsymbol{\sigma}_u + d\boldsymbol{\sigma}_v = \mathbf{t} = \dot{\boldsymbol{\gamma}}(t_0) = \dot{u}(t_0)\boldsymbol{\sigma}_u + \dot{v}(t_0)\boldsymbol{\sigma}_v,$$

i.e. $\dot{u}(t_0) = c$, $\dot{v}(t_0) = d$. Thus, finding a (unit-speed) geodesic $\boldsymbol{\gamma}$ passing through P at $t = t_0$ and having tangent vector $\mathbf{t}$ there is equivalent to solving the geodesic equations subject to the initial conditions (6). But we have said above that this problem has a unique solution. □

Example 8.5

We already know that all straight lines in a plane are geodesics. Since there is a straight line in the plane through any given point of the plane in any given direction parallel to the plane, it follows from Corollary 8.1 that there are no other geodesics.

Example 8.6

Similarly, on a sphere, the great circles are the only geodesics, for there is clearly a great circle passing through any given point of the sphere in any given direction tangent to the sphere. (If $\mathbf{p}$ is the position vector of the point and $\mathbf{t}$ the tangent direction, let Π be the plane passing through the origin and parallel to $\mathbf{p}$ and $\mathbf{t}$ (i.e. with normal $\mathbf{p} \times \mathbf{t}$); then take the intersection of the sphere with Π.)

The following consequence of Theorem 8.1 can also be used in some cases to find geodesics without solving the geodesic equations.

Corollary 8.2

An isometry between two surfaces takes the geodesics of one surface to the geodesics of the other.

Proof 8.2

Let $\mathcal{S}_1$ and $\mathcal{S}_2$ be the two surfaces, let $f : \mathcal{S}_1 \to \mathcal{S}_2$ be the isometry, and let $\boldsymbol{\gamma}$ be a geodesic in $\mathcal{S}_1$. Let $\boldsymbol{\sigma}(u, v)$ be a patch in $\mathcal{S}_1$ and suppose that the part of the geodesic lying in this patch is given by $\boldsymbol{\gamma}(t) = \boldsymbol{\sigma}(u(t), v(t))$. Then, u and v satisfy the geodesic equations (2), with E, F and G being the coefficients of the first fundamental form of $\boldsymbol{\sigma}$. By Theorem 5.1, $f \circ \boldsymbol{\sigma}$ is a patch of $\mathcal{S}_2$ with the *same* first fundamental form as $\boldsymbol{\sigma}$. Hence, by Theorem 8.1, $t \mapsto f(\boldsymbol{\sigma}(u(t), v(t)))$ is a geodesic on $\mathcal{S}_2$, in other words, $f \circ \boldsymbol{\gamma}$ is a geodesic. □

Example 8.7

On the circular cylinder $x^2 + y^2 = 1$, we know that the circles obtained by intersecting the cylinder with planes parallel to the xy-plane are geodesics (since they are normal sections). We also know that the straight lines on the cylinder parallel to the z-axis are geodesics. However, these are certainly not the only geodesics, for there is only one geodesic of each of the two types passing through each point of the cylinder (whereas we know that there is a geodesic passing through each point in *any given tangent direction*).

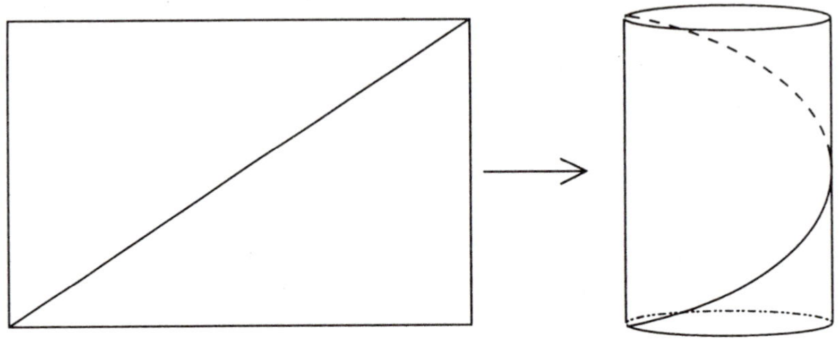

To find the missing geodesics, we recall that the cylinder is isometric to the plane (see Example 5.5). In fact, the isometry takes the point $(u, v, 0)$ of the xy-plane to the point $(\cos u, \sin u, v)$ of the cylinder. By Corollary 8.2, this map takes geodesics on the plane (i.e. straight lines) to geodesics on the cylinder, and vice versa. So to find all the geodesics on the cylinder, we have only to find the images under the isometry of all the straight lines in the plane. Any line not parallel to the y-axis has equation $y = mx + c$, where m and c are constants. Parametrising this line by $x = u$, $y = mu + c$, we see that its image is the curve

$$\boldsymbol{\gamma}(u) = (\cos u, \sin u, mu + c)$$

on the cylinder. Comparing with Example 2.1, we see that this is a *circular helix* of radius one and pitch $2\pi|m|$ (adding c to the z-coordinate just translates the helix vertically). Note that if $m = 0$ we get the circular geodesics that we already know. Finally, any straight line in the xy-plane parallel to the y-axis is mapped by the isometry to a straight line on the cylinder parallel to the z-axis, giving the other family of geodesics that we already know.

EXERCISES

8.6 Show that, if P and Q are distinct points of a circular cylinder, there are either two or infinitely many geodesics on the cylinder joining P and Q. Which pairs P, Q have the former property?

8.7 Use Corollary 8.2 to find all the geodesics on a circular cone. (Use Exercise 5.5.)

8.8 Use Corollary 8.2 to show that the geodesics on a generalised cylinder are exactly those constant-speed curves on the cylinder whose tangent vector makes a constant angle with the rulings of the cylinder.

8.9 Find the geodesics on a circular cylinder by solving the geodesic equations.

8.10 Let $\boldsymbol{\gamma}(t)$ be a unit-speed curve on the helicoid

$$\boldsymbol{\sigma}(u,v) = (u\cos v, u\sin v, v).$$

Show that

$$\dot{u}^2 + (1+u^2)\dot{v}^2 = 1$$

(a dot denotes d/dt). Show also that, if $\boldsymbol{\gamma}$ is a geodesic on $\boldsymbol{\sigma}$, then

$$\dot{v} = \frac{a}{1+u^2},$$

where a is a constant. Find the geodesics corresponding to $a = 0$ and $a = 1$.

8.11 Show that, if $\mathbf{N}$ is the standard unit normal of a surface patch $\boldsymbol{\sigma}(u,v)$ with first fundamental form $Edu^2 + 2Fdudv + Gdv^2$, then

$$\mathbf{N}\times\boldsymbol{\sigma}_u = \frac{E\boldsymbol{\sigma}_v - F\boldsymbol{\sigma}_u}{\sqrt{EG-F^2}}, \quad \mathbf{N}\times\boldsymbol{\sigma}_v = \frac{F\boldsymbol{\sigma}_v - G\boldsymbol{\sigma}_u}{\sqrt{EG-F^2}}.$$

(Use Proposition 5.2.) Deduce that, if $\boldsymbol{\sigma}(u(t),v(t))$ is a unit-speed curve on $\boldsymbol{\sigma}$, its geodesic curvature

$$\kappa_g = (\ddot{v}\dot{u} - \dot{v}\ddot{u})\sqrt{EG-F^2} + A\dot{u}^3 + B\dot{u}^2\dot{v} + C\dot{u}\dot{v}^2 + D\dot{v}^3,$$

where A, B, C and D can be expressed in terms of E, F, G and their derivatives. (Use the method of proof of Theorem 8.1 to calculate dot products such as $\boldsymbol{\sigma}_u.\boldsymbol{\sigma}_{vv}$.) This gives another proof of Corollary 8.2.

8.12 Show directly that the parameter of any curve satisfying the geodesic equations (2) is proportional to arc-length.

8.3. Geodesics on Surfaces of Revolution

It turns out that, although the geodesic equations for a surface of revolution cannot usually be solved explicitly, they can be used to get a good *qualitative* understanding of the geodesics on such a surface.

We parametrise the surface of revolution in the usual way

$$\boldsymbol{\sigma}(u,v) = (f(u)\cos v, f(u)\sin v, g(u)),$$

where we assume that $f > 0$ and $\left(\frac{df}{du}\right)^2 + \left(\frac{dg}{du}\right)^2 = 1$ (see Examples 4.13 and 6.2 – note that in these examples we used a dot to denote d/du, but now a dot is reserved for d/dt, where t is the parameter along a geodesic). We found in Example 6.2 that the first fundamental form of $\boldsymbol{\sigma}$ is $du^2 + f(u)^2 dv^2$. Referring to Eq. (2), we see that the geodesic equations are

$$\ddot{u} = f(u)\frac{df}{du}\dot{v}^2, \quad \frac{d}{dt}(f(u)^2\dot{v}) = 0. \tag{7}$$

We might as well consider unit-speed geodesics, so that

$$\dot{u}^2 + f(u)^2\dot{v}^2 = 1. \tag{8}$$

From this, we make the following easy deductions:

Proposition 8.5

On the surface of revolution

$$\boldsymbol{\sigma}(u, v) = (f(u)\cos v, f(u)\sin v, g(u)),$$

(i) every meridian is a geodesic;
(ii) a parallel $u = u_0$ (say) is a geodesic if and only if $df/du = 0$ when $u = u_0$, i.e. u_0 is a stationary point of f.

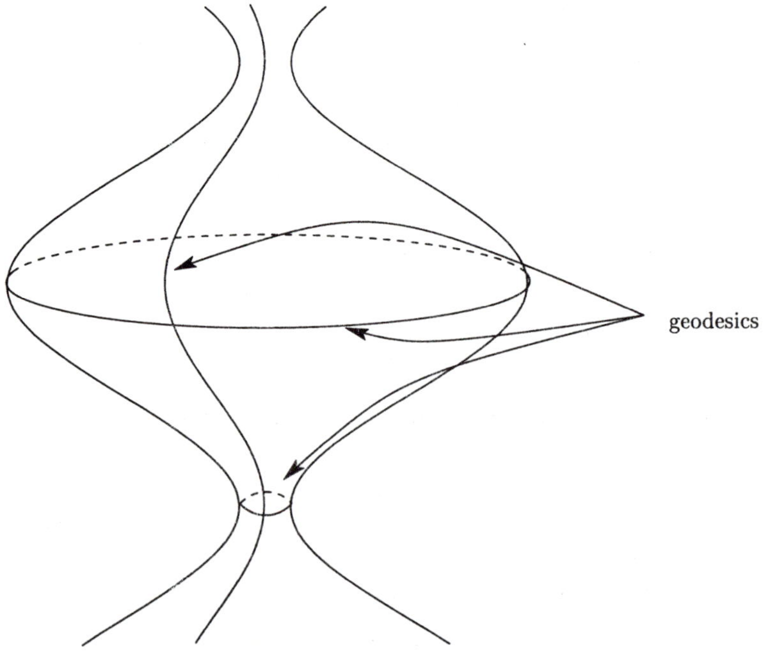

Proof 8.5

On a meridian, we have $v =$ constant so the second equation in (7) is obviously satisfied. Equation (8) gives $\dot{u} = \pm 1$, so $\dot{u}$ is constant and the first equation in (7) is also satisfied.

For (ii), note that if $u = u_0$ is constant, then by Eq. (8), $\dot{v} = \pm 1/f(u_0)$ is non-zero, so the first equation in (7) holds only if $df/du = 0$. Conversely, if $df/du = 0$ when $u = u_0$, the first equation in (7) obviously holds, and the second holds because $\dot{v} = \pm 1/f(u_0)$ and $f(u) = f(u_0)$ are constant. □

Of course, this proposition only gives some of the geodesics on a surface of revolution. The following result is very helpful in understanding the remaining geodesics.

Proposition 8.6 (Clairaut's Theorem)

Let $\boldsymbol{\gamma}$ be a geodesic on a surface of revolution $\mathcal{S}$, let ρ be the distance of a point of $\mathcal{S}$ from the axis of rotation, and let ψ be the angle between $\dot{\boldsymbol{\gamma}}$ and the meridians of $\mathcal{S}$. Then, $\rho \sin\psi$ is constant along $\boldsymbol{\gamma}$.

Conversely, if $\rho \sin\psi$ is constant along some curve $\boldsymbol{\gamma}$ in the surface, and if no part of $\boldsymbol{\gamma}$ is part of some parallel of $\mathcal{S}$, then $\boldsymbol{\gamma}$ is a geodesic.

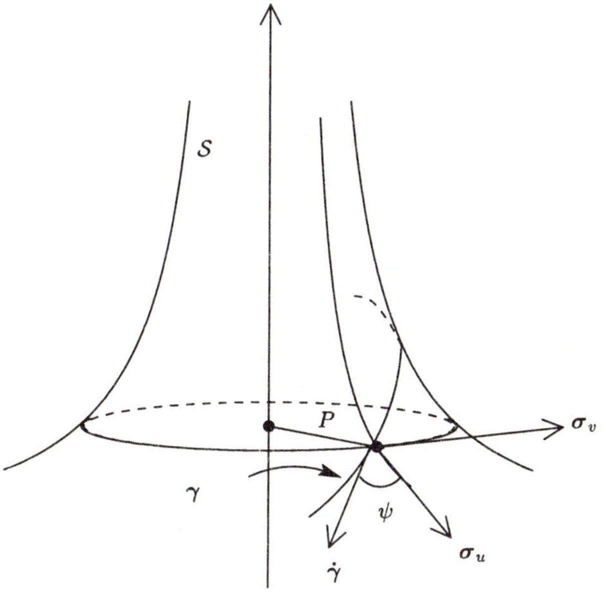

In the second paragraph of the proposition, by a 'part' of $\boldsymbol{\gamma}$ we mean $\boldsymbol{\gamma}(J)$, where J is an open interval. The hypothesis there cannot be relaxed, for on a parallel $\psi = \pi/2$, so $\rho \sin\psi$ is certainly constant. But parallels are not geodesics in general, as Proposition 8.5(ii) shows.

Proof 8.6

Parametrising $\mathcal{S}$ as in Proposition 8.5, we have $\rho = f(u)$. Note that $\boldsymbol{\sigma}_u / \parallel \boldsymbol{\sigma}_u \parallel = \boldsymbol{\sigma}_u$ and $\boldsymbol{\sigma}_v / \parallel \boldsymbol{\sigma}_v \parallel = \rho^{-1}\boldsymbol{\sigma}_v$ are unit vectors tangent to the parallels and the meridians, respectively, and that they are perpendicular since $F = 0$. Assuming that $\boldsymbol{\gamma}(t) = \boldsymbol{\sigma}(u(t), v(t))$ is unit-speed, we have

$$\dot{\boldsymbol{\gamma}} = \cos\psi\,\boldsymbol{\sigma}_u + \rho^{-1}\sin\psi\,\boldsymbol{\sigma}_v$$

(this equation actually serves to define the sign of ψ, which is left ambiguous in the statement of Clairaut's Theorem). Hence,

$$\boldsymbol{\sigma}_u \times \dot{\boldsymbol{\gamma}} = \rho^{-1}\sin\psi\,\boldsymbol{\sigma}_u \times \boldsymbol{\sigma}_v.$$

Since $\dot{\boldsymbol{\gamma}} = \dot{u}\boldsymbol{\sigma}_u + \dot{v}\boldsymbol{\sigma}_v$, this gives

$$\dot{v}\,\boldsymbol{\sigma}_u \times \boldsymbol{\sigma}_v = \rho^{-1}\sin\psi\,\boldsymbol{\sigma}_u \times \boldsymbol{\sigma}_v,$$
$$\therefore \quad \rho\dot{v} = \sin\psi.$$

Hence,

$$\rho\sin\psi = \rho^2\dot{v}.$$

But the second equation in (7) shows that this is a constant, say Ω, along the geodesic.

For the converse, if $\rho\sin\psi$ is a constant Ω along a unit-speed curve $\boldsymbol{\gamma}$ in $\mathcal{S}$, the above argument shows that the second equation in (7) is satisfied, and we must show that the first equation in (7) is satisfied too. Since

$$\dot{v} = \frac{\sin\psi}{\rho} = \frac{\Omega}{\rho^2}, \tag{9}$$

Eq. (8) gives

$$\dot{u}^2 = 1 - \frac{\Omega^2}{\rho^2}. \tag{10}$$

Differentiating both sides with respect to t gives

$$2\dot{u}\ddot{u} = \frac{2\Omega^2}{\rho^3}\dot{\rho} = \frac{2\Omega^2}{\rho^3}\frac{d\rho}{du}\dot{u},$$
$$\therefore \quad \dot{u}\left(\ddot{u} - \rho\frac{d\rho}{du}\dot{v}^2\right) = 0.$$

If the term in brackets does not vanish at some point of the curve, say at $\boldsymbol{\gamma}(t_0) = \boldsymbol{\sigma}(u_0, v_0)$, there will be a number $\epsilon > 0$ such that it does not vanish for $|t - t_0| < \epsilon$. But then $\dot{u} = 0$ for $|t - t_0| < \epsilon$, so $\boldsymbol{\gamma}$ coincides with the parallel

$u = u_0$ when $|t - t_0| < \epsilon$, contrary to our assumption. Hence, the term in brackets must vanish everywhere on $\boldsymbol{\gamma}$, i.e.

$$\ddot{u} = \rho \frac{d\rho}{du} \dot{v}^2,$$

showing that the first equation in (7) is indeed satisfied. □

Clairaut's Theorem has a simple mechanical interpretation. Recall that the geodesics on a surface $\mathcal{S}$ are the curves traced on $\mathcal{S}$ by a particle subject to no forces except a force normal to $\mathcal{S}$ that constrains it to move on $\mathcal{S}$. When $\mathcal{S}$ is a surface of revolution, the force at a point P of $\mathcal{S}$ lies in the plane containing the axis of revolution and P, and so has no moment about the axis. It follows that the angular momentum Ω of the particle about the axis is constant. But, if the particle moves along a unit-speed geodesic, the component of its velocity along the parallel through P is $\sin\psi$, so its angular momentum about the axis is proportional to $\rho \sin\psi$.

Example 8.8

We use Clairaut's Theorem to determine the geodesics on the pseudosphere (Section 7.2):

$$\boldsymbol{\sigma}(u, v) = (e^u \cos v, e^u \sin v, \sqrt{1 - e^{2u}} - \cosh^{-1}(e^{-u})).$$

We found there that its first fundamental form is

$$du^2 + e^{2u} dv^2.$$

It is convenient to reparametrise by setting $w = e^{-u}$. The reparametrised surface is

$$\tilde{\boldsymbol{\sigma}}(v, w) = \left(\frac{1}{w} \cos v, \frac{1}{w} \sin v, \sqrt{1 - \frac{1}{w^2}} - \cosh^{-1} w \right)$$

and its first fundamental form is

$$\frac{dv^2 + dw^2}{w^2}.$$

We must have $w > 1$ for $\tilde{\boldsymbol{\sigma}}$ to be well defined and smooth.

If $\boldsymbol{\gamma}(t) = \tilde{\boldsymbol{\sigma}}(w(t), v(t))$ is a unit-speed geodesic, the unit-speed condition gives

$$\dot{v}^2 + \dot{w}^2 = w^2, \tag{11}$$

and Clairaut's Theorem gives

$$\frac{1}{w} \sin\psi = \frac{1}{w^2} \dot{v} = \Omega, \tag{12}$$

where Ω is a constant, since $\rho = 1/w$. Thus, $\dot{v} = \Omega w^2$. If $\Omega = 0$ we get a meridian $v =$ constant. Assuming now that $\Omega \neq 0$ and substituting in Eq. (11) gives

$$\dot{w} = \pm w\sqrt{1 - \Omega^2 w^2}.$$

Hence, along the geodesic,

$$\frac{dv}{dw} = \frac{\dot{v}}{\dot{w}} = \pm\frac{\Omega w}{\sqrt{1 - \Omega^2 w^2}},$$

$$\therefore \quad (v - v_0) = \mp\frac{1}{\Omega}\sqrt{1 - \Omega^2 w^2},$$

$$\therefore \quad (v - v_0)^2 + w^2 = \frac{1}{\Omega^2}, \tag{13}$$

where v_0 is a constant. So the geodesics are the images under $\tilde{\boldsymbol{\sigma}}$ of the parts of the circles in the vw-plane given by Eq. (13) and lying in the region $w > 1$. Note that these circles all have centre on the v-axis, and so intersect the v-axis perpendicularly. The meridians correspond to straight lines perpendicular to the v-axis.

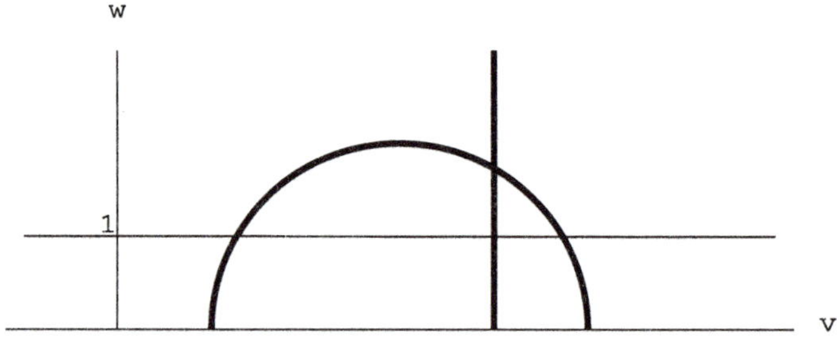

Since $w > 1$, any geodesic other than a meridian has a maximum value of w, which it attains, and a maximum and minimum value of v, which it approaches arbitrarily closely but does not attain (see the diagram below). This shows that the pseudosphere is 'incomplete', i.e. a geodesic on the pseudosphere cannot be continued indefinitely (in one direction if it is a meridian, in both directions otherwise).

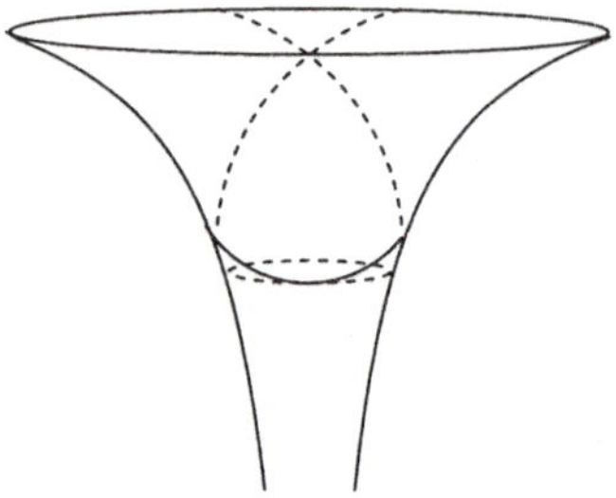

Returning now to an arbitrary surface of revolution $\mathcal{S}$, we describe how Clairaut's Theorem allows us to describe the *qualitative* behaviour of the geodesics on $\mathcal{S}$.

Note first that, in general, there are two geodesics passing through any given point P of $\mathcal{S}$ with a given angular momentum Ω, for $\dot{v}$ is determined by Eq. (9) and $\dot{u}$ up to sign by Eq. (10). In fact, one geodesic is obtained from the other by reflecting in the plane through P containing the axis of rotation (which changes Ω to $-\Omega$) followed by changing the parameter t of the geodesic to $-t$ (which changes the angular momentum back to Ω again).

The discussion in the preceding paragraph shows that we may as well assume that $\Omega > 0$, which we do from now on. Then, Eq. (10) shows that *the geodesic is confined to the part of $\mathcal{S}$ which is at a distance $\geq \Omega$ from the axis.*

If all of $\mathcal{S}$ is a distance $> \Omega$ from the axis, the geodesic will cross every parallel of $\mathcal{S}$. For otherwise, u would be bounded above or below on $\mathcal{S}$, say the former. Let u_0 be the least upper bound of u on the geodesic, and let $\Omega + 2\epsilon$, where $\epsilon > 0$, be the radius of the parallel $u = u_0$. If u is sufficiently close to u_0, the radius of the corresponding parallel will be $\geq \Omega + \epsilon$, and on the part of the geodesic lying in this region we shall have

$$|\dot{u}| \geq \sqrt{1 - \left(\frac{\Omega}{\Omega + \epsilon}\right)^2} > 0$$

by Eq. (10). But this clearly implies that the geodesic will cross $u = u_0$, contradicting our assumption.

Thus, the interesting case is that in which part of $\mathcal{S}$ is within a distance Ω of the axis. The discussion of this case will be clearer if we consider a concrete example whose geodesics nevertheless exhibit essentially all possible forms of behaviour.

Example 8.9

We consider the hyperboloid of one sheet obtained by rotating the hyperbola

$$x^2 - z^2 = 1, \quad x > 0,$$

in the xz-plane around the z-axis. Since all of the surface is at a distance ≥ 1 from the z-axis, we have seen above that, if $0 \leq \Omega < 1$, a geodesic with angular momentum Ω crosses every parallel of the hyperboloid and so extends from $z = -\infty$ to $z = \infty$.

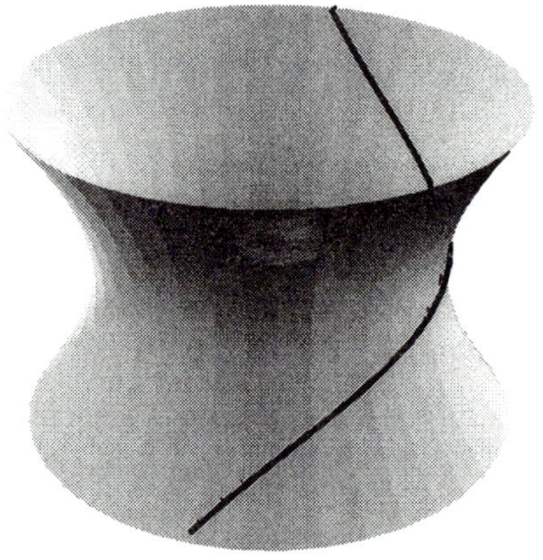

$0 < \Omega < 1$

Suppose now that $\Omega > 1$. Then the geodesic is confined to one of the two regions

$$z \geq \sqrt{\Omega^2 - 1}, \quad z \leq -\sqrt{\Omega^2 - 1},$$

which are bounded by circles Γ^+ and Γ^-, respectively, of radius Ω. Let P be a point on Γ^-, and consider the geodesic $\mathcal{C}$ that passes through P and is tangent to Γ^- there. Then, $\psi = \pi/2$ and $\rho = \Omega$ at P, so $\mathcal{C}$ has angular momentum Ω. Now $\mathcal{C}$ cannot be contained in Γ^-, since Γ^- is not a geodesic (by Proposition 8.5(ii)), so $\mathcal{C}$ must head into the region below Γ^- as it leaves P. Moreover, $\mathcal{C}$ must be symmetric about P, since reflection in the plane through P containing the z-axis takes $\mathcal{C}$ to another geodesic that also passes through P and is tangent to Γ^- there, and so must coincide with $\mathcal{C}$ by the uniqueness part of Corollary 8.1. Since $\dot{u} \neq 0$ in the region below Γ^- by Eq. (10), the geodesic crosses every parallel below Γ^- and $z \to -\infty$ as $t \to \pm\infty$.

Suppose now that $\tilde{\mathcal{C}}$ is *any* geodesic with angular momentum $\Omega > 1$ in the region below Γ^-. Then a suitable rotation around the z-axis will cause $\tilde{\mathcal{C}}$ to intersect $\mathcal{C}$, say at Q, and so to coincide with it (possibly after reflecting in the plane through Q containing the z-axis and changing t to $-t$). We have therefore described the behaviour of every geodesic with angular momentum $\Omega > 1$ that is confined to the region below Γ^-. Of course, the geodesics with angular momentum $\Omega > 1$ in the region above Γ^+ are obtained by reflecting those below Γ^- in the xy-plane.

Suppose finally that $\Omega = 1$. Let $\mathcal{C}$ be a geodesic with angular momentum

1 passing through a point P. If P is on the waist Γ of the hyperboloid (i.e. the unit circle in the xy-plane), which is a geodesic by Proposition 8.5(ii), then $\rho = 1$ at P so $\psi = \pi/2$ and $\mathcal{C}$ is tangent to Γ at P. It must therefore coincide with Γ.

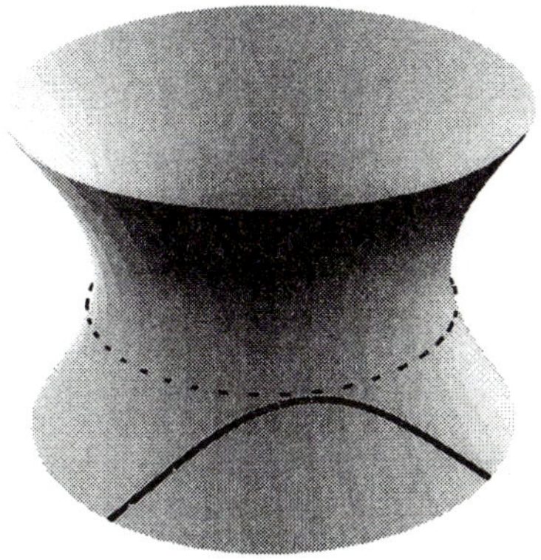

$\Omega > 1$

$\Omega = 1$

Assume now that P is in the region below Γ. Then $0 < \psi < \pi/2$ at P, so as it leaves P in one direction, $\mathcal{C}$ approaches Γ. It must in fact get arbitrarily close to Γ. For if it were to stay always below a parallel $\tilde{\Gamma}$ of radius $1 + \epsilon$, say (with $\epsilon > 0$), then we would have

$$|\dot{u}| \geq \sqrt{1 - \left(\frac{1}{1+\epsilon}\right)^2}$$

everywhere along $\mathcal{C}$ by Eq. (10), which clearly implies that $\mathcal{C}$ must cross every parallel, contradicting our assumption. So, if $\Omega = 1$ the geodesic spirals around the hyperboloid approaching, and getting arbitrarily close to, Γ but never quite reaching it.

EXERCISES

8.13 There is another way to see that all the meridians, and the parallels corresponding to the stationary points of f, are geodesics on a surface of revolution considered in this section. What is it?

8.14 A surface of revolution has the property that *every* parallel is a geodesic. What kind of surface is it?

8.15 Show that a geodesic on the pseudosphere with non-zero angular momentum Ω intersects itself if and only if $\Omega < (1+\pi^2)^{-1/2}$. How many self-intersections are there in that case? (The condition for a self-intersection is that, for some value of $w > 1$, the two values of v satisfying Eq. (13) should differ by an integer multiple of 2π.)

8.16 Let $f : \tilde{\boldsymbol{\sigma}}(v, w) \mapsto \tilde{\boldsymbol{\sigma}}(\tilde{v}, \tilde{w})$ be an isometry of the pseudosphere, where the parametrisation $\tilde{\boldsymbol{\sigma}}$ is that in Example 8.8.

(i) Show that f takes meridians to meridians, and deduce that $\tilde{u}$ does not depend on w. (Use the fact that meridians are the only geodesics on the pseudosphere that can be extended indefinitely in one direction.)

(ii) Deduce that f takes parallels to parallels. (Parallels are the curves that are perpendicular to every meridian.)

(iii) Deduce from (ii) and Exercise 7.11 that $\tilde{w} = w$.

(iv) Show that f is a rotation about the axis of the pseudosphere or a reflection in a plane containing the axis of rotation.

8.17 What do the geodesics on a pseudosphere correspond to in the disc model (Exercise 7.12(ii))? (Use the solution to Exercise 7.12 and the fact that the Möbius transformation $z \mapsto \frac{z-i}{z+i}$ is a conformal map from the (complex) plane to itself that takes lines and circles to lines and circles.)

8.18 Describe the geodesics on

(i) a spheroid, obtained by rotating an ellipse around one of its axes;

(ii) a torus (Exercise 4.10).

8.4. Geodesics as Shortest Paths

Everyone knows that the straight line segment joining two points P and Q in a plane is the shortest path between P and Q. It is almost as well known that great circles are the shortest paths on a sphere. And we have seen that the straight lines are the geodesics in a plane, and the great circles are the geodesics on a sphere.

To see the connection between geodesics and shortest paths on an arbitrary surface $\mathcal{S}$, we consider a unit-speed curve $\boldsymbol{\gamma}$ on $\mathcal{S}$ passing through two fixed points $\mathbf{p}$ and $\mathbf{q}$ on the surface. If $\boldsymbol{\gamma}$ is a shortest path on the surface from $\mathbf{p}$ to $\mathbf{q}$, then the part of $\boldsymbol{\gamma}$ contained in any surface patch $\boldsymbol{\sigma}$ of $\mathcal{S}$ must be the shortest path between any two of its points. For if $\mathbf{p}'$ and $\mathbf{q}'$ are any two points of $\boldsymbol{\gamma}$ in $\boldsymbol{\sigma}$, and if there were a shorter path in $\boldsymbol{\sigma}$ from $\mathbf{p}'$ to $\mathbf{q}'$ than $\boldsymbol{\gamma}$, we could replace the part of $\boldsymbol{\gamma}$ between $\mathbf{p}'$ and $\mathbf{q}'$ by this shorter path, thus giving a shorter path from $\mathbf{p}$ to $\mathbf{q}$ in $\mathcal{S}$.

We may therefore consider a path $\boldsymbol{\gamma}$ entirely contained in a surface patch $\boldsymbol{\sigma}$. To test whether $\boldsymbol{\gamma}$ has smaller length than any other path in $\boldsymbol{\sigma}$ passing through two fixed points $\mathbf{p}$ and $\mathbf{q}$ in $\boldsymbol{\sigma}$, we embed $\boldsymbol{\gamma}$ in a smooth family of curves on $\boldsymbol{\sigma}$ passing through $\mathbf{p}$ and $\mathbf{q}$. By such a family, we mean a curve $\boldsymbol{\gamma}^\tau$ on $\boldsymbol{\sigma}$, for each τ in an open interval $(-\delta, \delta)$, such that

(i) there is an $\epsilon > 0$ such that $\boldsymbol{\gamma}^\tau(t)$ is defined for all $t \in (-\epsilon, \epsilon)$ and all $\tau \in (-\delta, \delta)$;

(ii) for some a, b with $-\epsilon < a < b < \epsilon$, we have

$$\boldsymbol{\gamma}^\tau(a) = \mathbf{p} \quad \text{and} \quad \boldsymbol{\gamma}^\tau(b) = \mathbf{q} \quad \text{for all} \quad \tau \in (-\delta, \delta);$$

(iii) the map from the rectangle $(-\delta, \delta) \times (-\epsilon, \epsilon)$ into $\mathbf{R}^3$ given by

$$(\tau, t) \mapsto \boldsymbol{\gamma}^\tau(t)$$

is smooth;

(iv) $\boldsymbol{\gamma}^0 = \boldsymbol{\gamma}$.

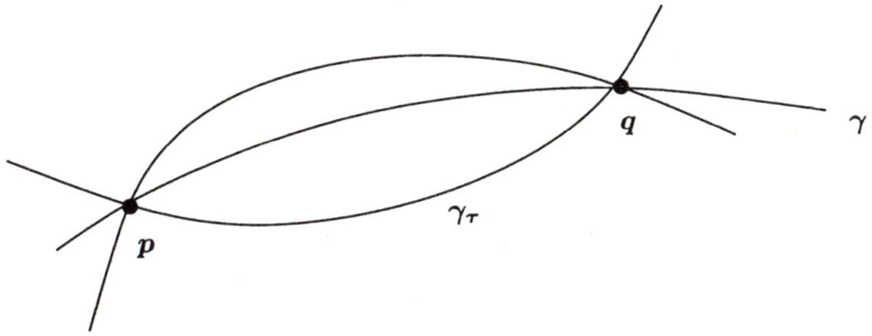

The length of the part of $\boldsymbol{\gamma}^\tau$ between $\mathbf{p}$ and $\mathbf{q}$ is

$$\mathcal{L}(\tau) = \int_a^b \| \dot{\boldsymbol{\gamma}}^\tau \| \, dt,$$

where a dot denotes d/dt.

Theorem 8.2

With the above notation, the unit-speed curve $\boldsymbol{\gamma}$ is a geodesic if and only if

$$\frac{d}{d\tau}\mathcal{L}(\tau) = 0 \quad \text{when } \tau = 0$$

for all families of curves $\boldsymbol{\gamma}^\tau$ with $\boldsymbol{\gamma}^0 = \boldsymbol{\gamma}$.

Note Although we assumed that $\boldsymbol{\gamma} = \boldsymbol{\gamma}^0$ is unit-speed, we *do not* assume that $\boldsymbol{\gamma}^\tau$ is unit-speed if $\tau \neq 0$.

Proof 8.2

We use the formula for 'differentiating under the integral sign': if $f(\tau, t)$ is smooth,

$$\frac{d}{d\tau}\int f(\tau,t)dt = \int \frac{\partial f}{\partial \tau}dt.$$

Thus,

$$\begin{aligned}
\frac{d}{d\tau}\mathcal{L}(\tau) &= \frac{d}{d\tau}\int_a^b \parallel \dot{\boldsymbol{\gamma}}^\tau \parallel \, dt \\
&= \frac{d}{d\tau}\int_a^b (E\dot{u}^2 + 2F\dot{u}\dot{v} + G\dot{v}^2)^{1/2}\, dt \\
&= \int_a^b \frac{\partial}{\partial \tau}(g(\tau,t)^{1/2})\, dt \\
&= \frac{1}{2}\int_a^b g(\tau,t)^{-1/2}\frac{\partial g}{\partial \tau}\, dt, \qquad (14)
\end{aligned}$$

where

$$g(\tau,t) = E\dot{u}^2 + 2F\dot{u}\dot{v} + G\dot{v}^2$$

and a dot denotes d/dt. Now,

$$\begin{aligned}
\frac{\partial g}{\partial \tau} &= \frac{\partial E}{\partial \tau}\dot{u}^2 + 2\frac{\partial F}{\partial \tau}\dot{u}\dot{v} + \frac{\partial G}{\partial \tau}\dot{v}^2 + 2E\dot{u}\frac{\partial \dot{u}}{\partial \tau} + 2F\left(\frac{\partial \dot{u}}{\partial \tau}\dot{v} + \dot{u}\frac{\partial \dot{v}}{\partial \tau}\right) + 2G\dot{v}\frac{\partial \dot{v}}{\partial \tau} \\
&= \left(E_u\frac{\partial u}{\partial \tau} + E_v\frac{\partial v}{\partial \tau}\right)\dot{u}^2 + 2\left(F_u\frac{\partial u}{\partial \tau} + F_v\frac{\partial v}{\partial \tau}\right)\dot{u}\dot{v} + \left(G_u\frac{\partial u}{\partial \tau} + G_v\frac{\partial v}{\partial \tau}\right)\dot{v}^2 \\
&\quad + 2E\dot{u}\frac{\partial^2 u}{\partial \tau \partial t} + 2F\left(\frac{\partial^2 u}{\partial \tau \partial t}\dot{v} + \dot{u}\frac{\partial^2 v}{\partial \tau \partial t}\right) + 2G\dot{v}\frac{\partial^2 v}{\partial \tau \partial t} \\
&= (E_u\dot{u}^2 + 2F_u\dot{u}\dot{v} + G_u\dot{v}^2)\frac{\partial u}{\partial \tau} + (E_v\dot{u}^2 + 2F_v\dot{u}\dot{v} + G_v\dot{v}^2)\frac{\partial v}{\partial \tau} \\
&\quad + 2(E\dot{u} + F\dot{v})\frac{\partial^2 u}{\partial \tau \partial t} + 2(F\dot{u} + G\dot{v})\frac{\partial^2 v}{\partial \tau \partial t}.
\end{aligned}$$

The contribution to the integral in Eq. (14) coming from the terms involving the second partial derivatives is

$$\begin{aligned}
&\int_a^b g^{-1/2}\left\{(E\dot{u} + F\dot{v})\frac{\partial^2 u}{\partial \tau \partial t} + (F\dot{u} + G\dot{v})\frac{\partial^2 v}{\partial \tau \partial t}\right\} dt \\
&\quad = g^{-1/2}\left\{(E\dot{u} + F\dot{v})\frac{\partial u}{\partial \tau} + (F\dot{u} + G\dot{v})\frac{\partial v}{\partial \tau}\right\}\Bigg|_{t=a}^{t=b} \\
&\qquad - \int_a^b \left(\frac{\partial}{\partial t}\left\{g^{-1/2}(E\dot{u} + F\dot{v})\right\}\frac{\partial u}{\partial \tau} + \frac{\partial}{\partial t}\left\{g^{-1/2}(F\dot{u} + G\dot{v})\right\}\frac{\partial v}{\partial \tau}\right) dt, \qquad (15)
\end{aligned}$$

using integration by parts. Now, since $\boldsymbol{\gamma}^\tau(a)$ and $\boldsymbol{\gamma}^\tau(b)$ are independent of τ (being equal to $\mathbf{p}$ and $\mathbf{q}$, respectively), we have

$$\frac{\partial \boldsymbol{\gamma}^\tau}{\partial \tau} = \mathbf{0} \quad \text{when } t = a \text{ or } b.$$

Since

$$\frac{\partial \boldsymbol{\gamma}^\tau}{\partial \tau} = \frac{\partial u}{\partial \tau}\boldsymbol{\sigma}_u + \frac{\partial v}{\partial \tau}\boldsymbol{\sigma}_v,$$

we see that

$$\frac{\partial u}{\partial \tau} = \frac{\partial v}{\partial \tau} = 0 \quad \text{when } t = a \text{ or } b.$$

Hence, the first term on the right-hand side of Eq. (15) is zero. Inserting the remaining terms in Eq. (15) back into Eq. (14), we get

$$\frac{d}{d\tau}\mathcal{L}(\tau) = \int_a^b \left(U\frac{\partial u}{\partial \tau} + V\frac{\partial v}{\partial \tau} \right) dt, \tag{16}$$

where

$$\begin{aligned} U(\tau, t) &= \frac{1}{2}g^{-1/2}(E_u\dot{u}^2 + 2F_u\dot{u}\dot{v} + G_u\dot{v}^2) - \frac{d}{dt}\left\{g^{-1/2}(E\dot{u} + F\dot{v})\right\}, \\ V(\tau, t) &= \frac{1}{2}g^{-1/2}(E_v\dot{u}^2 + 2F_v\dot{u}\dot{v} + G_v\dot{v}^2) - \frac{d}{dt}\left\{g^{-1/2}(F\dot{u} + G\dot{v})\right\}. \end{aligned} \tag{17}$$

Now $\boldsymbol{\gamma}^0 = \boldsymbol{\gamma}$ is unit-speed, so since $\| \dot{\boldsymbol{\gamma}}^\tau \|^2 = g(\tau, t)$, we have $g(\tau, t) = 1$ for all t when $\tau = 0$. Comparing Eqs. (17) with the geodesic equations in (2), we see that, if $\boldsymbol{\gamma}$ is a geodesic, then $U = V = 0$ when $\tau = 0$, and hence by Eq. (16),

$$\frac{d}{d\tau}\mathcal{L}(\tau) = 0 \quad \text{when } \tau = 0.$$

For the converse, we have to show that, if

$$\int_a^b \left(U\frac{\partial u}{\partial \tau} + V\frac{\partial v}{\partial \tau} \right) dt = 0 \quad \text{when } \tau = 0 \tag{18}$$

for *all* families of curves $\boldsymbol{\gamma}^\tau$, then $U = V = 0$ when $\tau = 0$ (since this will prove that $\boldsymbol{\gamma}$ satisfies the geodesic equations). Assume, then, that condition (18) holds, and suppose, for example, that $U \neq 0$ when $\tau = 0$. We will show that this leads to a contradiction.

Since $U \neq 0$ when $\tau = 0$, there is some $t_0 \in (a, b)$ such that $U(0, t_0) \neq 0$, say $U(0, t_0) > 0$. Since U is a continuous function, there exists $\eta > 0$ such that

$$U(0, t) > 0 \quad \text{if } t \in (t_0 - \eta, t_0 + \eta).$$

Let ϕ be a smooth function such that

$$\phi(t) > 0 \text{ if } t \in (t_0 - \eta, t_0 + \eta) \text{ and } \phi(t) = 0 \text{ if } t \notin (t_0 - \eta, t_0 + \eta). \tag{19}$$

(The construction of such a function ϕ is outlined in Exercise 8.20.) Suppose that $\gamma(t) = \boldsymbol{\sigma}(u(t), v(t))$, and consider the family of curves $\boldsymbol{\gamma}^\tau(t) = \boldsymbol{\sigma}(u(\tau, t), v(\tau, t))$, where

$$u(\tau, t) = u(t) + \tau\phi(t), \quad v(\tau, t) = v(t).$$

Then, $\partial u/\partial\tau = \phi$ and $\partial v/\partial\tau = 0$ for all τ and t, so Eq. (18) gives

$$0 = \int_a^b \left(U\frac{\partial u}{\partial \tau} + V\frac{\partial v}{\partial \tau} \right)\bigg|_{\tau=0} dt = \int_{t_0-\eta}^{t_0+\eta} U(0,t)\phi(t)\, dt. \tag{20}$$

But $U(0, t)$ and $\phi(t)$ are both > 0 for all $t \in (t_0 - \eta, t_0 + \eta)$, so the integral on the right-hand side of Eq. (20) is > 0. This contradiction proves that we must have $U(0, t) = 0$ for all $t \in (a, b)$. One proves similarly that $V(0, t) = 0$ for all $t \in (a, b)$. Together, these results prove that $\boldsymbol{\gamma}$ satisfies the geodesic equations. □

It is worth making several comments on Theorem 8.2 to be clear about what it says, and also what it does not say.

Firstly, if $\boldsymbol{\gamma}$ is a shortest path on $\boldsymbol{\sigma}$ from $\mathbf{p}$ to $\mathbf{q}$, then $\mathcal{L}(\tau)$ must have an absolute minimum when $\tau = 0$. This implies that $\frac{d}{d\tau}\mathcal{L}(\tau) = 0$ when $\tau = 0$, and hence by Theorem 8.2 that $\boldsymbol{\gamma}$ is a geodesic.

Secondly, if $\boldsymbol{\gamma}$ is a geodesic on $\boldsymbol{\sigma}$ passing through $\mathbf{p}$ and $\mathbf{q}$, then $\mathcal{L}(\tau)$ has a stationary point (extremum) when $\tau = 0$, but this need not be an absolute minimum, or even a local minimum, so $\boldsymbol{\gamma}$ need not be a shortest path from $\mathbf{p}$

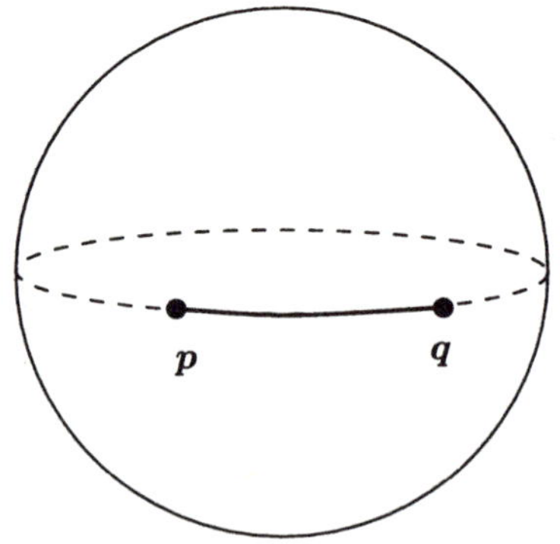

to $\mathbf{q}$. For example, if $\mathbf{p}$ and $\mathbf{q}$ are two nearby points on a sphere, the short great circle arc joining $\mathbf{p}$ and $\mathbf{q}$ is the shortest path from $\mathbf{p}$ to $\mathbf{q}$ (this is not quite obvious – see below), but the long great circle arc joining $\mathbf{p}$ and $\mathbf{q}$ is also a geodesic.

Thirdly, in general, a shortest path joining two points on a surface may not exist. For example, consider the surface $\mathcal{S}$ consisting of the xy-plane with the origin removed. This is a perfectly good surface, but there is *no* shortest path on the surface from the point $\mathbf{p} = (-1, 0)$ to the point $\mathbf{q} = (1, 0)$. Of course, the shortest path should be the straight line segment joining the two points, but this does not lie entirely on the surface, since it passes through the origin which is not part of the surface. For a 'real life' analogy, imagine trying to walk from $\mathbf{p}$ to $\mathbf{q}$ but finding that there is a deep hole in the ground at the origin. The solution might be to walk in a straight line as long as possible, and then skirt around the hole at the last minute, say taking something like the following route:

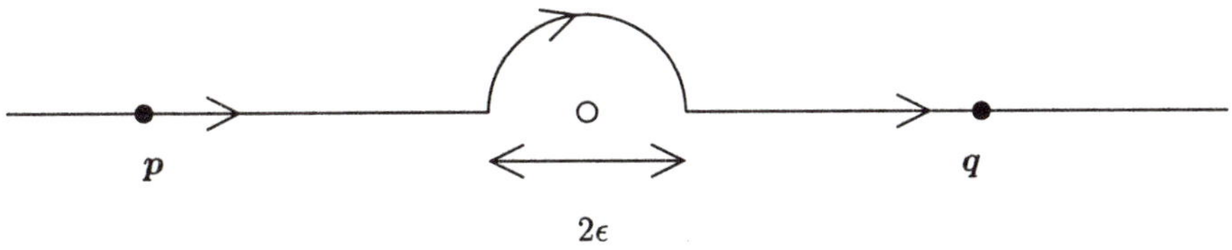

This path consists of two straight line segments of length $1 - \epsilon$, together with a semicircle of radius ϵ, so its total length is

$$2(1-\epsilon) + \pi\epsilon = 2 + (\pi - 2)\epsilon.$$

Of course, this is greater than the straight line distance 2, but it can be made as close as we like to 2 by taking ϵ sufficiently small. In the language of real analysis, the greatest lower bound of the lengths of curves on the surface joining $\mathbf{p}$ and $\mathbf{q}$ is 2, but there is no curve from $\mathbf{p}$ to $\mathbf{q}$ *in the surface* whose length is equal to this lower bound.

Finally, it can be proved that if a surface $\mathcal{S}$ is a *closed* subset of $\mathbf{R}^3$ (i.e. if the set of points of $\mathbf{R}^3$ that are *not* in $\mathcal{S}$ is an open subset of $\mathbf{R}^3$), and if there is *some* path in $\mathcal{S}$ joining any two points of $\mathcal{S}$, then there is always a shortest path joining any two points of $\mathcal{S}$. For example, a plane is a closed subset of $\mathbf{R}^3$, so there is a shortest path joining any two points. This path must be a straight line, for by the first remark above it is a geodesic, and we know that the only geodesics on a plane are the straight lines. Similarly, a sphere is a closed subset of $\mathbf{R}^3$, and it follows that the short great circle arc joining two points on the sphere is the shortest path joining them. But the surface $\mathcal{S}$ considered above is *not* a closed subset of $\mathbf{R}^3$, for $(0, 0)$ is a point not in $\mathcal{S}$, but any open ball containing $(0, 0)$ must clearly contain points of $\mathcal{S}$, so the set of points not in $\mathcal{S}$ is not open.

Another property of surfaces that are closed subsets of $\mathbf{R}^3$ (that we shall also not prove) is that geodesics on such surfaces can be extended indefinitely, i.e. they can be defined on the whole of $\mathbf{R}$. This is clear for straight lines in the plane, for example, and for great circles on the sphere (although in the latter case the geodesics 'close up' after an increment in the unit-speed parameter equal to the circumference of the sphere). But, for the straight line $\boldsymbol{\gamma}(t) = (t-1, 0)$ on the surface $\mathcal{S}$ defined above, which passes through $\mathbf{p}$ when $t = 0$, the largest interval containing $t = 0$ on which it is defined as a curve *in the surface* is $(-\infty, 1)$. We encountered a less artificial example of this 'incompleteness' in Example 8.8: the pseudosphere considered there fails to be a closed subset of $\mathbf{R}^3$ because the points of its boundary circle in the xy-plane are not in the surface.

EXERCISES

8.19 The geodesics on a circular (half) cone were determined in Exercise 8.7. Interpreting 'line' as 'geodesic', which of the following (true) statements in plane euclidean geometry are true for the cone?

(i) There is a line passing through any two points.

(ii) There is a unique line passing through any two points.

(iii) Any two distinct lines intersect in at most one point.

(iv) There are lines that do not intersect each other.

(v) Any line can be continued indefinitely.

(vi) A line defines the shortest distance between any two of its points.

(vii) A line cannot intersect itself transversely (i.e. with two non-parallel tangent vectors at the point of intersection).

8.20 Construct a smooth function with the properties in (19) in the following steps:

(i) Show that, for all integers n (positive and negative), $t^n e^{-1/t^2}$ tends to 0 as t tends to 0. (Use L'Hopital's rule.)

(ii) Deduce from (i) that the function

$$\theta(t) = \begin{cases} e^{-1/t^2} & \text{if } t \geq 0, \\ 0 & \text{if } t \leq 0 \end{cases}$$

is smooth everywhere.

(iii) Show that the function

$$\psi(t) = \theta(1+t)\theta(1-t)$$

is smooth everywhere, that $\psi(t) > 0$ if $-1 < t < 1$, and that $\psi(t) = 0$ otherwise.

(iv) Show that the function

$$\phi(t) = \psi\left(\frac{t - t_0}{\eta}\right)$$

has the properties we want.

8.5. Geodesic Coordinates

The existence of geodesics on a surface $\mathcal{S}$ allows us to construct a very useful atlas for $\mathcal{S}$.

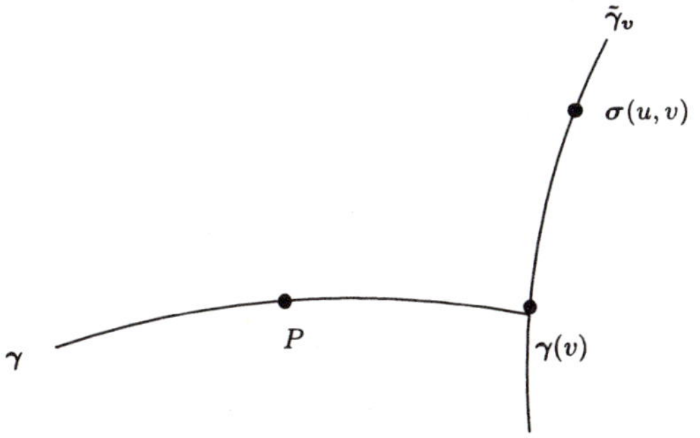

For this, let P be a point of $\mathcal{S}$ and let $\boldsymbol{\gamma}$, with parameter v say, be a unit-speed geodesic on $\mathcal{S}$ with $\boldsymbol{\gamma}(0) = P$. For any value of v, there is a unique unit-speed geodesic $\tilde{\boldsymbol{\gamma}}^v$, with parameter u say, such that $\tilde{\boldsymbol{\gamma}}^v(0) = \boldsymbol{\gamma}(v)$ and which is perpendicular to $\boldsymbol{\gamma}$ at $\boldsymbol{\gamma}(v)$. We define $\boldsymbol{\sigma}(u, v) = \tilde{\boldsymbol{\gamma}}^v(u)$.

Proposition 8.7

With the above notation, there is an open subset U of $\mathbf{R}^2$ containing $(0, 0)$ such that $\boldsymbol{\sigma} : U \to \mathbf{R}^3$ is an allowable surface patch for $\mathcal{S}$. Moreover, the first fundamental form of $\boldsymbol{\sigma}$ is

$$du^2 + G(u, v)dv^2,$$

where G is a smooth function on U such that

$$G(0, v) = 1, \quad G_u(0, v) = 0,$$

whenever $(0, v) \in U$.

Proof 8.7

The proof that $\boldsymbol{\sigma}$ is (for a suitable open set U) an allowable surface patch makes use of the inverse function theorem, and is similar to the proof of Proposition 4.1 (see Section 4.6).

Note first that, for any value of v,

$$\boldsymbol{\sigma}_u(0,v) = \frac{d}{du}\tilde{\boldsymbol{\gamma}}^v(u)\bigg|_{u=0}, \quad \boldsymbol{\sigma}_v(0,v) = \frac{d}{dv}\tilde{\boldsymbol{\gamma}}^v(0) = \frac{d}{dv}\boldsymbol{\gamma}(v),$$

and that these are perpendicular unit vectors by construction. If $\boldsymbol{\sigma}(u,v) = (f(u,v), g(u,v), h(u,v))$, it follows that the jacobian matrix

$$\begin{pmatrix} f_u & f_v \\ g_u & g_v \\ h_u & h_v \end{pmatrix}$$

has rank 2 when $u = v = 0$. Hence, at least one of its three 2×2 submatrices is invertible at $(0,0)$, say

$$\begin{pmatrix} f_u & f_v \\ g_u & g_v \end{pmatrix}. \tag{21}$$

By the inverse function theorem 4.2, there is an open subset U of $\mathbf{R}^2$ such that the map given by

$$F(u,v) = (f(u,v), g(u,v))$$

is a bijection from U to an open subset $F(U)$ of $\mathbf{R}^2$, and such that its inverse map $F(U) \to U$ is also smooth. The matrix (21) is then invertible for all $(u,v) \in U$, so $\boldsymbol{\sigma}_u$ and $\boldsymbol{\sigma}_v$ are linearly independent for $(u,v) \in U$. It follows that $\boldsymbol{\sigma} : U \to \mathbf{R}^3$ is an allowable surface patch.

As to the first fundamental form of $\boldsymbol{\sigma}$, note first that

$$E = \| \boldsymbol{\sigma}_u \|^2 = \| \frac{d}{du}\tilde{\boldsymbol{\gamma}}^v(u) \|^2 = 1$$

because $\tilde{\boldsymbol{\gamma}}^v$ is a unit-speed curve. Next, we apply the second of the geodesic equations (2) to $\tilde{\boldsymbol{\gamma}}^v$. The unit-speed parameter is u and v is constant, so we get $F_u = 0$. But when $u = 0$, we have already seen that $\boldsymbol{\sigma}_u$ and $\boldsymbol{\sigma}_v$ are perpendicular, so $F = 0$. It follows that $F = 0$ everywhere. Hence, the first fundamental form of $\boldsymbol{\sigma}$ is

$$du^2 + G(u,v)dv^2.$$

We have

$$G(0,v) = \| \boldsymbol{\sigma}_v(0,v) \|^2 = \| \frac{d\boldsymbol{\gamma}}{dv} \|^2 = 1$$

because $\boldsymbol{\gamma}$ is unit-speed. Finally, from the first geodesic equation in (2) applied to the geodesic $\boldsymbol{\gamma}$, for which $u = 0$ and v is the unit-speed parameter, we get $G_u(0,v) = 0$. □

A surface patch $\boldsymbol{\sigma}$ constructed as above is called a *geodesic patch*, and u and v are called *geodesic coordinates*.

Example 8.10

If P is a point on the equator of the unit sphere S^2, take $\boldsymbol{\gamma}$ to be the equator with parameter the longitude φ, and let $\tilde{\boldsymbol{\gamma}}^{\varphi}$ be the meridian parametrised by latitude θ and passing through the point on the equator with longitude φ. The corresponding geodesic patch is the usual latitude longitude patch, for which the first fundamental form is

$$d\theta^2 + \cos^2\theta\, d\varphi^2,$$

in accordance with Proposition 8.7.

We shall give an application of geodesic coordinates in the proof of Theorem 10.4.

EXERCISES

8.21 Let P be a point of a surface $\mathcal{S}$ and let $\mathbf{v}$ be a unit tangent vector to $\mathcal{S}$ at P. Let $\boldsymbol{\gamma}^{\theta}(r)$ be the unit-speed geodesic on $\mathcal{S}$ passing through P when $r = 0$ and with tangent vector $\mathbf{v}$ there. Let $\boldsymbol{\sigma}(r,\theta) = \boldsymbol{\gamma}^{\theta}(r)$. It can be shown that $\boldsymbol{\sigma}$ is smooth for $-\epsilon < r < \epsilon$ and all values of θ, where ϵ is some positive number, and that it is an allowable surface patch for $\mathcal{S}$ defined for $0 < r < \epsilon$ and for θ in any open interval of length $\leq 2\pi$. This is called a *geodesic polar patch* on $\mathcal{S}$. Show that, if $0 < R < \epsilon$,

$$\int_0^R \parallel \frac{d\boldsymbol{\gamma}^{\theta}}{dr} \parallel^2 dr = R.$$

By differentiating both sides with respect to θ, prove that

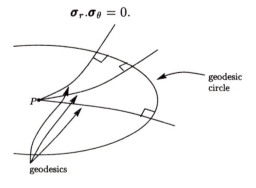

$$\boldsymbol{\sigma}_r.\boldsymbol{\sigma}_{\theta} = 0.$$

This is called *Gauss's Lemma* – geometrically, it means that the parameter curve $r = R$, called the *geodesic circle* with centre P and radius R, is perpendicular to each of its radii, i.e. the geodesics passing through P.

Deduce that the first fundamental form of $\boldsymbol{\sigma}$ is

$$dr^2 + G(r, \theta)d\theta^2,$$

for some smooth function $G(r, \theta)$.

9
Minimal Surfaces

In Section 8.4 we considered the problem of finding the shortest paths between two points on a surface. We now consider the analogous problem in one higher dimension, that of finding a surface of minimal area with a fixed curve as its boundary. This is called *Plateau's Problem.* The solutions to Plateau's problem turn out to be surfaces whose mean curvature vanishes everywhere. The study of these so-called minimal surfaces was initiated by Euler and Lagrange in the mid-18th century, but new examples of minimal surfaces have been discovered quite recently.

9.1. Plateau's Problem

In Section 8.4, we found the condition for a curve on a surface to minimise distance between its endpoints by embedding the given curve in a family of curves passing through the same two points, and studying how the length of the curve varies as the curve varies through the family. Accordingly, we shall now study a family of surface patches $\boldsymbol{\sigma}^\tau : U \to \mathbf{R}^3$, where U is an open subset of $\mathbf{R}^2$ independent of τ, and τ lies in some open interval $(-\delta, \delta)$, for some $\delta > 0$. Let $\boldsymbol{\sigma} = \boldsymbol{\sigma}^0$. The family is required to be *smooth*, in the sense that the map $(u, v, \tau) \mapsto \boldsymbol{\sigma}^\tau(u, v)$ from the open subset $\{(u, v, \tau) \mid (u, v) \in U,\ \tau \in (-\delta, \delta)\}$ of $\mathbf{R}^3$ to $\mathbf{R}^3$ is smooth. The *surface variation* of the family is the function $\boldsymbol{\varphi} : U \to \mathbf{R}^3$ given by

$$\boldsymbol{\varphi} = \dot{\boldsymbol{\sigma}}^\tau|_{\tau=0} ,$$

where here and elsewhere in this section, a dot denotes $d/d\tau$.

Let $\boldsymbol{\pi}$ be a simple closed curve that is contained, along with its interior $\mathrm{int}(\boldsymbol{\pi})$, in U (see Section 3.1). Then $\boldsymbol{\pi}$ corresponds to a simple closed curve $\boldsymbol{\gamma}^\tau = \boldsymbol{\sigma}^\tau \circ \boldsymbol{\pi}$ in the surface patch $\boldsymbol{\sigma}^\tau$, and we define the area function $\mathcal{A}(\tau)$ to be the area of the part of $\boldsymbol{\sigma}^\tau$ inside $\boldsymbol{\gamma}^\tau$:

$$\mathcal{A}(\tau) = \iint_{\mathrm{int}(\boldsymbol{\pi})} d\mathcal{A}_{\boldsymbol{\sigma}^\tau}.$$

Note that, if we are considering a family of surfaces with a *fixed* boundary curve $\boldsymbol{\gamma}$, then $\boldsymbol{\gamma}^\tau = \boldsymbol{\gamma}$ for all τ, and hence $\boldsymbol{\varphi}^\tau(u,v) = \mathbf{0}$ when (u,v) is a point on the curve $\boldsymbol{\pi}$.

Theorem 9.1

With the above notation, assume that the surface variation $\boldsymbol{\varphi}^\tau$ vanishes along the boundary curve $\boldsymbol{\pi}$. Then,

$$\dot{\mathcal{A}}(0) = -2\iint_{\mathrm{int}(\boldsymbol{\pi})} H(EG - F^2)^{1/2}\alpha\, dudv, \tag{1}$$

where H is the mean curvature of $\boldsymbol{\sigma}$, E, F and G are the coefficients of its first fundamental form, $\alpha = \boldsymbol{\varphi}.\mathbf{N}$, and $\mathbf{N}$ is the standard unit normal of $\boldsymbol{\sigma}$.

We defer the proof of this theorem to the end of this section.

If $\boldsymbol{\sigma}$ has the smallest area among all surfaces with the given boundary curve $\boldsymbol{\gamma}$, then $\mathcal{A}$ must have an absolute minimum at $\tau = 0$, so $\dot{\mathcal{A}}(0) = 0$ for *all* smooth families of surfaces as above. This means that the integral in (1) must vanish for all smooth functions $\alpha : U \to \mathbf{R}$. As in the proof of Theorem 8.2, this can happen only if the term that multiplies α in the integrand vanishes, in other words only if $H = 0$. This suggests the following definition.

Definition 9.1

A *minimal surface* is a surface whose mean curvature is zero everywhere.

Theorem 9.1 and the preceding discussion then give

Corollary 9.1

If a surface $\mathcal{S}$ has least area among all surfaces with the same boundary curve, then $\mathcal{S}$ is a minimal surface.

Minimal surfaces have an interesting physical interpretation as the shapes taken up by soap films. A soap film has energy by virtue of surface tension, and this energy is proportional to its area. A soap film spanning a wire in the shape of a curve $\mathcal{C}$ should therefore adopt the shape of a surface of least area with boundary $\mathcal{C}$. By Corollary 9.1, this will be a minimal surface.

More generally, if the soap film separates two regions of different pressure, the film will adopt the shape of a surface of *constant* mean curvature. This is the case for a soap bubble, for example, for which the air pressure inside the bubble is greater than the pressure outside. To see this, we apply the principle of 'virtual work'. This tells us that, if the soap film is in equilibrium, and we imagine a ('virtual') change in the surface, the change in the energy of the film must be the same as the work done by the film against the air pressure. If p is the pressure difference, the force exerted by the air on a small piece of the surface of area $\Delta\mathcal{A}$ is $p\Delta\mathcal{A}$, so the work done when it moves a small distance α perpendicular to itself is $\alpha p\Delta\mathcal{A}$. On the other hand, the formula in Theorem 9.1 shows that the change in area of the surface is proportional to $\alpha H\Delta\mathcal{A}$ (note that α is the component of the variation $\boldsymbol{\varphi}$ perpendicular to the surface). So p is proportional to H. Since the pressure difference must be the same across the whole surface, so must the mean curvature H. Surfaces of constant non-zero mean curvature were discussed in Section 7.4.

For the moment, we give only one example of a minimal surface; others will be given in the next section. This example already shows, however, that the converse of Corollary 9.1 is false.

Example 9.1

A catenoid is the surface obtained by revolving the curve $x = \frac{1}{a}\cosh az$ in the xz-plane around the z-axis, where a is a non-zero constant (a picture of the catenoid can be found in Example 9.2). We take $a = 1$ for simplicity. The catenoid can be parametrised by

$$\boldsymbol{\sigma}(u, v) = (\cosh u \cos v, \cosh u \sin v, u).$$

Then,

$$\begin{aligned}
\boldsymbol{\sigma}_u = (\sinh u \cos v, \sinh u \sin v, 1), \quad \boldsymbol{\sigma}_v &= (-\cosh u \sin v, \cosh u \cos v, 0),\\
\boldsymbol{\sigma}_u \times \boldsymbol{\sigma}_v &= (-\cosh u \cos v, -\cosh u \sin v, \sinh u \cosh u),\\
\mathbf{N} &= (-\mathrm{sech}\, u \cos v, -\mathrm{sech}\, u \sin v, \mathrm{tanh} u),\\
\boldsymbol{\sigma}_{uu} &= (\cosh u \cos v, \cosh u \sin v, 0),\\
\boldsymbol{\sigma}_{uv} &= (-\sinh u \sin v, \sinh u \cos v, 0),\\
\boldsymbol{\sigma}_{vv} &= (-\cosh u \cos v, -\cosh u \sin v, 0).
\end{aligned}$$

This gives the coefficients of the first and second fundamental forms of $\boldsymbol{\sigma}$ as

$$E = G = \cosh^2 u, \ F = 0, \quad L = -1, \ M = 0, \ N = 1.$$

The first three of these equations show that the parametrisation $\boldsymbol{\sigma}$ is conformal, and Proposition 7.1(ii) gives

$$H = \frac{LG - 2MF + NE}{2(EG - F^2)} = 0,$$

showing that the catenoid is a minimal surface.

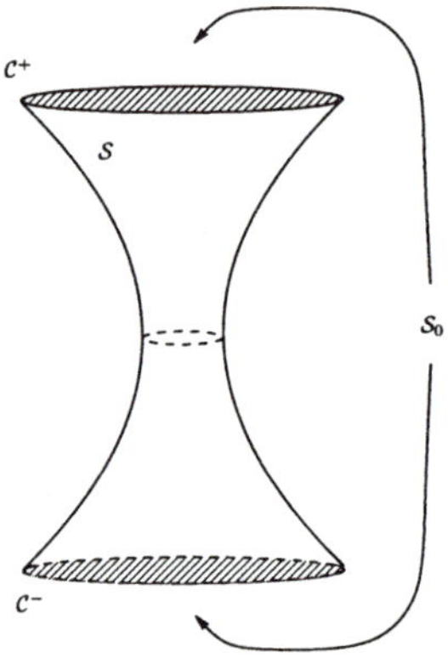

Fix $a > 0$, and let $b = \cosh a$. The surface $\mathcal{S}$ consisting of the part of the catenoid with $|z| < a$ has boundary the two circles $\mathcal{C}^{\pm}$ of radius b in the planes $z = \pm a$ with centres on the z-axis. Another surface spanning the same two circles is, of course, the surface $\mathcal{S}_0$ consisting of the two discs $x^2 + y^2 \leq b^2$ in the planes $z = \pm a$. The area of $\mathcal{S}$ is, by Proposition 5.2,

$$\int_0^{2\pi}\int_{-a}^{a}(EG - F^2)^{1/2}dudv = \int_0^{2\pi}\int_{-a}^{a}\cosh^2 u\, dudv = 2\pi(a + \sinh a \cosh a).$$

The area of $\mathcal{S}_0$ is, of course, $2\pi b^2 = 2\pi\cosh^2 a$. So the minimal surface $\mathcal{S}$ will *not* minimise the area among all surfaces with boundary the two circles $\mathcal{C}^{\pm}$ if $\cosh^2 a < a + \sinh a \cosh a$, i.e. if

$$1 + e^{-2a} < 2a. \tag{2}$$

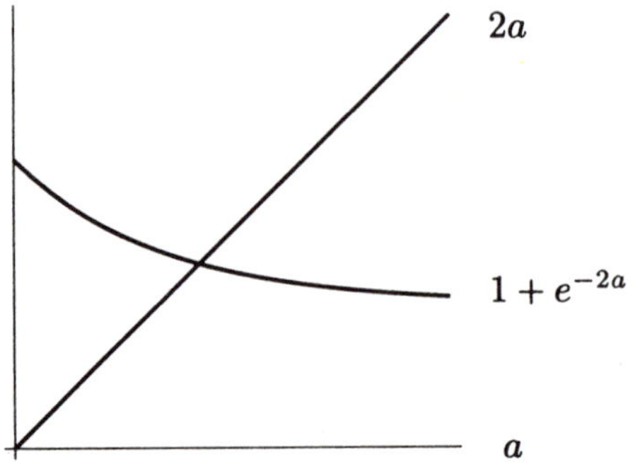

The graphs of $1+e^{-2a}$ and $2a$ as functions of a clearly intersect in exactly one point $a = a_0$, say, and the inequality (2) holds if $a > a_0$. If this condition is satisfied, the catenoid is not area minimising.

It can be shown that if $a < a_0$ the catenoid does have least area among all surfaces spanning the circles $\mathcal{C}^+$ and $\mathcal{C}^-$.

It is time to prove Theorem 9.1.

Proof 9.1

Let $\boldsymbol{\varphi}^\tau = \dot{\boldsymbol{\sigma}}^\tau$, so that $\boldsymbol{\varphi}^0 = \boldsymbol{\varphi}$, and let $\mathbf{N}^\tau$ be the standard unit normal of $\boldsymbol{\sigma}^\tau$. There are smooth functions α^τ, β^τ and γ^τ of (u, v, τ) such that

$$\boldsymbol{\varphi}^\tau = \alpha^\tau \mathbf{N}^\tau + \beta^\tau \boldsymbol{\sigma}_u^\tau + \gamma^\tau \boldsymbol{\sigma}_v^\tau,$$

so that $\alpha = \alpha^0$. To simplify the notation, we drop the superscript τ for the rest of the proof; at the end of the proof we put $\tau = 0$.

We have

$$\mathcal{A}(\tau) = \iint_{\mathrm{int}(\boldsymbol{\pi})} \parallel \boldsymbol{\sigma}_u \times \boldsymbol{\sigma}_v \parallel dudv = \iint_{\mathrm{int}(\boldsymbol{\pi})} \mathbf{N}.(\boldsymbol{\sigma}_u \times \boldsymbol{\sigma}_v)\, dudv,$$

so

$$\dot{\mathcal{A}} = \iint_{\mathrm{int}(\boldsymbol{\pi})} \frac{\partial}{\partial \tau} (\mathbf{N}.(\boldsymbol{\sigma}_u \times \boldsymbol{\sigma}_v))\, dudv. \tag{3}$$

Now,

$$\frac{\partial}{\partial \tau}(\mathbf{N}.(\boldsymbol{\sigma}_u \times \boldsymbol{\sigma}_v)) = \dot{\mathbf{N}}.(\boldsymbol{\sigma}_u \times \boldsymbol{\sigma}_v) + \mathbf{N}.(\dot{\boldsymbol{\sigma}}_u \times \boldsymbol{\sigma}_v) + \mathbf{N}.(\boldsymbol{\sigma}_u \times \dot{\boldsymbol{\sigma}}_v). \tag{4}$$

Since $\mathbf{N}$ is a unit vector,

$$\dot{\mathbf{N}}.(\boldsymbol{\sigma}_u \times \boldsymbol{\sigma}_v) = \dot{\mathbf{N}}.\mathbf{N} \parallel \boldsymbol{\sigma}_u \times \boldsymbol{\sigma}_v \parallel = 0.$$

On the other hand,

$$\begin{aligned}\mathbf{N}.(\dot{\boldsymbol{\sigma}}_u \times \boldsymbol{\sigma}_v) &= \frac{(\boldsymbol{\sigma}_u \times \boldsymbol{\sigma}_v).(\dot{\boldsymbol{\sigma}}_u \times \boldsymbol{\sigma}_v)}{\parallel \boldsymbol{\sigma}_u \times \boldsymbol{\sigma}_v \parallel} \\ &= \frac{(\boldsymbol{\sigma}_u.\dot{\boldsymbol{\sigma}}_u)(\boldsymbol{\sigma}_v.\boldsymbol{\sigma}_v) - (\boldsymbol{\sigma}_u.\boldsymbol{\sigma}_v)(\boldsymbol{\sigma}_v.\dot{\boldsymbol{\sigma}}_u)}{\parallel \boldsymbol{\sigma}_u \times \boldsymbol{\sigma}_v \parallel} \\ &= \frac{G(\boldsymbol{\sigma}_u.\dot{\boldsymbol{\sigma}}_u) - F(\boldsymbol{\sigma}_v.\dot{\boldsymbol{\sigma}}_u)}{(EG - F^2)^{1/2}},\end{aligned}$$

using Proposition 5.2. Similarly,

$$\mathbf{N}.(\boldsymbol{\sigma}_u \times \dot{\boldsymbol{\sigma}}_v) = \frac{E(\boldsymbol{\sigma}_v.\dot{\boldsymbol{\sigma}}_v) - F(\boldsymbol{\sigma}_u.\dot{\boldsymbol{\sigma}}_v)}{(EG - F^2)^{1/2}}.$$

Substituting these results into Eq. (4), we get

$$\frac{\partial}{\partial\tau}(\mathbf{N}.(\boldsymbol{\sigma}_u \times \boldsymbol{\sigma}_v)) = \frac{E(\boldsymbol{\sigma}_v.\dot{\boldsymbol{\sigma}}_v) - F(\dot{\boldsymbol{\sigma}}_u.\boldsymbol{\sigma}_v + \boldsymbol{\sigma}_u.\dot{\boldsymbol{\sigma}}_v) + G(\boldsymbol{\sigma}_u.\dot{\boldsymbol{\sigma}}_u)}{(EG-F^2)^{1/2}}. \tag{5}$$

Now

$$\dot{\boldsymbol{\sigma}}_u = \boldsymbol{\varphi}_u = \alpha_u\mathbf{N} + \beta_u\boldsymbol{\sigma}_u + \gamma_u\boldsymbol{\sigma}_v + \alpha\mathbf{N}_u + \beta\boldsymbol{\sigma}_{uu} + \gamma\boldsymbol{\sigma}_{uv},$$

$$\therefore\quad \boldsymbol{\sigma}_u.\dot{\boldsymbol{\sigma}}_u = E\beta_u + F\gamma_u + (\boldsymbol{\sigma}_u.\mathbf{N}_u)\alpha + (\boldsymbol{\sigma}_u.\boldsymbol{\sigma}_{uu})\beta + (\boldsymbol{\sigma}_u.\boldsymbol{\sigma}_{uv})\gamma.$$

Since $\boldsymbol{\sigma}_u.\mathbf{N}_u = -\boldsymbol{\sigma}_{uu}.\mathbf{N} = -L$, $\boldsymbol{\sigma}_u.\boldsymbol{\sigma}_{uu} = \frac{1}{2}E_u$ and $\boldsymbol{\sigma}_u.\boldsymbol{\sigma}_{uv} = \frac{1}{2}E_v$, we get

$$\boldsymbol{\sigma}_u.\dot{\boldsymbol{\sigma}}_u = E\beta_u + F\gamma_u - L\alpha + \frac{1}{2}E_u\beta + \frac{1}{2}E_v\gamma.$$

Similarly,

$$\begin{aligned}
\boldsymbol{\sigma}_v.\dot{\boldsymbol{\sigma}}_u &= F\beta_u + G\gamma_u - M\alpha + (F_u - \frac{1}{2}E_v)\beta + \frac{1}{2}G_u\gamma,\\
\boldsymbol{\sigma}_u.\dot{\boldsymbol{\sigma}}_v &= E\beta_v + F\gamma_v - M\alpha + \frac{1}{2}E_v\beta + (F_v - \frac{1}{2}G_u)\gamma,\\
\boldsymbol{\sigma}_v.\dot{\boldsymbol{\sigma}}_v &= F\beta_v + G\gamma_v - N\alpha + \frac{1}{2}G_u\beta + \frac{1}{2}G_v\gamma.
\end{aligned}$$

Substituting these last four equations into the right-hand side of Eq. (5), simplifying, and using the formula for H in Proposition 7.1(ii), we find that

$$\frac{\partial}{\partial\tau}(\mathbf{N}.(\boldsymbol{\sigma}_u \times \boldsymbol{\sigma}_v)) = \left(\beta(EG-F^2)^{1/2}\right)_u + \left(\gamma(EG-F^2)^{1/2}\right)_v - 2\alpha H(EG-F^2)^{1/2}. \tag{6}$$

Comparing with Eq. (3), and reinstating the superscripts, we see that we must prove that

$$\int\!\!\int_{\text{int}(\boldsymbol{\pi})} \left\{ \left(\beta^0(EG-F^2)^{1/2}\right)_u + \left(\gamma^0(EG-F^2)^{1/2}\right)_v \right\} dudv = 0. \tag{7}$$

But by Green's Theorem (see Section 3.1), this integral is equal to

$$\int_{\boldsymbol{\pi}} (EG-F^2)^{1/2}(\beta^0 dv - \gamma^0 du),$$

and this obviously vanishes because $\beta^0 = \gamma^0 = 0$ along the boundary curve $\boldsymbol{\pi}$. This completes the proof of Theorem 9.1. □

Note that we did not quite use the full force of the assumptions in Theorem 9.1, since they imply that α^0 $(= \alpha)$ vanishes along the boundary curve, and this was not used in the proof. So *Eq. (1) holds provided the surface variation $\boldsymbol{\varphi}$ is normal to the surface along the boundary curve.*

Note also that Theorem 9.1 is intuitively obvious for variations $\boldsymbol{\varphi}$ that are parallel to the surface, i.e. those for which $\alpha = 0$ everywhere on the surface, since such a parallel variation causes the surface to slide along itself and will not change the shape, and in particular the area, of the surface. Thus, the main point is to prove Theorem 9.1 for normal variations, i.e. those for which $\beta = \gamma = 0$ everywhere on the surface. Making this restriction simplifies the above proof considerably.

EXERCISES

9.1 Show that any rigid motion of $\mathbf{R}^3$ takes a minimal surface to another minimal surface, as does any dilation $(x, y, z) \mapsto a(x, y, z)$, where a is a non-zero constant.

9.2 Show that $z = f(x, y)$, where f is a smooth function of two variables, is a minimal surface if and only if

$$(1 + f_y^2)f_{xx} - 2f_x f_y f_{xy} + (1 + f_x^2)f_{yy} = 0.$$

9.3 Show that every umbilic on a minimal surface is a planar point (see Proposition 6.3).

9.4 Show that the gaussian curvature of a minimal surface is ≤ 0 everywhere, and that it is zero everywhere if and only if the surface is part of a plane. (Use Proposition 6.5.) We shall obtain a much more precise result in Corollary 9.2.

9.5 Show that there is no compact minimal surface. (Use Proposition 7.6 and Exercise 9.4.)

9.2. Examples of Minimal Surfaces

The simplest minimal surface is, of course, the plane, for which both principal curvatures are zero everywhere. Apart from this, the first minimal surfaces to be discovered were those in the following two examples.

Example 9.2

A *catenoid* is obtained by rotating a curve $x = \frac{1}{a}\cosh az$ in the xz-plane around the z-axis, where $a > 0$ is a constant. We showed in Example 9.1 that this is a minimal surface (we only dealt there with the case $a = 1$, but the general case follows from it by using Exercise 9.1).

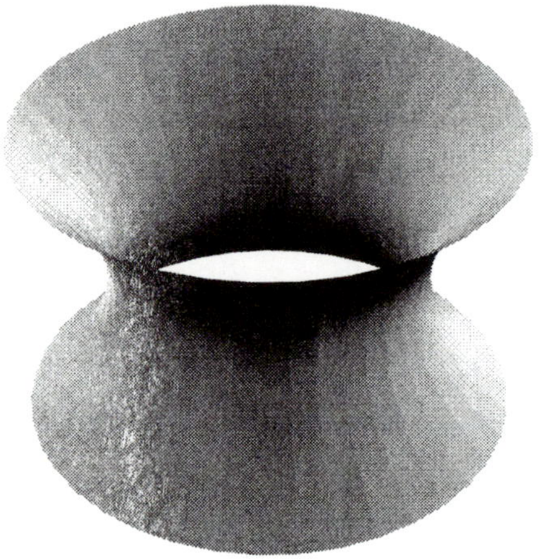

The catenoid is a surface of revolution. In fact, apart from the plane it is the only minimal surface of revolution:

Proposition 9.1

Any minimal surface of revolution is either part of a plane or can be obtained by applying a rigid motion to part of a catenoid.

Proof 9.1

By applying a rigid motion, we can assume that the axis of the surface $\mathcal{S}$ is the z-axis and the profile curve lies in the xz-plane. We parametrise $\mathcal{S}$ in the usual way (see Example 6.2):

$$\boldsymbol{\sigma}(u, v) = (f(u)\cos v, f(u)\sin v, g(u)),$$

where the profile curve $u \mapsto (f(u), 0, g(u))$ is assumed to be unit-speed and $f > 0$. From Example 6.2, the first and second fundamental forms are

$$du^2 + f(u)^2 dv^2 \quad \text{and} \quad (\dot{f}\ddot{g} - \ddot{f}\dot{g})du^2 + f\dot{g}dv^2,$$

respectively, a dot denoting d/du. By Proposition 7.1(ii), the mean curvature is

$$H = \frac{1}{2}\left(\dot{f}\ddot{g} - \ddot{f}\dot{g} + \frac{\dot{g}}{f}\right).$$

We suppose now that, for some value of u, say $u = u_0$, we have $\dot{g}(u_0) \neq 0$. We shall then have $\dot{g}(u) \neq 0$ for u in some open interval containing u_0. Let (α, β) be the largest such interval. Supposing now that $u \in (\alpha, \beta)$, the unit-speed condition $\dot{f}^2 + \dot{g}^2 = 1$ gives (as in Example 7.2)

$$\dot{f}\ddot{g} - \ddot{f}\dot{g} = -\frac{\ddot{f}}{\dot{g}},$$

so we get

$$H = \frac{1}{2}\left(\frac{\dot{g}}{f} - \frac{\ddot{f}}{\dot{g}}\right).$$

Since $\dot{g}^2 = 1 - \dot{f}^2$, $\mathcal{S}$ is minimal if and only if

$$f\ddot{f} = 1 - \dot{f}^2. \tag{8}$$

To solve the differential equation (8), put $h = \dot{f}$, and note that

$$\ddot{f} = \frac{dh}{dt} = \frac{dh}{df}\frac{df}{dt} = h\frac{dh}{df}.$$

Hence, Eq. (8) becomes

$$fh\frac{dh}{df} = 1 - h^2.$$

Note that, since $\dot{g} \neq 0$, we have $h^2 \neq 1$, so we can integrate this equation as follows:

$$\int \frac{hdh}{1 - h^2} = \int \frac{df}{f},$$

$$\therefore \quad \frac{1}{\sqrt{1 - h^2}} = af,$$

$$\therefore \quad h = \frac{\sqrt{a^2f^2 - 1}}{af},$$

where a is a non-zero constant. (We have omitted a $\pm$, but the sign can be changed by replacing u by $-u$ if necessary.) Writing $h = df/du$ and integrating again,

$$\int \frac{af\,df}{\sqrt{a^2f^2 - 1}} = \int du,$$

$$\therefore \quad f = \frac{1}{a}\sqrt{1 + a^2(u + b)^2},$$

where b is a constant. By a change of parameter $u \mapsto u + b$, we can assume that $b = 0$. So

$$f = \frac{1}{a}\sqrt{1 + a^2u^2}.$$

To compute g, we have

$$\dot{g}^2 = 1 - \dot{f}^2 = 1 - h^2 = \frac{1}{a^2 f^2},$$

$$\therefore \quad \frac{dg}{du} = \pm \frac{1}{\sqrt{1 + a^2 u^2}},$$

$$\therefore \quad g = \pm \frac{1}{a} \sinh^{-1}(au) + c \qquad \text{(where } c \text{ is a constant)},$$

$$\therefore \quad au = \pm \sinh(a(g - c)),$$

$$\therefore \quad f = \frac{1}{a} \cosh(a(g - c)).$$

Thus, the profile curve of $\mathcal{S}$ is

$$x = \frac{1}{a} \cosh(a(z - c)).$$

By a translation along the z-axis, we can assume that $c = 0$, so we have a catenoid.

We are not quite finished, however. So far, we have only shown that the part of $\mathcal{S}$ corresponding to $u \in (\alpha, \beta)$ is part of the catenoid, for in the proof we used in an essential way that $\dot{g} \neq 0$. This is why the proof has so far excluded the possibility that $\mathcal{S}$ is a plane. To complete the proof, we argue as follows. Suppose that $\beta < \infty$. Then, if the profile curve is defined for values of $u \geq \beta$, we must have $\dot{g}(\beta) = 0$, for otherwise $\dot{g}$ would be non-zero on an open interval containing β, which would contradict our assumption that (α, β) is the *largest* open interval containing u_0 on which $\dot{g} \neq 0$. But the formulas above show that

$$\dot{g}^2 = \frac{1}{1 + a^2 u^2} \quad \text{if } u \in (\alpha, \beta),$$

so, since $\dot{g}$ is a continuous function of u, $\dot{g}(\beta) = \pm(1 + a^2\beta^2)^{-1/2} \neq 0$. This contradiction shows that the profile curve is not defined for values of $u \geq \beta$. Of course, this also holds trivially if $\beta = \infty$. A similar argument applies to α, and shows that (α, β) is the entire domain of definition of the profile curve. Hence, the whole of $\mathcal{S}$ is part of a catenoid.

The only remaining case to consider is that in which $\dot{g}(u) = 0$ for all values of u for which the profile curve is defined. But then $g(u)$ is a constant, say d, and $\mathcal{S}$ is part of the plane $z = d$. □

Example 9.3

A *helicoid* is a ruled surface swept out by a straight line that rotates at constant speed about an axis perpendicular to the line while simultaneously moving at constant speed along the axis. We can take the axis to be the z-axis. Let ω be the angular velocity of the rotating line and α its speed along the z-axis. If the line starts along the x-axis, at time v the centre of the line is at $(0,0,\alpha v)$ and it has rotated by an angle ωv. Hence, the point of the line initially at $(u,0,0)$ is now at the point with position vector

$$\boldsymbol{\sigma}(u,v) = (u\cos\omega v, u\sin\omega v, \alpha v).$$

We leave it to Exercise 9.6 to check that this is a minimal surface.

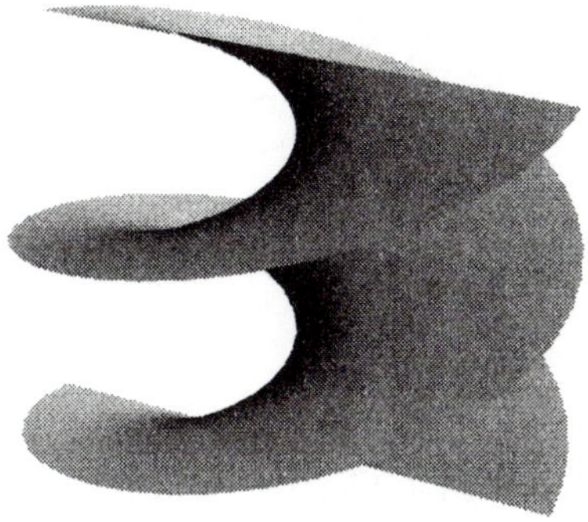

We have the following analogue of Proposition 9.1.

Proposition 9.2

Any ruled minimal surface is part of a plane or part of a helicoid.

Proof 9.2

We take the usual parametrisation

$$\boldsymbol{\sigma}(u,v) = \boldsymbol{\gamma}(u) + v\boldsymbol{\delta}(u)$$

(see Example 4.12), where $\boldsymbol{\gamma}$ is a curve that meets each of the rulings and $\boldsymbol{\delta}(u)$ is a vector parallel to the ruling through $\boldsymbol{\gamma}(u)$. We begin the proof by making some simplifications to the parametrisation.

First, we can certainly assume that $\parallel \boldsymbol{\delta}(u) \parallel = 1$ for all values of u. We assume also that $\dot{\boldsymbol{\delta}}$ *is never zero*, where the dot denotes d/du. (We shall consider later what happens if $\dot{\boldsymbol{\delta}}(u) = \mathbf{0}$ for some values of u.) We can then assume that $\boldsymbol{\delta}$ is a unit-speed curve (we do *not* assume that $\boldsymbol{\gamma}$ is unit-speed). These assumptions imply that $\boldsymbol{\delta}.\dot{\boldsymbol{\delta}} = \dot{\boldsymbol{\delta}}.\ddot{\boldsymbol{\delta}} = 0$. Now we consider the curve

$$\tilde{\boldsymbol{\gamma}}(u) = \boldsymbol{\gamma}(u) - (\dot{\boldsymbol{\gamma}}.\dot{\boldsymbol{\delta}})\boldsymbol{\delta}(u).$$

If $\tilde{v} = v + \dot{\boldsymbol{\gamma}}.\dot{\boldsymbol{\delta}}$, the surface can be reparametrised using $\tilde{\boldsymbol{\gamma}}$ and the parameters u and $\tilde{v}$, namely

$$\boldsymbol{\sigma}(u,\tilde{v}) = \tilde{\boldsymbol{\gamma}}(u) + \tilde{v}\boldsymbol{\delta}(u),$$

but now we also have

$$\dot{\tilde{\boldsymbol{\gamma}}}.\dot{\boldsymbol{\delta}} = \left(\dot{\boldsymbol{\gamma}} - \frac{d}{du}(\boldsymbol{\gamma}.\dot{\boldsymbol{\delta}})\boldsymbol{\delta} - (\dot{\boldsymbol{\gamma}}.\dot{\boldsymbol{\delta}})\boldsymbol{\delta}\right).\dot{\boldsymbol{\delta}} = 0,$$

since $\boldsymbol{\delta}.\dot{\boldsymbol{\delta}} = 0$ and $\dot{\boldsymbol{\delta}}.\dot{\boldsymbol{\delta}} = 1$. This means that we could have assumed that $\dot{\boldsymbol{\gamma}}.\dot{\boldsymbol{\delta}} = 0$ at the beginning, and we make this assumption from now on.

We have

$$\boldsymbol{\sigma}_u = \dot{\boldsymbol{\gamma}} + v\dot{\boldsymbol{\delta}}, \quad \boldsymbol{\sigma}_v = \boldsymbol{\delta},$$
$$\therefore \quad E = \| \dot{\boldsymbol{\gamma}} + v\dot{\boldsymbol{\delta}} \|^2, \quad F = (\dot{\boldsymbol{\gamma}} + v\dot{\boldsymbol{\delta}}).\boldsymbol{\delta} = \dot{\boldsymbol{\gamma}}.\boldsymbol{\delta}, \quad G = 1.$$

Let $A = (EG - F^2)^{1/2}$. Then,

$$\mathbf{N} = A^{-1}(\dot{\boldsymbol{\gamma}} + v\dot{\boldsymbol{\delta}}) \times \boldsymbol{\delta}.$$

Next, we have

$$\begin{aligned}
&\boldsymbol{\sigma}_{uu} = \ddot{\boldsymbol{\gamma}} + v\ddot{\boldsymbol{\delta}}, \quad \boldsymbol{\sigma}_{uv} = \dot{\boldsymbol{\delta}}, \quad \boldsymbol{\sigma}_{vv} = \mathbf{0},\\
\therefore \quad &L = A^{-1}(\ddot{\boldsymbol{\gamma}} + v\ddot{\boldsymbol{\delta}}).((\dot{\boldsymbol{\gamma}} + v\dot{\boldsymbol{\delta}}) \times \boldsymbol{\delta}),\\
&M = A^{-1}\dot{\boldsymbol{\delta}}.((\dot{\boldsymbol{\gamma}} + v\dot{\boldsymbol{\delta}}) \times \boldsymbol{\delta}) = A^{-1}\dot{\boldsymbol{\delta}}.(\dot{\boldsymbol{\gamma}} \times \boldsymbol{\delta}),\\
&N = 0.
\end{aligned}$$

Hence, the minimal surface condition

$$H = \frac{LG - 2MF + NE}{2A^2} = 0$$

gives

$$(\ddot{\boldsymbol{\gamma}} + v\ddot{\boldsymbol{\delta}}).((\dot{\boldsymbol{\gamma}} + v\dot{\boldsymbol{\delta}}) \times \boldsymbol{\delta}) = 2(\boldsymbol{\delta}.\dot{\boldsymbol{\gamma}})(\dot{\boldsymbol{\delta}}.(\dot{\boldsymbol{\gamma}} \times \boldsymbol{\delta})).$$

This equation must hold for all values of (u, v). Equating coefficients of powers of v gives

$$\ddot{\boldsymbol{\gamma}}.(\dot{\boldsymbol{\gamma}} \times \boldsymbol{\delta}) = 2(\boldsymbol{\delta}.\dot{\boldsymbol{\gamma}})(\dot{\boldsymbol{\delta}}.(\dot{\boldsymbol{\gamma}} \times \boldsymbol{\delta})), \tag{9}$$
$$\ddot{\boldsymbol{\gamma}}.(\dot{\boldsymbol{\delta}} \times \boldsymbol{\delta}) + \ddot{\boldsymbol{\delta}}.(\dot{\boldsymbol{\gamma}} \times \boldsymbol{\delta}) = 0, \tag{10}$$
$$\ddot{\boldsymbol{\delta}}.(\dot{\boldsymbol{\delta}} \times \boldsymbol{\delta}) = 0. \tag{11}$$

Equation (11) shows that $\boldsymbol{\delta}, \dot{\boldsymbol{\delta}}$ and $\ddot{\boldsymbol{\delta}}$ are linearly dependent. Since $\boldsymbol{\delta}$ and $\dot{\boldsymbol{\delta}}$ are perpendicular unit vectors, there are smooth functions $\alpha(u)$ and $\beta(u)$ such that

$$\ddot{\boldsymbol{\delta}} = \alpha\boldsymbol{\delta} + \beta\dot{\boldsymbol{\delta}}.$$

But, since $\boldsymbol{\delta}$ is unit-speed, $\dot{\boldsymbol{\delta}}.\ddot{\boldsymbol{\delta}} = 0$. Also, differentiating $\boldsymbol{\delta}.\dot{\boldsymbol{\delta}} = 0$ gives $\boldsymbol{\delta}.\ddot{\boldsymbol{\delta}} = -\dot{\boldsymbol{\delta}}.\dot{\boldsymbol{\delta}} = -1$. Hence, $\alpha = -1$ and $\beta = 0$, so

$$\ddot{\boldsymbol{\delta}} = -\boldsymbol{\delta}. \tag{12}$$

Equation (12) shows that the curvature of the curve $\boldsymbol{\delta}$ is 1, and that its principal normal is $-\boldsymbol{\delta}$. Hence, its binormal is $\dot{\boldsymbol{\delta}} \times (-\boldsymbol{\delta})$, and since

$$\frac{d}{du}(\dot{\boldsymbol{\delta}} \times \boldsymbol{\delta}) = \ddot{\boldsymbol{\delta}} \times \boldsymbol{\delta} + \dot{\boldsymbol{\delta}} \times \dot{\boldsymbol{\delta}} = -\boldsymbol{\delta} \times \boldsymbol{\delta} = \mathbf{0},$$

it follows that the torsion of $\boldsymbol{\delta}$ is zero. Hence, $\boldsymbol{\delta}$ parametrises a circle of radius 1 (see Proposition 1.5). By applying a rigid motion, we can assume that $\boldsymbol{\delta}$ is the circle with radius 1 and centre the origin in the xy-plane, so that

$$\boldsymbol{\delta}(u) = (\cos u, \sin u, 0).$$

From Eq. (12), we get $\ddot{\boldsymbol{\delta}}.(\dot{\boldsymbol{\gamma}} \times \boldsymbol{\delta}) = -\boldsymbol{\delta}.(\dot{\boldsymbol{\gamma}} \times \boldsymbol{\delta}) = 0$, so by Eq. (10),

$$\ddot{\boldsymbol{\gamma}}.(\dot{\boldsymbol{\delta}} \times \boldsymbol{\delta}) = 0.$$

It follows that $\ddot{\boldsymbol{\gamma}}$ is parallel to the xy-plane, and hence that

$$\boldsymbol{\gamma}(u) = (f(u), g(u), au + b),$$

where f and g are smooth functions and a and b are constants. If $a = 0$, the surface is part of the plane $z = b$. Otherwise, Eq. (9) gives

$$\ddot{g}\cos u - \ddot{f}\sin u = 2(\dot{f}\cos u + \dot{g}\sin u). \tag{13}$$

We finally make use of the condition $\dot{\boldsymbol{\gamma}}.\dot{\boldsymbol{\delta}} = 0$, which gives

$$\dot{f}\sin u = \dot{g}\cos u. \tag{14}$$

Differentiating this gives

$$\ddot{f}\sin u + \dot{f}\cos u = \ddot{g}\cos u - \dot{g}\sin u. \tag{15}$$

Equations (13) and (15) together give

$$\dot{f}\cos u + \dot{g}\sin u = 0.$$

and using Eq. (14) we get $\dot{f} = \dot{g} = 0$. Thus, f and g are constants. By a translation of the surface, we can assume that the constants f, g and b are zero, so that

$$\boldsymbol{\gamma}(u) = (0, 0, au)$$

and

$$\boldsymbol{\sigma}(u, v) = (v\cos u, v\sin u, au),$$

which is a helicoid.

We assumed at the beginning that $\dot{\boldsymbol{\delta}}$ is never zero. If $\dot{\boldsymbol{\delta}}$ is *always* zero, then $\boldsymbol{\delta}$ is a constant vector and the surface is a generalised cylinder. But in fact a generalised cylinder is a minimal surface only if the cylinder is part of a plane (Exercise 9.8). The proof is now completed by an argument similar to that used at the end of the proof of Proposition 9.1, which shows that the whole surface is either part of a plane or part of a helicoid. □

After the catenoid and helicoid, the next minimal surfaces to be discovered were the following two.

Example 9.4

Enneper's minimal surface is

$$\boldsymbol{\sigma}(u,v) = \left(u - \frac{1}{3}u^3 + uv^2, v - \frac{1}{3}v^3 + vu^2, u^2 - v^2\right).$$

It was shown in Exercise 7.15 that this is a minimal surface.

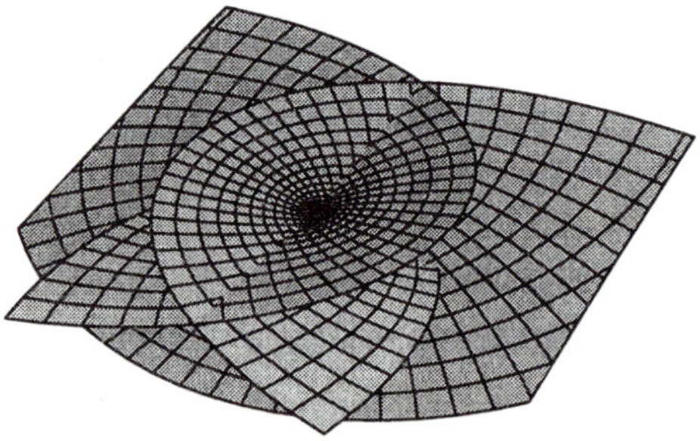

Strictly speaking, this is not a surface patch in the sense used in this book as it is not injective. The self-intersections are clearly visible in the picture above. However, if we restrict (u, v) to lie in sufficiently small open sets, $\boldsymbol{\sigma}$ will be injective by the inverse function theorem.

Example 9.5

Scherk's minimal surface is the surface with cartesian equation

$$z = \ln\left(\frac{\cos y}{\cos x}\right).$$

It was shown in Exercise 7.16 that this is a minimal surface. Note that the surface exists only when $\cos x$ and $\cos y$ are both > 0 or both < 0, in other words in the interiors of the white squares of the following chess board pattern, in which the squares have vertices at the points $(\pi/2 + m\pi, \pi/2 + n\pi)$, where m and n are integers, no two squares with a common edge have the same colour, and the square containing the origin is white:

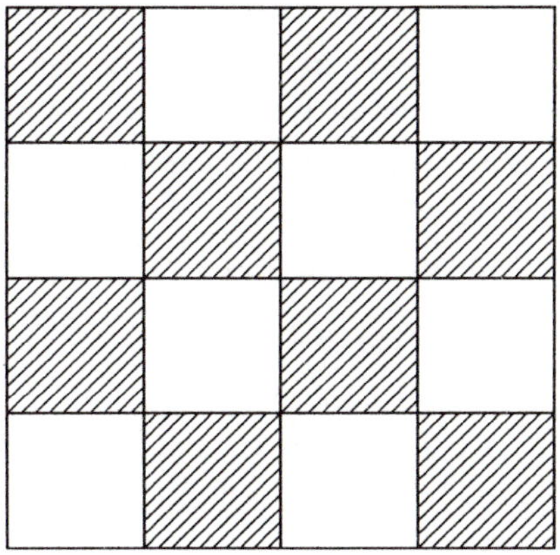

The white squares have centres of the form $(m\pi, n\pi)$, where m and n are integers with $m+n$ even. Since, for such m, n,

$$\frac{\cos(y+n\pi)}{\cos(x+m\pi)} = \frac{\cos y}{\cos x},$$

it follows that the part of the surface over the square with centre $(m\pi, n\pi)$ is obtained from the part over the square with centre $(0,0)$ by the translation $(x, y, z) \mapsto (x + m\pi, y + n\pi, z)$. So it suffices to exhibit the part of the surface over a single square:

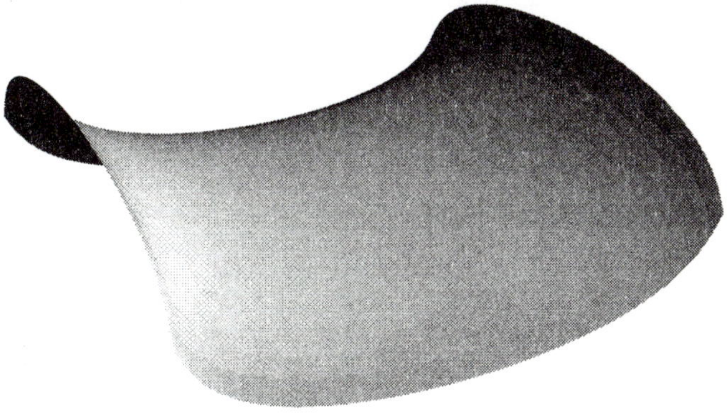

EXERCISES

9.6 Show that the helicoid is a minimal surface.

9.7 Show that the surfaces $\boldsymbol{\sigma}^t$ in the isometric deformation of the helicoid into the catenoid given in Exercise 5.8 are minimal surfaces.

9.8 Show that a generalised cylinder is a minimal surface only when the cylinder is part of a plane.

9.9 A *translation surface* is a surface of the form

$$z = f(x) + g(y),$$

where f and g are smooth functions. (It is obtained by moving the curve $u \mapsto (u, 0, f(u))$ parallel to itself along the curve $v \mapsto (0, v, g(v))$.) Use Exercise 9.2 to show that this is a minimal surface if and only if

$$\frac{d^2f/dx^2}{1 + (df/dx)^2} = -\frac{d^2g/dy^2}{1 + (dg/dy)^2}.$$

Deduce that any minimal translation surface is part of a plane or can be transformed into part of Scherk's surface in Example 9.5 by a translation and a dilation $(x, y, z) \mapsto a(x, y, z)$ for some non-zero constant a.

9.10 Verify that *Catalan's surface*

$$\boldsymbol{\sigma}(u, v) = \left(u - \sin u \cosh v, 1 - \cos u \cosh v, -4 \sin \frac{u}{2} \sinh \frac{v}{2}\right)$$

is a conformally parametrised minimal surface. (As in the case of Enneper's surface, Catalan's surface has self-intersections, so it is only a surface if we restrict (u, v) to sufficiently small open sets.)

Show that

(i) the parameter curve on the surface given by $u = 0$ is a straight line;

(ii) the parameter curve $u = \pi$ is a parabola;
(iii) the parameter curve $v = 0$ is a cycloid (see Exercise 1.7).
Show also that each of these curves, when suitably parametrised, is a geodesic on Catalan's surface.

9.3. Gauss Map of a Minimal Surface

Recall from Section 7.3 that the Gauss map of a surface patch $\boldsymbol{\sigma} : U \to \mathbf{R}^3$ associates to a point $\boldsymbol{\sigma}(u, v)$ of the surface the standard unit normal $\mathbf{N}(u, v)$ regarded as a point of the unit sphere S^2. By Eq. (15) in Chapter 7,

$$\mathbf{N}_u \times \mathbf{N}_v = K\,\boldsymbol{\sigma}_u \times \boldsymbol{\sigma}_v,$$

where K is the gaussian curvature of $\boldsymbol{\sigma}$, so $\mathbf{N}$ will be regular provided K is nowhere zero, and we assume this for the remainder of this section.

Proposition 9.3

Let $\boldsymbol{\sigma}(u, v)$ be a minimal surface patch with nowhere vanishing gaussian curvature. Then, the Gauss map is a conformal *map from $\boldsymbol{\sigma}$ to part of the unit sphere.*

We should really be a little more careful in the statement of this proposition, since for us conformal maps are always diffeomorphisms (see Section 5.3). However, even if $\mathbf{N}_u \times \mathbf{N}_v$ is never zero, it does not follow that the map $\boldsymbol{\sigma}(u, v) \mapsto \mathbf{N}(u, v)$ is *injective* (see Exercise 9.12(ii)). Nevertheless, the inverse function theorem tells us that, if $(u_0, v_0) \in \mathbf{R}^2$ is a point where $\boldsymbol{\sigma}$ (and hence $\mathbf{N}$) is defined, there is an open set U containing (u_0, v_0) on which $\boldsymbol{\sigma}$ is defined and on which $\mathbf{N}$ *is* injective. Then, $\mathbf{N} : U \to S^2$ is an allowable surface patch on the unit sphere S^2, and the Gauss map *is* a diffeomorphism from $\boldsymbol{\sigma}(U)$ to $\mathbf{N}(U)$.

Proof 9.3

By Theorem 5.2, we have to show that the first fundamental form

$$(\mathbf{N}_u.\mathbf{N}_u)du^2 + 2(\mathbf{N}_u.\mathbf{N}_v)du dv + (\mathbf{N}_v.\mathbf{N}_v)dv^2$$

of $\mathbf{N}$ is proportional to that of $\boldsymbol{\sigma}$. Form the symmetric 2×2 matrix

$$\mathcal{F}_{III} = \begin{pmatrix} \mathbf{N}_u.\mathbf{N}_u & \mathbf{N}_u.\mathbf{N}_v \\ \mathbf{N}_u.\mathbf{N}_v & \mathbf{N}_v.\mathbf{N}_v \end{pmatrix}$$

in the same way as we associated symmetric 2×2 matrices $\mathcal{F}_I$ and $\mathcal{F}_{II}$ to the first and second fundamental forms of $\boldsymbol{\sigma}$ in Section 6.3. Then, we have to show that

$$\mathcal{F}_{III} = \lambda \mathcal{F}_I, \tag{16}$$

for some scalar λ.

By Proposition 6.4,

$$\mathbf{N}_u.\mathbf{N}_u = a^2 \boldsymbol{\sigma}_u.\boldsymbol{\sigma}_u + 2ab\boldsymbol{\sigma}_u.\boldsymbol{\sigma}_v + b^2 \boldsymbol{\sigma}_v.\boldsymbol{\sigma}_v = a^2E + 2abF + b^2G,$$

where $\begin{pmatrix} a & c \\ b & d \end{pmatrix} = -\mathcal{W}$ and $\mathcal{W} = \mathcal{F}_I^{-1}\mathcal{F}_{II}$ is the Weingarten matrix. Computing $\mathbf{N}_u.\mathbf{N}_v$ and $\mathbf{N}_v.\mathbf{N}_v$ in the same way gives

$$\begin{aligned}
\mathcal{F}_{III} &= \begin{pmatrix} a^2E + 2abF + b^2G & acE + (ad+bc)F + bdG \\ acE + (ad+bc)F + bdG & c^2E + 2cdF + d^2G \end{pmatrix} \\
&= \begin{pmatrix} a & b \\ c & d \end{pmatrix} \begin{pmatrix} E & F \\ F & G \end{pmatrix} \begin{pmatrix} a & c \\ b & d \end{pmatrix} \\
&= (-\mathcal{W})^t \mathcal{F}_I (-\mathcal{W}) \\
&= (-\mathcal{F}_I^{-1}\mathcal{F}_{II})^t \mathcal{F}_I (-\mathcal{F}_I^{-1}\mathcal{F}_{II}) \\
&= \mathcal{F}_{II}\mathcal{F}_I^{-1}\mathcal{F}_I\mathcal{F}_I^{-1}\mathcal{F}_{II} \\
&= \mathcal{F}_{II}\mathcal{F}_I^{-1}\mathcal{F}_{II}.
\end{aligned}$$

Hence, Eq. (16) is equivalent to

$$\mathcal{F}_I^{-1}\mathcal{F}_{II}\mathcal{F}_I^{-1}\mathcal{F}_{II} = \lambda I,$$
$$\text{i.e.} \quad \mathcal{W}^2 = \lambda I.$$

But,

$$\mathcal{W}^2 = \begin{pmatrix} a & c \\ b & d \end{pmatrix}^2 = \begin{pmatrix} a^2 + bc & c(a+d) \\ b(a+d) & d^2 + bc \end{pmatrix}.$$

Now, recall from Section 6.3 that the principal curvatures κ_1 and κ_2 are the eigenvalues of $\mathcal{W}$. Since the sum of the eigenvalues of a matrix is equal to the sum of its diagonal entries,

$$\kappa_1 + \kappa_2 = -(a+d).$$

If $\boldsymbol{\sigma}$ is minimal, the mean curvature $H = \frac{1}{2}(\kappa_1 + \kappa_2)$ vanishes, so $a + d = 0$ and hence

$$\mathcal{W}^2 = (a^2 + bc)I,$$

as we want. □

We saw in Exercise 5.14 that a conformal parametrisation of the plane is necessarily holomorphic or anti-holomorphic, so this proposition strongly suggests a connection between minimal surfaces and holomorphic functions. This connection turns out to be very extensive, and we shall give an introduction to it in the next section.

EXERCISES

9.11 Show that the scalar λ appearing in the proof of Proposition 9.3 is equal to $-K$, where K is the gaussian curvature of the surface.

9.12 Show that
(i) the Gauss map of the catenoid is injective and its image is the whole of the unit sphere except for the north and south poles;
(ii) the image of the Gauss map of the helicoid is the same as that of the catenoid, but that infinitely many points on the helicoid are sent by the Gauss map to any given point in its image.

9.4. Minimal Surfaces and Holomorphic Functions

In this section, we shall make use of certain elementary properties of holomorphic functions. Readers without the necessary background in complex analysis may safely omit this section, whose results are not used anywhere else in the book.

We shall need to make use of special surface patches on a minimal surface. Recall from Section 5.3 that a surface patch $\boldsymbol{\sigma} : U \to \mathbf{R}^3$ is called *conformal* if its first fundamental form is equal to $E(du^2 + dv^2)$ for some positive smooth function E on U.

Proposition 9.4

Every surface has an atlas consisting of conformal surface patches.

We shall accept this result without proof (the proof is non-trivial).

Let $\boldsymbol{\sigma} : U \to \mathbf{R}^3$ be a conformal surface patch. We introduce complex coordinates in the plane in which U lies by setting

$$\zeta = u + iv \quad \text{for } (u, v) \in U,$$

and we define

$$\boldsymbol{\varphi}(\zeta) = \boldsymbol{\sigma}_u - i\boldsymbol{\sigma}_v. \tag{17}$$

Thus, $\boldsymbol{\varphi} = (\varphi_1, \varphi_2, \varphi_3)$ has three components, each of which is a complex-valued function of (u, v), i.e. of ζ. The basic result which establishes the connection between minimal surfaces and holomorphic functions is

Proposition 9.5

Let $\boldsymbol{\sigma} : U \to \mathbf{R}^3$ be a conformal surface patch. Then $\boldsymbol{\sigma}$ is minimal if and only if the function $\boldsymbol{\varphi}$ defined in Eq. (17) is holomorphic on U.

Saying that $\boldsymbol{\varphi}$ is holomorphic means that each of its components φ_1, φ_2 and φ_3 is holomorphic.

Proof 9.5

Let $\varphi(u, v)$ be a complex-valued smooth function, and let α and β be its real and imaginary parts, so that $\varphi = \alpha + i\beta$. The Cauchy–Riemann equations

$$\alpha_u = \beta_v \quad \text{and} \quad \alpha_v = -\beta_u$$

are the necessary and sufficient conditions for φ to be holomorphic. Applying this to each of the components of $\boldsymbol{\varphi}$, we see that $\boldsymbol{\varphi}$ is holomorphic if and only if

$$(\boldsymbol{\sigma}_u)_u = (-\boldsymbol{\sigma}_v)_v \quad \text{and} \quad (\boldsymbol{\sigma}_u)_v = -(-\boldsymbol{\sigma}_v)_u.$$

The second equation imposes no condition on $\boldsymbol{\sigma}$, and the first is equivalent to $\boldsymbol{\sigma}_{uu} + \boldsymbol{\sigma}_{vv} = \mathbf{0}$. So we have to show that $\boldsymbol{\sigma}$ is minimal if and only if the laplacian $\Delta\boldsymbol{\sigma} = \boldsymbol{\sigma}_{uu} + \boldsymbol{\sigma}_{vv}$ is zero.

By Proposition 7.2(ii) and the fact that $\boldsymbol{\sigma}$ is conformal, the mean curvature of $\boldsymbol{\sigma}$ is given by

$$H = \frac{L + N}{2E},$$

so $\boldsymbol{\sigma}$ is minimal if and only if $L + N = 0$, i.e.

$$(\boldsymbol{\sigma}_{uu} + \boldsymbol{\sigma}_{vv}).\mathbf{N} = 0. \tag{18}$$

Obviously, then, $\boldsymbol{\sigma}$ is minimal if $\Delta\boldsymbol{\sigma} = \mathbf{0}$. For the converse, we have to show that $\Delta\boldsymbol{\sigma} = \mathbf{0}$ if Eq. (18) holds. It is enough to prove that $\Delta\boldsymbol{\sigma}.\boldsymbol{\sigma}_u = \Delta\boldsymbol{\sigma}.\boldsymbol{\sigma}_v = 0$, since $\{\boldsymbol{\sigma}_u, \boldsymbol{\sigma}_v, \mathbf{N}\}$ is a basis of $\mathbf{R}^3$.

We compute

$$\begin{aligned}\Delta\boldsymbol{\sigma}.\boldsymbol{\sigma}_u &= \boldsymbol{\sigma}_{uu}.\boldsymbol{\sigma}_u + \boldsymbol{\sigma}_{vv}.\boldsymbol{\sigma}_u \\ &= \frac{1}{2}(\boldsymbol{\sigma}_u.\boldsymbol{\sigma}_u)_u + (\boldsymbol{\sigma}_v.\boldsymbol{\sigma}_u)_v - (\boldsymbol{\sigma}_v.\boldsymbol{\sigma}_{uv}) \\ &= \frac{1}{2}(\boldsymbol{\sigma}_u.\boldsymbol{\sigma}_u - \boldsymbol{\sigma}_v.\boldsymbol{\sigma}_v)_u + (\boldsymbol{\sigma}_v.\boldsymbol{\sigma}_u)_v.\end{aligned}$$

But, since $\boldsymbol{\sigma}$ is conformal, $\boldsymbol{\sigma}_u.\boldsymbol{\sigma}_u = \boldsymbol{\sigma}_v.\boldsymbol{\sigma}_v$ and $\boldsymbol{\sigma}_u.\boldsymbol{\sigma}_v = 0$. Hence, $\Delta\boldsymbol{\sigma}.\boldsymbol{\sigma}_u = 0$. Similarly, $\Delta\boldsymbol{\sigma}.\boldsymbol{\sigma}_v = 0$. □

The holomorphic function $\boldsymbol{\varphi}$ associated to a minimal surface $\boldsymbol{\sigma}$ is not arbitrary, however:

Theorem 9.2

If $\boldsymbol{\sigma} : U \to \mathbf{R}^3$ is a conformally parametrised minimal surface, the vector-valued holomorphic function $\boldsymbol{\varphi} = (\varphi_1, \varphi_2, \varphi_3)$ defined in Eq. (17) satisfies the following conditions:
(i) $\varphi_1^2 + \varphi_2^2 + \varphi_3^2 = 0$;
(ii) $\boldsymbol{\varphi}$ *is nowhere zero.*

Conversely, if U is simply-connected, and if φ_1, φ_2 and φ_3 are holomorphic functions on U satisfying conditions (i) and (ii) above, there is a conformally parametrised minimal surface $\boldsymbol{\sigma} : U \to \mathbf{R}^3$ such that $\boldsymbol{\varphi} = (\varphi_1, \varphi_2, \varphi_3)$ satisfies Eq. (17). Moreover, $\boldsymbol{\sigma}$ is uniquely determined by φ_1, φ_2 and φ_3 up to a translation.

An open subset U of $\mathbf{R}^2$ is said to be *simply-connected* if every simple closed curve in U can be shrunk to a point staying inside U. Intuitively, this means that U has no 'holes'.

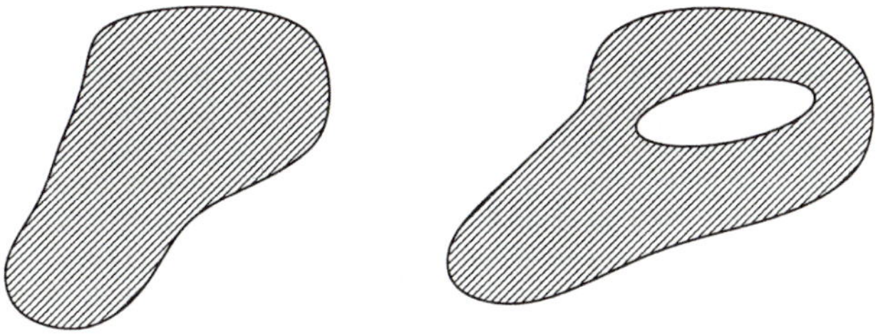

simply-connected not simply-connected

In the course of the following proof, and in the proof of Proposition 9.7 below, we shall need to recall that, if F is a holomorphic function of $\zeta = u + iv$, then

$$F_u = F', \quad F_v = iF', \quad (\overline{F})_u = \overline{F'}, \quad (\overline{F})_v = -i\overline{F'},$$

where $F' = dF/d\zeta$ is the complex derivative of F, and the bar denotes complex-conjugate.

Proof 9.2

Suppose first that $\boldsymbol{\sigma} = (\sigma^1, \sigma^2, \sigma^3)$ is minimal, where $\sigma^k : U \to \mathbf{R}$ for $k = 1, 2, 3$. We have to show that $\boldsymbol{\varphi} = (\varphi_1, \varphi_2, \varphi_3)$ satisfies conditions (i) and (ii). Since $\varphi_k = \sigma_u^k - i\sigma_v^k$ for $k = 1, 2, 3$,

$$\sum_{k=1}^{3} \varphi_k^2 = \sum_{k=1}^{3} \left((\sigma_u^k)^2 - (\sigma_v^k)^2 - 2i\sigma_u^k\sigma_v^k\right) = \parallel \boldsymbol{\sigma}_u \parallel^2 - \parallel \boldsymbol{\sigma}_v \parallel^2 - 2i\boldsymbol{\sigma}_u.\boldsymbol{\sigma}_v, \qquad (19)$$

which vanishes since $\boldsymbol{\sigma}$ is conformal. Finally, $\boldsymbol{\varphi} = \mathbf{0}$ if and only if $\boldsymbol{\sigma}_u = \boldsymbol{\sigma}_v = \mathbf{0}$, and this is impossible since $\boldsymbol{\sigma}$ is regular.

For the converse, take $\boldsymbol{\varphi}$ satisfying conditions (i) and (ii). We must show that $\boldsymbol{\varphi}$ arises from a minimal surface as above, and that this minimal surface is unique up to a translation of $\mathbf{R}^3$. Fix $(u_0, v_0) \in U$ and define $\boldsymbol{\sigma}$ as the real part of a complex line integral:

$$\boldsymbol{\sigma}(u, v) = \mathfrak{Re} \int_{\boldsymbol{\pi}} \boldsymbol{\varphi}(\xi) d\xi,$$

where $\boldsymbol{\pi}$ is any curve in U from (u_0, v_0) to $(u, v) \in U$. The fact that U is simply-connected implies, by virtue of Cauchy's Theorem, that $\int_{\boldsymbol{\pi}} \boldsymbol{\varphi}(\xi) d\xi$ is independent of the path $\boldsymbol{\pi}$ chosen, and hence so is $\boldsymbol{\sigma}(u, v)$. Now, $\boldsymbol{\Phi}(\zeta) = \int_{\boldsymbol{\pi}} \boldsymbol{\varphi}(\xi) d\xi$ is a holomorphic function of $\zeta = u + iv$, and $\boldsymbol{\Phi}'(\zeta) = \boldsymbol{\varphi}(\zeta)$. Hence, by the facts stated just before the beginning of the proof,

$$\begin{aligned} \boldsymbol{\sigma}_u &= \mathfrak{Re}(\boldsymbol{\Phi}_u) = \mathfrak{Re}(\boldsymbol{\Phi}') = \mathfrak{Re}(\boldsymbol{\varphi}), \\ \boldsymbol{\sigma}_v &= \mathfrak{Re}(\boldsymbol{\Phi}_v) = \mathfrak{Re}(i\boldsymbol{\Phi}') = -\mathfrak{Im}(\boldsymbol{\varphi}), \end{aligned} \qquad (20)$$

so $\boldsymbol{\varphi} = \boldsymbol{\sigma}_u - i\boldsymbol{\sigma}_v$.

To complete the proof, we have to show that $\boldsymbol{\sigma}$ is a conformal surface patch. But, condition (ii) and Eq. (20) show that $\boldsymbol{\sigma}_u$ and $\boldsymbol{\sigma}_v$ are not both zero. By condition (i) and Eq. (19), $\parallel \boldsymbol{\sigma}_u \parallel = \parallel \boldsymbol{\sigma}_v \parallel$ and $\boldsymbol{\sigma}_u.\boldsymbol{\sigma}_v = 0$. Since $\boldsymbol{\sigma}_u$ and $\boldsymbol{\sigma}_v$ are not both zero, this proves that $\boldsymbol{\sigma}_u$ and $\boldsymbol{\sigma}_v$ are both non-zero and perpendicular, hence linearly independent, so that $\boldsymbol{\sigma}$ is a regular surface patch; it also proves that $\boldsymbol{\sigma}$ is conformal.

If another conformal minimal surface $\tilde{\boldsymbol{\sigma}}$ corresponds to the same holomorphic function $\boldsymbol{\varphi}$ as $\boldsymbol{\sigma}$, then $\tilde{\boldsymbol{\sigma}}_u = \boldsymbol{\sigma}_u$ and $\tilde{\boldsymbol{\sigma}}_v = \boldsymbol{\sigma}_v$ everywhere on U, which implies that $\tilde{\boldsymbol{\sigma}} - \boldsymbol{\sigma}$ is a constant, say $\mathbf{a}$, so that $\tilde{\boldsymbol{\sigma}}$ is obtained from $\boldsymbol{\sigma}$ by translating by the vector $\mathbf{a}$. □

Before giving some examples, we observe that, if a holomorphic function $\boldsymbol{\varphi}$ satisfies the conditions in Theorem 9.2, so does $i\boldsymbol{\varphi}$. If $\boldsymbol{\varphi}$ is the holomorphic function corresponding to a minimal surface $\mathcal{S}$, the minimal surface to which $i\boldsymbol{\varphi}$ corresponds is called the *conjugate* of $\mathcal{S}$. It is well defined by $\mathcal{S}$ up to a translation.

Example 9.6

The parametrisation

$$\boldsymbol{\sigma}(u, v) = (\cosh u \cos v, \cosh u \sin v, u)$$

of the catenoid is conformal (see Example 9.1). The associated holomorphic function is

$$\begin{aligned}
\boldsymbol{\varphi}(\zeta) &= \boldsymbol{\sigma}_u - i\boldsymbol{\sigma}_v \\
&= (\sinh u \cos v + i \cosh u \sin v, \sinh u \sin v - i \cosh u \cos v, 1) \\
&= (\sinh(u + iv), -i \cosh(u + iv), 1) \\
&= (\sinh \zeta, -i \cosh \zeta, 1).
\end{aligned}$$

Note that conditions (i) and (ii) in Theorem 9.2 are satisfied, since $\boldsymbol{\varphi}$ is clearly never zero and the sum of the squares of its components is

$$\sinh^2 \zeta - \cosh^2 \zeta + 1 = 0.$$

Let us determine the conjugate minimal surface $\tilde{\boldsymbol{\sigma}}$ of the catenoid. From the proof of Theorem 9.2,

$$\begin{aligned}
\tilde{\boldsymbol{\sigma}}(u, v) &= \mathfrak{Re} \int_{\boldsymbol{\pi}} (i \sinh \xi, \cosh \xi, i)\, d\xi \\
&= \mathfrak{Re}(i \cosh \zeta, \sinh \zeta, i\zeta) \\
&= (-\sinh u \sin v, \sinh u \cos v, -v),
\end{aligned}$$

up to a translation. If we reparametrise by defining $\tilde{u} = \sinh u, \tilde{v} = v + \pi/2$, we get the surface

$$(\tilde{u}, \tilde{v}) \mapsto (\tilde{u} \cos \tilde{v}, \tilde{u} \sin \tilde{v}, -\tilde{v}),$$

after translating by $(0, 0, -\pi/2)$, which is obtained from the helicoid in Exercise 4.14 by reflecting in the z-axis.

Note that the parametrisation of the helicoid given in Example 4.14 is not conformal, so the constructions in this section cannot be applied to it.

It is actually possible to 'solve' the conditions on $\boldsymbol{\varphi}$ in Theorem 9.2.

Proposition 9.6

Let $f(\zeta)$ be a holomorphic function on an open set U in the complex plane, not identically zero, and let $g(\zeta)$ be a meromorphic function on U such that, if $\zeta_0 \in U$ is a pole of g of order $m \geq 1$, say, then ζ_0 is also a zero of f of order $\geq 2m$. Then,

$$\boldsymbol{\varphi} = \left(\frac{1}{2}f(1-g^2), \frac{i}{2}f(1+g^2), fg\right) \tag{21}$$

satisfies conditions (i) and (ii) in Theorem 9.2, and conversely every holomorphic function $\boldsymbol{\varphi}$ satisfying these conditions arises in this way.

The correspondence given by Theorem 9.2 and Proposition 9.6 between pairs of functions f and g and minimal surfaces is called *Weierstrass's representation.*

Proof 9.6

Suppose that f and g are as in the statement of the proposition. If g has a pole of order $m \geq 1$ at $\zeta_0 \in U$, and f has a zero of order $n \geq 2m$ at ζ_0, then the Laurent expansions of f and g about ζ_0 are of the form

$$f(\zeta) = a(\zeta - \zeta_0)^n + \cdots \quad \text{and} \quad g(\zeta) = \frac{b}{(\zeta - \zeta_0)^m} + \cdots,$$

where a and b are non-zero complex numbers and the $\cdots$ indicates terms involving higher powers of $\zeta - \zeta_0$. Then,

$$f(1 \pm g^2) = \pm ab^2(\zeta - \zeta_0)^{n-2m} + \cdots \quad \text{and} \quad fg = ab(\zeta - \zeta_0)^{n-m} + \cdots$$

involve only non-negative powers of $\zeta - \zeta_0$, so $\boldsymbol{\varphi}$ is holomorphic near ζ_0. Since it is clear that $\boldsymbol{\varphi}$ is holomorphic wherever g is holomorphic, it follows that the function $\boldsymbol{\varphi}$ defined by Eq. (21) is holomorphic everywhere on U. It is clear that $\boldsymbol{\varphi}$ is identically zero only if f is identically zero, and simple algebra shows that $\boldsymbol{\varphi}$ satisfies condition (i) in Theorem 9.2.

Conversely, suppose that $\boldsymbol{\varphi} = (\varphi_1, \varphi_2, \varphi_3)$ is a holomorphic function satisfying conditions (i) and (ii) in Theorem 9.2. If $\varphi_1 - i\varphi_2$ is not identically zero, define

$$f = \varphi_1 - i\varphi_2, \quad g = \frac{\varphi_3}{\varphi_1 - i\varphi_2}. \tag{22}$$

Since $\boldsymbol{\varphi}$ is holomorphic, f is holomorphic and g is meromorphic. Condition (i) implies that $(\varphi_1 + i\varphi_2)(\varphi_1 - i\varphi_2) = -\varphi_3^2$, and hence that

$$\varphi_1 + i\varphi_2 = -fg^2. \tag{23}$$

Simple algebra shows that Eqs. (22) and (23) imply Eq. (21). Equation (23) implies that fg^2 is holomorphic, and the argument with Laurent expansions in the first part of the proof now gives the condition on the zeros and poles of f and g. Finally, if $\varphi_1 - i\varphi_2 = 0$, we repeat the above argument replacing $\varphi_1 \pm i\varphi_2$ by $\varphi_1 \mp i\varphi_2$ (note that $\varphi_1 - i\varphi_2$ and $\varphi_1 + i\varphi_2$ cannot both be zero, for if they were we would have $\varphi_1 = \varphi_2 = 0$, hence $\varphi_3 = 0$ by condition (i), and this would violate condition (ii)). □

We give only one application of Weierstrass's representation.

Proposition 9.7

The gaussian curvature of the minimal surface corresponding to the functions f and g in Weierstrass's representation is

$$K = \frac{-16|dg/d\zeta|^2}{|f|^2(1+|g|^2)^4}.$$

Proof 9.7

This is a straightforward, if tedious, computation, and we shall omit many of the details. Define $\overline{\boldsymbol{\varphi}}$ by taking the complex-conjugate of each component of $\boldsymbol{\varphi}$. Then, $\boldsymbol{\sigma}_u = \frac{1}{2}(\boldsymbol{\varphi} + \overline{\boldsymbol{\varphi}})$, $\boldsymbol{\sigma}_v = \frac{1}{2i}(\overline{\boldsymbol{\varphi}} - \boldsymbol{\varphi})$. Since $\boldsymbol{\varphi}.\boldsymbol{\varphi} = \overline{\boldsymbol{\varphi}}.\overline{\boldsymbol{\varphi}} = 0$, the first fundamental form is $\frac{1}{2}\boldsymbol{\varphi}.\overline{\boldsymbol{\varphi}}(du^2 + dv^2)$. Substituting the formula for $\boldsymbol{\varphi}$ into Eq. (14) and simplifying, we find that the first fundamental form is

$$\frac{1}{4}|f|^2(1+|g|^2)^2(du^2+dv^2). \tag{24}$$

Next,

$$\begin{aligned}
\boldsymbol{\sigma}_u \times \boldsymbol{\sigma}_v &= \frac{1}{4i}(\boldsymbol{\varphi}+\overline{\boldsymbol{\varphi}}) \times (\overline{\boldsymbol{\varphi}} - \boldsymbol{\varphi}) = \frac{1}{2i}\boldsymbol{\varphi} \times \overline{\boldsymbol{\varphi}}, \\
\therefore \quad \| \boldsymbol{\sigma}_u \times \boldsymbol{\sigma}_v \|^2 &= -\frac{1}{4}(\boldsymbol{\varphi} \times \overline{\boldsymbol{\varphi}}).(\boldsymbol{\varphi} \times \overline{\boldsymbol{\varphi}}), \\
&= -\frac{1}{4}((\boldsymbol{\varphi}.\boldsymbol{\varphi})(\overline{\boldsymbol{\varphi}}.\overline{\boldsymbol{\varphi}}) - (\boldsymbol{\varphi}.\overline{\boldsymbol{\varphi}})^2), \\
&= \frac{1}{4}(\boldsymbol{\varphi}.\overline{\boldsymbol{\varphi}})^2, \\
\therefore \quad \mathbf{N} &= i\frac{\overline{\boldsymbol{\varphi}} \times \boldsymbol{\varphi}}{\boldsymbol{\varphi}.\overline{\boldsymbol{\varphi}}}.
\end{aligned}$$

In terms of f and g, this becomes

$$\mathbf{N} = \frac{1}{1+|g|^2}\left(g+\overline{g}, -i(g-\overline{g}), |g|^2 - 1\right). \tag{25}$$

Using the remarks preceding the proof of Theorem 9.2 and the formulas

$$L = -\boldsymbol{\sigma}_u.\mathbf{N}_u, \quad M = -\boldsymbol{\sigma}_u.\mathbf{N}_v, \quad N = -\boldsymbol{\sigma}_v.\mathbf{N}_v$$

(which follow by differentiating $\boldsymbol{\sigma}_u.\mathbf{N} = \boldsymbol{\sigma}_v.\mathbf{N} = 0$), we find that the second fundamental form is

$$-\frac{1}{2}\left((fg' + \overline{fg'})(du^2 + dv^2) + 2i(fg' - \overline{fg'})dudv\right). \tag{26}$$

Combining Eqs. (24), (25) and (26), and using the formula for the gaussian curvature K in Proposition 7.1(i), we finally obtain the formula in the statement of the proposition. □

Corollary 9.2

Let $\mathcal{S}$ be a minimal surface that is not part of a plane. Then, the zeros of the gaussian curvature of $\mathcal{S}$ are isolated.

This means that, if the gaussian curvature K vanishes at a point P of $\mathcal{S}$, then K does not vanish at any other point of $\mathcal{S}$ sufficiently near to P. More precisely, if P lies in a surface patch $\boldsymbol{\sigma}$ of $\mathcal{S}$, say $P = \boldsymbol{\sigma}(u_0, v_0)$, there is a number $\epsilon > 0$ such that K does not vanish at the point $\boldsymbol{\sigma}(u, v)$ of $\mathcal{S}$ if $0 < (u - u_0)^2 + (v - v_0)^2 < \epsilon^2$.

Proof 9.5

From the formula for K in Proposition 9.8, K vanishes exactly where the meromorphic function g' vanishes. If g' is zero everywhere, so is K and $\mathcal{S}$ is part of a plane (this was shown in Proposition 6.5, but follows immediately from Eq. (18) which shows that $\mathbf{N}$ is constant if g is constant). But it is a standard result of complex analysis that the zeros of a non-zero meromorphic function are isolated, so if K is not identically zero its zeros must be isolated. □

EXERCISES

9.13 Find the holomorphic function $\boldsymbol{\varphi}$ corresponding to the plane passing through the origin with unit normal $\mathbf{a}$. What is its conjugate surface?

9.14 If a minimal surface $\mathcal{S}$ corresponds to a pair of functions f and g in Weierstrass's representation, to which pair of functions does the conjugate minimal surface of $\mathcal{S}$ correspond?

9.15 Calculate the functions f and g in Weierstrass's representation for the catenoid and the helicoid.

9.16 Find the holomorphic function $\boldsymbol{\varphi}$ corresponding to Enneper's minimal surface given in Example 9.4. Show that its conjugate minimal surface coincides with a reparametrisation of the same surface rotated by $\pi/4$ around the z-axis.

9.17 Find a parametrisation of *Henneberg's surface*, the minimal surface corresponding to the functions $f(\zeta) = 1 - \zeta^{-4}$, $g(\zeta) = \zeta$ in Weierstrass's representation. (Reparametrise by putting $\zeta = e^{\tilde{\zeta}}$.) The following are a 'close up' view and a 'large scale' view of this surface.

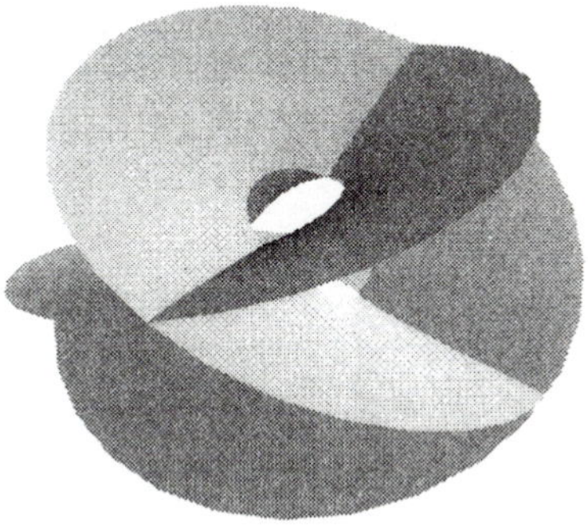

Henneberg: close up

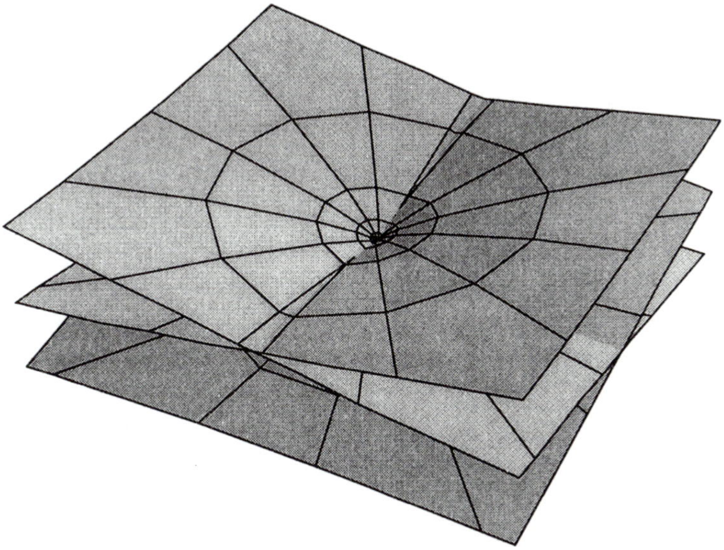

Henneberg: large scale

9.18 Let $\mathcal{G}$ be the Gauss map of a minimal surface in Weierstrass's representation. Let $\pi : S^2\backslash(0,0,1) \to \mathbf{R}^2$ be the stereographic projection map defined in Example 5.7, and identify the point $(u, v) \in \mathbf{R}^2$ with the complex number $u + iv$ as usual. Show that $\pi \circ \mathcal{G}$ is the function g.

10
Gauss's Theorema Egregium

One of Gauss's most important discoveries about surfaces is that the gaussian curvature is unchanged when the surface is bent without stretching. Gauss called this result 'egregium', and the Latin word for 'remarkable' has remained attached to his theorem ever since.

10.1. Gauss's Remarkable Theorem

The purpose of this section is to prove the following important theorem and to study its consequences.

Theorem 10.1

The gaussian curvature of a surface is preserved by isometries.

This means, more precisely, that if $\mathcal{S}_1$ and $\mathcal{S}_2$ are two surfaces and if $f : \mathcal{S}_1 \to \mathcal{S}_2$ is an isometry between them, then for any point P in $\mathcal{S}_1$ the gaussian curvature of $\mathcal{S}_1$ at P is equal to that of $\mathcal{S}_2$ at $f(P)$. The theorem is sometimes expressed by saying that the gaussian curvature is an *intrinsic* property of a surface, for it implies that the gaussian curvature could be measured by a bug living in the surface.

In proving the theorem, it is enough by Theorem 5.1 to consider the case of a surface patch $\boldsymbol{\sigma}$ on $\mathcal{S}_1$, and to prove that, if $\boldsymbol{\sigma}$ and $f \circ \boldsymbol{\sigma}$ have the same first fundamental forms, then they have the same gaussian curvature. This is far from obvious, for the formula

$$K = \frac{LN - M^2}{EG - F^2}$$

that we obtained for the gaussian curvature K in Proposition 7.1(i) depends on the coefficients L, N and M of the *second* fundamental form, as well as the coefficients E, F and G of the first fundamental form. Hence, the theorem is telling us that $LN - M^2$ can be expressed in terms of E, F and G (although it is *not* saying that L, M and N individually can be so expressed).

To prove the theorem, we shall make use of a smooth *orthonormal* basis $\{\mathbf{e}', \mathbf{e}''\}$ of the tangent plane at each point of the surface patch, where 'smooth' means that $\mathbf{e}'$ and $\mathbf{e}''$ are smooth functions of the surface parameters (u, v). Then, $\{\mathbf{e}', \mathbf{e}'', \mathbf{N}\}$ is an orthonormal basis of $\mathbf{R}^3$ ($\mathbf{N}$ being the standard unit normal of $\boldsymbol{\sigma}$), and we shall assume that it is *right-handed*, i.e. that $\mathbf{N} = \mathbf{e}' \times \mathbf{e}''$. This can always be achieved by interchanging $\mathbf{e}'$ and $\mathbf{e}''$ if necessary. Note that the dashes on $\mathbf{e}'$ and $\mathbf{e}''$ have nothing to do with derivatives.

We can express the partial derivatives of $\mathbf{e}'$ and $\mathbf{e}''$ with respect to u and v in terms of the orthonormal basis $\{\mathbf{e}', \mathbf{e}'', \mathbf{N}\}$. Since both partial derivatives of $\mathbf{e}'$ are perpendicular to $\mathbf{e}'$, the $\mathbf{e}'$ components of $\mathbf{e}'_u$ and $\mathbf{e}'_v$ are zero (and similarly for $\mathbf{e}''$). Thus,

$$\begin{aligned}
\mathbf{e}'_u &= \phantom{-\alpha'\mathbf{e}'} \alpha\mathbf{e}'' + \lambda'\mathbf{N}, \\
\mathbf{e}'_v &= \phantom{-\alpha'\mathbf{e}'} \beta\mathbf{e}'' + \mu'\mathbf{N}, \\
\mathbf{e}''_u &= -\alpha'\mathbf{e}' \phantom{+\alpha\mathbf{e}''} + \lambda''\mathbf{N}, \\
\mathbf{e}''_v &= -\beta'\mathbf{e}' \phantom{+\alpha\mathbf{e}''} + \mu''\mathbf{N},
\end{aligned}$$

for some scalars $\alpha, \beta, \alpha', \beta', \lambda', \mu', \lambda'', \mu''$ (which may depend on u and v). Moreover, by differentiating the equation $\mathbf{e}'.\mathbf{e}'' = 0$ with respect to u, we see that $\mathbf{e}'_u.\mathbf{e}'' = -\mathbf{e}'.\mathbf{e}''_u$. i.e. $\alpha' = \alpha$ (and similarly $\beta' = \beta$). Thus,

$$\begin{aligned}
\mathbf{e}'_u &= \phantom{-\alpha\mathbf{e}'} \alpha\mathbf{e}'' + \lambda'\mathbf{N}, \\
\mathbf{e}'_v &= \phantom{-\alpha\mathbf{e}'} \beta\mathbf{e}'' + \mu'\mathbf{N}, \\
\mathbf{e}''_u &= -\alpha\mathbf{e}' \phantom{+\alpha\mathbf{e}''} + \lambda''\mathbf{N}, \\
\mathbf{e}''_v &= -\beta\mathbf{e}' \phantom{+\alpha\mathbf{e}''} + \mu''\mathbf{N}.
\end{aligned} \tag{1}$$

The following lemma is the crucial step in the proof of Theorem 10.1.

Lemma 10.1

With the above notation, we have

$$\mathbf{e}'_u.\mathbf{e}''_v - \mathbf{e}''_u.\mathbf{e}'_v = \lambda'\mu'' - \lambda''\mu' \quad (2)$$

$$= \alpha_v - \beta_u \quad (3)$$

$$= \frac{LN - M^2}{(EG - F^2)^{1/2}}. \quad (4)$$

Proof 10.1

Equation (2) follows immediately from Eq. (1), since $\mathbf{e}'$, $\mathbf{e}''$ and $\mathbf{N}$ are perpendicular unit vectors.

Next, we compute

$$\begin{aligned}\alpha_v - \beta_u &= \frac{\partial}{\partial u}(\mathbf{e}'.\mathbf{e}''_v) - \frac{\partial}{\partial v}(\mathbf{e}'.\mathbf{e}''_u) \quad \text{(by Eq. (1))}\\ &= \mathbf{e}'_u.\mathbf{e}''_v + \mathbf{e}'.\mathbf{e}''_{uv} - \mathbf{e}'_v.\mathbf{e}''_u - \mathbf{e}'.\mathbf{e}''_{uv}\\ &= \mathbf{e}'_u.\mathbf{e}''_v - \mathbf{e}'_v.\mathbf{e}''_u.\end{aligned}$$

This proves Eq. (3).

To prove Eq. (4), we use the formula

$$\mathbf{N}_u \times \mathbf{N}_v = K\boldsymbol{\sigma}_u \times \boldsymbol{\sigma}_v$$

(see Eq. (15) in Chapter 7). Combining this with the formulas

$$\mathbf{N} = \frac{\boldsymbol{\sigma}_u \times \boldsymbol{\sigma}_v}{\parallel \boldsymbol{\sigma}_u \times \boldsymbol{\sigma}_v \parallel}, \quad \parallel \boldsymbol{\sigma}_u \times \boldsymbol{\sigma}_v \parallel = (EG - F^2)^{1/2}$$

(see Proposition 5.2), we get

$$\mathbf{N}_u \times \mathbf{N}_v = \frac{LN - M^2}{(EG - F^2)^{1/2}}\mathbf{N},$$

and hence

$$(\mathbf{N}_u \times \mathbf{N}_v).\mathbf{N} = \frac{LN - M^2}{(EG - F^2)^{1/2}}. \quad (5)$$

Since $\mathbf{N} = \mathbf{e}' \times \mathbf{e}''$, we get

$$\begin{aligned}(\mathbf{N}_u \times \mathbf{N}_v).\mathbf{N} &= (\mathbf{N}_u \times \mathbf{N}_v).(\mathbf{e}' \times \mathbf{e}''),\\ &= (\mathbf{N}_u.\mathbf{e}')(\mathbf{N}_v.\mathbf{e}'') - (\mathbf{N}_u.\mathbf{e}'')(\mathbf{N}_v.\mathbf{e}')\\ &= (\mathbf{N}.\mathbf{e}'_u)(\mathbf{N}.\mathbf{e}''_v) - (\mathbf{N}.\mathbf{e}''_u)(\mathbf{N}.\mathbf{e}'_v)\\ &= \lambda'\mu'' - \lambda''\mu' \quad \text{(by Eq. (1))}, \quad (6)\end{aligned}$$

where in passing from the second line to the third we used the equations

$$\mathbf{N}_u.\mathbf{e}' = -\mathbf{N}.\mathbf{e}'_u, \ \ \mathbf{N}_u.\mathbf{e}'' = -\mathbf{N}.\mathbf{e}''_u,$$

$$\mathbf{N}_v.\mathbf{e}' = -\mathbf{N}.\mathbf{e}'_v, \ \ \mathbf{N}_v.\mathbf{e}'' = -\mathbf{N}.\mathbf{e}''_v,$$

which follow by differentiating $\mathbf{N}.\mathbf{e}' = 0 = \mathbf{N}.\mathbf{e}''$ with respect to u and v. Putting Eqs. (5) and (6) together shows that the right-hand sides of Eqs. (2) and (4) are equal. Since Eq. (2) has already been established, this proves Eq. (4). □

Now we can prove Theorem 10.1.

Proof 10.1

Combining Eqs. (3) and (4), we get

$$K = \frac{\alpha_v - \beta_u}{(EG - F^2)^{1/2}}, \tag{7}$$

so to prove the theorem it suffices to show that, *for a suitable choice of* $\{\mathbf{e}', \mathbf{e}''\}$, *the scalars* α *and* β *depend only on* E, F *and* G. We shall construct $\{\mathbf{e}', \mathbf{e}''\}$ by applying the Gram–Schmidt process to the basis $\{\boldsymbol{\sigma}_u, \boldsymbol{\sigma}_v\}$ of the tangent plane, and will then show that they have the desired property.

So we first define

$$\mathbf{e}' = \frac{\boldsymbol{\sigma}_u}{\| \boldsymbol{\sigma}_u \|} = \epsilon \boldsymbol{\sigma}_u,$$

where $\epsilon = E^{-1/2}$. Now we look for a vector $\mathbf{e}'' = \gamma \boldsymbol{\sigma}_u + \delta \boldsymbol{\sigma}_v$, for some scalars γ, δ, such that $\mathbf{e}''$ is a unit vector perpendicular to $\mathbf{e}'$. These conditions give

$$E^{-1/2}(\gamma E + \delta F) = 0, \quad \gamma^2 E + 2\gamma\delta F + \delta^2 G = 1.$$

The first equation gives $\gamma = -\delta F/E$, and substituting in the second equation then gives

$$\delta^2 \left(\frac{F^2}{E} - 2\frac{F^2}{E} + G \right) = 1,$$

$$\therefore \quad \delta = \frac{E^{1/2}}{(EG - F^2)^{1/2}}, \quad \gamma = -\frac{FE^{-1/2}}{(EG - F^2)^{1/2}}, \quad \epsilon = E^{-1/2}. \tag{8}$$

(We could change the sign of δ, and hence also that of γ, but it would make no difference in the end.) Thus,

$$\mathbf{e}' = \epsilon \boldsymbol{\sigma}_u, \quad \mathbf{e}'' = \gamma \boldsymbol{\sigma}_u + \delta \boldsymbol{\sigma}_v, \tag{9}$$

where γ, δ and ϵ depend only on E, F and G.

We now compute α and β. First,

$$\begin{aligned}\alpha &= \mathbf{e}'_u.\mathbf{e}'' \quad \text{(by Eq. (1))}\\ &= (\epsilon_u\boldsymbol{\sigma}_u + \epsilon\boldsymbol{\sigma}_{uu}).(\gamma\boldsymbol{\sigma}_u + \delta\boldsymbol{\sigma}_v) \quad \text{(by Eq. (9))}\\ &= \frac{\epsilon_u}{\epsilon}(\epsilon\boldsymbol{\sigma}_u).(\gamma\boldsymbol{\sigma}_u + \delta\boldsymbol{\sigma}_v) + \epsilon\gamma\boldsymbol{\sigma}_{uu}.\boldsymbol{\sigma}_u + \epsilon\delta\boldsymbol{\sigma}_{uu}.\boldsymbol{\sigma}_v\\ &= \frac{\epsilon_u}{\epsilon}\mathbf{e}'.\mathbf{e}'' + \frac{1}{2}\epsilon\gamma(\boldsymbol{\sigma}_u.\boldsymbol{\sigma}_u)_u + \epsilon\delta((\boldsymbol{\sigma}_u.\boldsymbol{\sigma}_v)_u - \boldsymbol{\sigma}_u.\boldsymbol{\sigma}_{uv})\\ &= \frac{1}{2}\epsilon\gamma E_u + \epsilon\delta(F_u - \frac{1}{2}E_v) \quad (\text{since } \mathbf{e}'.\mathbf{e}'' = 0), \qquad (10)\end{aligned}$$

which does indeed depend only on E, F and G (because the same is true for γ, δ and ϵ). And finally,

$$\begin{aligned}\beta &= \mathbf{e}'_v.\mathbf{e}''\\ &= (\epsilon_v\boldsymbol{\sigma}_u + \epsilon\boldsymbol{\sigma}_{uv}).(\gamma\boldsymbol{\sigma}_u + \delta\boldsymbol{\sigma}_v)\\ &= \frac{\epsilon_v}{\epsilon}\mathbf{e}'.\mathbf{e}'' + \epsilon\gamma\boldsymbol{\sigma}_{uv}.\boldsymbol{\sigma}_u + \epsilon\delta\boldsymbol{\sigma}_{uv}.\boldsymbol{\sigma}_v\\ &= \frac{1}{2}\epsilon\gamma E_v + \frac{1}{2}\epsilon\delta G_u, \qquad (11)\end{aligned}$$

which also depends only on E, F and G.

This completes the proof of Gauss's Theorem. □

By substituting the actual values of γ, δ and ϵ into these formulas for α and β, and then using Eq. (7), we get an *explicit* formula for K in terms of E, F and G. Here is the result:

Corollary 10.1

The gaussian curvature is given by

$$K = \frac{\begin{vmatrix} -\frac{1}{2}E_{vv} + F_{uv} - \frac{1}{2}G_{uu} & \frac{1}{2}E_u & F_u - \frac{1}{2}E_v \\ F_v - \frac{1}{2}G_u & E & F \\ \frac{1}{2}G_v & F & G \end{vmatrix} - \begin{vmatrix} 0 & \frac{1}{2}E_v & \frac{1}{2}G_u \\ \frac{1}{2}E_v & E & F \\ \frac{1}{2}G_u & F & G \end{vmatrix}}{(EG - F^2)^2}.$$

We shall not go through the details of this calculation, partly because the proof is very tedious, and partly because the following special cases are often all that is needed:

Corollary 10.2

(i) If $F = 0$, we have

$$K = -\frac{1}{2\sqrt{EG}}\left\{\frac{\partial}{\partial u}\left(\frac{G_u}{\sqrt{EG}}\right) + \frac{\partial}{\partial v}\left(\frac{E_v}{\sqrt{EG}}\right)\right\}.$$

(ii) If $E = 1$ and $F = 0$, we have

$$K = -\frac{1}{\sqrt{G}}\frac{\partial^2\sqrt{G}}{\partial u^2}.$$

Proof 10.2

If $F = 0$, Eq. (8) gives

$$\gamma = 0, \quad \delta = G^{-1/2}, \quad \epsilon = E^{-1/2}.$$

Substituting in Eqs. (10) and (11), we get

$$\alpha = -\frac{1}{2}(EG)^{-1/2}E_v, \quad \beta = \frac{1}{2}(EG)^{-1/2}G_u.$$

Hence,

$$K = \frac{\alpha_v - \beta_u}{(EG)^{1/2}} = -\frac{1}{2\sqrt{EG}}\left\{\frac{\partial}{\partial v}\left(\frac{E_v}{\sqrt{EG}}\right) + \frac{\partial}{\partial u}\left(\frac{G_u}{\sqrt{EG}}\right)\right\}, \tag{12}$$

proving the formula in (i).

If, in addition, $E = 1$, the first term on the right-hand side of (12) vanishes, so

$$K = -\frac{1}{2\sqrt{G}}\frac{\partial}{\partial u}\left(\frac{G_u}{\sqrt{G}}\right) = -\frac{1}{\sqrt{G}}\frac{\partial^2\sqrt{G}}{\partial u^2},$$

proving the formula in (ii). □

Example 10.1

For the surface of revolution

$$\boldsymbol{\sigma}(u, v) = (f(u)\cos v, f(u)\sin v, g(u)),$$

where $f > 0$ and $\dot{f}^2 + \dot{g}^2 = 1$ (a dot denoting d/du), we found in Example 6.2 that $E = 1$, $F = 0$, $G = f(u)^2$. Hence, Corollary 10.2(ii) applies and gives

$$K = -\frac{1}{\sqrt{G}}\frac{\partial^2\sqrt{G}}{\partial u^2} = -\frac{\ddot{f}}{f},$$

in agreement with Eq. (2) of Chapter 7.

We are now in a position to give the application of geodesic coordinates that we promised in Section 8.5.

Theorem 10.2

Any point of a surface of constant gaussian curvature is contained in a patch that is isometric to part of a plane, a sphere or a pseudosphere.

Proof 10.2

Let P be a point of a surface $\mathcal{S}$ with constant gaussian curvature K. By applying a dilation of $\mathbf{R}^3$ (see Exercise 7.6), we need only consider the cases $K = 0, 1$ and -1.

We take a geodesic patch $\boldsymbol{\sigma}(u, v)$ with $\boldsymbol{\sigma}(0,0) = P$. Writing $g = \sqrt{G}$, the first fundamental form is

$$du^2 + g(u, v)^2 dv^2.$$

By Corollary 10.2(ii),

$$\frac{\partial^2 g}{\partial u^2} + Kg = 0. \tag{13}$$

Note that

$$g(0, v) = 1, \quad g_u(0, v) = 0, \tag{14}$$

by Proposition 8.7.

If $K = 0$, the solution of Eq. (13) is $g(u, v) = \alpha u + \beta$, where α and β are smooth functions of v only. The boundary conditions (14) give $\alpha = 0$, $\beta = 1$, so $g = 1$ and the first fundamental form of $\boldsymbol{\sigma}$ is

$$du^2 + dv^2.$$

This is the same as the first fundamental form of the usual parametrisation of the plane (see Example 5.1), and Theorem 5.1 now shows that $\boldsymbol{\sigma}$ is isometric to part of the plane.

If $K = 1$, the general solution of Eq. (13) is $g = \alpha \cos u + \beta \sin u$, where α and β only depend on v. The boundary conditions (14) give $\alpha = 1$, $\beta = 0$, and the first fundamental form of $\boldsymbol{\sigma}$ is

$$du^2 + \cos^2 u \, dv^2.$$

This is the first fundamental form of the unit sphere, with u and v being latitude and longitude, respectively (see Example 5.2). Hence, $\boldsymbol{\sigma}$ is isometric to part of the unit sphere.

Finally, if $K = -1$, we find in the same way that the first fundamental form of $\boldsymbol{\sigma}$ is

$$du^2 + \cosh^2 u\, dv^2.$$

We have not encountered this first fundamental form before. However, let us reparametrise $\boldsymbol{\sigma}$ by defining

$$V = e^v \tanh u, \quad W = e^v \operatorname{sech} u.$$

We then find, using the formulas in Exercise 5.4, for example, that the first fundamental form becomes

$$\frac{dV^2 + dW^2}{W^2}.$$

Comparing with Example 8.8, we see that this is the first fundamental form of the pseudosphere. □

EXERCISES

10.1 If a surface patch has first fundamental form $e^\lambda(du^2 + dv^2)$, where λ is a smooth function of u and v, show that its gaussian curvature K satisfies

$$\Delta\lambda + 2Ke^\lambda = 0,$$

where Δ denotes the laplacian $\partial^2/\partial u^2 + \partial^2/\partial v^2$.

10.2 With the notation of Exercise 8.21, define $u = r\cos\theta$, $v = r\sin\theta$, and let $\tilde{\boldsymbol{\sigma}}(u, v)$ be the corresponding reparametrisation of $\boldsymbol{\sigma}$. It can be shown that $\tilde{\boldsymbol{\sigma}}$ is an allowable surface patch for $\mathcal{S}$ defined on the open set $u^2 + v^2 < \epsilon^2$. (Note that this is not quite obvious because $\boldsymbol{\sigma}$ is not allowable when $r = 0$.) Show that the first fundamental form of $\tilde{\boldsymbol{\sigma}}$ is $\tilde{E}du^2 + 2\tilde{F}dudv + \tilde{G}dv^2$, where

$$\tilde{E} = \frac{u^2}{r^2} + \frac{Gv^2}{r^4}, \quad \tilde{F} = \left(1 - \frac{G}{r^2}\right)\frac{uv}{r^2}, \quad \tilde{G} = \frac{v^2}{r^2} + \frac{Gu^2}{r^4}$$

(use Exercise 5.4). Show that $u^2(\tilde{E} - 1) = v^2(\tilde{G} - 1)$, and by considering the Taylor expansions of $\tilde{E}$ and $\tilde{G}$ about $u = v = 0$, deduce that

$$G(r, \theta) = r^2 + kr^4 + \text{remainder}$$

for some constant k, where remainder$/r^4$ tends to zero as r tends to zero. Show finally that $k = -K(P)/3$, where $K(P)$ is the value of the gaussian curvature of $\mathcal{S}$ at P (use Corollary 10.2(ii)).

10.3 With the notation of Exercises 8.21 and 10.2, show that

(i) the circumference of the geodesic circle with centre P and radius R is

$$C_R = 2\pi R\left(1 - \frac{K(P)}{6}R^2 + \text{remainder}\right),$$

where remainder$/R^2$ tends to zero as R tends to zero;

(ii) the area inside the geodesic circle in (i) is

$$A_R = \pi R^2\left(1 - \frac{K(P)}{12}R^2 + \text{remainder}\right),$$

where the remainder satisfies the same condition as in (i).

Calculate C_R and A_R exactly when $\mathcal{S}$ is the unit sphere, and verify that the results agree with those in (i) and (ii) above.

10.4 Let ABC be a triangle on a surface $\boldsymbol{\sigma}$ whose sides are arcs of geodesics. Assume that ABC is contained in a geodesic patch $\boldsymbol{\sigma}$ as in Exercise 8.21 with $P = A$. Thus, with the notation in that exercise, if we take $\mathbf{v}$ to be parallel to AB at A, then AB and AC are the parameter curves $\theta = 0$ and $\theta = \angle A$, and BC can be parametrised by $\boldsymbol{\gamma}(\theta) = \boldsymbol{\sigma}(f(\theta), \theta)$ for some smooth function f and $0 \le \theta \le \angle A$.

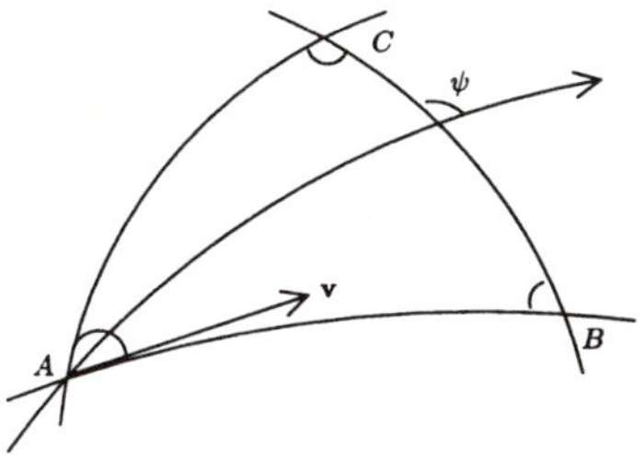

(i) Use the geodesic equations (Chapter 8, Eqs. (2)) to show that

$$f'' - \frac{f'\lambda'}{\lambda^2} = \frac{1}{2}\frac{\partial G}{\partial r},$$

where a dash denotes $d/d\theta$ and $\lambda = \parallel \boldsymbol{\gamma}' \parallel$.

(ii) Show that, if $\psi(\theta)$ is the angle between $\boldsymbol{\sigma}_r$ and the tangent vector to BC at $\boldsymbol{\gamma}(\theta)$, then

$$\psi'(\theta) = -\frac{\partial\sqrt{G}}{\partial r}(f(\theta), \theta).$$

(iii) Show that, if K is the gaussian curvature of $\mathcal{S}$,

$$\iint_{ABC} K\, d\mathcal{A}_{\boldsymbol{\sigma}} = \angle A + \angle B + \angle C - \pi.$$

This result will be generalised in Chapter 11.

10.2. Isometries of Surfaces

The Theorema Egregium provides a necessary condition for the existence of an isometry between surfaces: if such an isometry exists, the gaussian curvature must be the same at corresponding points of the two surfaces. We give two examples of this idea; others can be found in the exercises.

Our first result shows that it is impossible to draw a 'perfect' map of the Earth (which is why cartography is an interesting subject).

Proposition 10.1

Any map of any region of the earth's surface must distort distances.

Proof 10.1

A map of a region of the earth's surface which did *not* distort distances would be a diffeomorphism from this region of a sphere to a region in a plane (the map) which multiplied all distances by the same constant factor, say C. We might as well assume that the plane passes through the origin. Then, by composing this map with the map from the plane to itself which takes a point with position vector $\mathbf{r}$ to the point with position vector $C^{-1}\mathbf{r}$, we would get an *isometry* between this region of the sphere and some region of a plane. This would imply, by the Theorema Egregium, that these regions of the sphere and the plane have the same gaussian curvature. But we know that a plane has gaussian curvature zero everywhere, and a sphere has constant positive gaussian curvature everywhere (if the sphere has radius R, the gaussian curvature is $1/R^2$). So no such isometry can exist. □

Note, on the other hand, that it *is* possible to draw a map of the Earth that correctly represents angles, for we saw in Example 5.7 and Exercise 5.11 that the stereographic and Mercator projections are conformal, and Archimedes's Theorem 5.4 shows that there is a map that correctly represents areas.

Our next example shows how the Theorema Egregium can sometimes be used to determine all the isometries of a surface.

Proposition 10.2

The only isometries of a helicoid

$$\boldsymbol{\sigma}(u,v) = (u\cos v, u\sin v, v)$$

are S_λ, $R_x \circ S_\lambda$, $R_y \circ S_\lambda$ and $R_z \circ S_\lambda$ for some value of λ, where S_λ is the screwing motion $\boldsymbol{\sigma}(u,v) \mapsto \boldsymbol{\sigma}(u, v+\lambda)$, and R_x, R_y and R_z are rotations by π around the x, y and z-axes.

Proof 10.2

Suppose that an isometry of the helicoid takes $\boldsymbol{\sigma}(u,v)$ to $\boldsymbol{\sigma}(\tilde{u},\tilde{v})$, where $\tilde{u}$ and $\tilde{v}$ are smooth functions of u and v. Since the gaussian curvature at $\boldsymbol{\sigma}(u,v)$ is $-1/(1+u^2)^2$ (see Exercise 7.2), the Theorema Egregium tells us that

$$\frac{-1}{(1+u^2)^2} = \frac{-1}{(1+\tilde{u}^2)^2},$$

so $\tilde{u} = \pm u$. Applying a rotation R_z by π around the z-axis changes u to $-u$ (and fixes v), so we assume that $\tilde{u} = u$. Let $\tilde{v} = f(u,v)$. By Theorem 5.1, the patches $\boldsymbol{\sigma}(u,v)$ and $\tilde{\boldsymbol{\sigma}}(u,v) = \boldsymbol{\sigma}(u, f(u,v))$ have the same first fundamental form. That of $\boldsymbol{\sigma}$ is $du^2 + (1+u^2)dv^2$, and that of $\tilde{\boldsymbol{\sigma}}$ is found to be

$$(1 + (1+u^2)f_u^2)du^2 + 2(1+u^2)f_u f_v du dv + (1+u^2)f_v^2 dv^2.$$

Equating these, we find that $f_u = 0$ and $f_v = \pm 1$. Hence,

$$\tilde{v} = f(u,v) = \pm v + \lambda,$$

where λ is a constant. A rotation R_x by π around the x-axis changes v to $-v$ (and fixes u), so we take the + sign. This gives the isometry

$$S_\lambda : \boldsymbol{\sigma}(u,v) \mapsto \boldsymbol{\sigma}(u, v+\lambda).$$

This proves the proposition (the isometry $R_y \circ S_\lambda$ arises because $R_y = R_x \circ R_z$.) □

EXERCISES

10.5 Show that there is no isometry between any region of a sphere and any region of a (generalised) cylinder or a (generalised) cone. (Use Proposition 10.1 and Exercise 5.7.)

10.6 Show that the gaussian curvature of the Möbius band in Example 4.9 is equal to $-1/4$ everywhere along its median circle. Deduce that this Möbius band *cannot* be constructed by taking a strip of paper and joining the ends together with a half-twist. (The analytic description of the 'cut and paste' Möbius band is more complicated than the version in Example 4.9.)

10.7 Consider the surface patches

$$\boldsymbol{\sigma}(u,v) = (u\cos v, u\sin v, \ln u), \quad \tilde{\boldsymbol{\sigma}}(u,v) = (u\cos v, u\sin v, v).$$

Prove that the gaussian curvature of $\boldsymbol{\sigma}$ at $\boldsymbol{\sigma}(u,v)$ is the same as that of $\tilde{\boldsymbol{\sigma}}$ at $\tilde{\boldsymbol{\sigma}}(u,v)$, but that the map from $\boldsymbol{\sigma}$ to $\tilde{\boldsymbol{\sigma}}$ which takes $\boldsymbol{\sigma}(u,v)$ to $\tilde{\boldsymbol{\sigma}}(u,v)$ is *not* an isometry. Prove that, in fact, there is *no* isometry from $\boldsymbol{\sigma}$ to $\tilde{\boldsymbol{\sigma}}$.

10.8 Show that the only isometries from the catenoid to itself are products of rotations around its axis, reflections in planes containing the axis, and reflection in the plane containing the waist of the catenoid.

10.3. The Codazzi–Mainardi Equations

Gauss's Theorema Egregium shows that the coefficients of the first and second fundamental forms of a surface cannot be arbitrary smooth functions, for it shows that $LN - M^2$ can be expressed in terms of E, F and G. It is natural to ask if there are any further relations between these coefficients. In this section, we find that there are indeed some additional relations, and we show that, in a sense we shall explain, there are no others.

We begin with a computation similar to that in Lemma 10.1.

Proposition 10.3 (Gauss Equations)

Let $\boldsymbol{\sigma}(u,v)$ be a surface patch. Then,

$$\begin{aligned}
\boldsymbol{\sigma}_{uu} &= \Gamma^1_{11}\boldsymbol{\sigma}_u + \Gamma^2_{11}\boldsymbol{\sigma}_v + L\mathbf{N},\\
\boldsymbol{\sigma}_{uv} &= \Gamma^1_{12}\boldsymbol{\sigma}_u + \Gamma^2_{12}\boldsymbol{\sigma}_v + M\mathbf{N},\\
\boldsymbol{\sigma}_{vv} &= \Gamma^1_{22}\boldsymbol{\sigma}_u + \Gamma^2_{22}\boldsymbol{\sigma}_v + N\mathbf{N},
\end{aligned}$$

where

$$\Gamma^1_{11} = \frac{GE_u - 2FF_u + FE_v}{2(EG - F^2)}, \quad \Gamma^2_{11} = \frac{2EF_u - EE_v - FE_u}{2(EG - F^2)},$$

$$\Gamma^1_{12} = \frac{GE_v - FG_u}{2(EG - F^2)}, \quad \Gamma^2_{12} = \frac{EG_u - FE_v}{2(EG - F^2)},$$

$$\Gamma^1_{22} = \frac{2GF_v - GG_u - FG_u}{2(EG - F^2)}, \quad \Gamma^2_{22} = \frac{EG_v - 2FF_v + FG_u}{2(EG - F^2)}.$$

The six Γ coefficients in these formulas are called *Christoffel symbols.*

Proof 10.3

Since $\{\boldsymbol{\sigma}_u, \boldsymbol{\sigma}_v, \mathbf{N}\}$ is a basis of $\mathbf{R}^3$, scalar functions $\alpha_1, \ldots, \gamma_3$ satisfying

$$\begin{aligned} \boldsymbol{\sigma}_{uu} &= \alpha_1 \boldsymbol{\sigma}_u + \alpha_2 \boldsymbol{\sigma}_v + \alpha_3 \mathbf{N}, \\ \boldsymbol{\sigma}_{uv} &= \beta_1 \boldsymbol{\sigma}_u + \beta_2 \boldsymbol{\sigma}_v + \beta_3 \mathbf{N}, \\ \boldsymbol{\sigma}_{vv} &= \gamma_1 \boldsymbol{\sigma}_u + \gamma_2 \boldsymbol{\sigma}_v + \gamma_3 \mathbf{N}, \end{aligned} \tag{15}$$

certainly exist. Taking the dot product of each equation with $\mathbf{N}$ gives

$$\alpha_3 = L, \quad \beta_3 = M, \quad \gamma_3 = N.$$

Now we take the dot product of each equation in (15) with $\boldsymbol{\sigma}_u$ and $\boldsymbol{\sigma}_v$. This gives six scalar equations from which we determine the remaining six coefficients. For example, taking the dot product of the first equation in (15) with $\boldsymbol{\sigma}_u$ and $\boldsymbol{\sigma}_v$ gives the two equations

$$\begin{aligned} E\alpha_1 + F\alpha_2 &= \boldsymbol{\sigma}_{uu}.\boldsymbol{\sigma}_u = \frac{1}{2}E_u, \\ F\alpha_1 + G\alpha_2 &= \boldsymbol{\sigma}_{uu}.\boldsymbol{\sigma}_v = (\boldsymbol{\sigma}_u.\boldsymbol{\sigma}_v)_u - \boldsymbol{\sigma}_u.\boldsymbol{\sigma}_{uv} = F_u - \frac{1}{2}E_v. \end{aligned}$$

Solving these equations gives $\alpha_1 = \Gamma_{11}^1, \alpha_2 = \Gamma_{11}^2$; similarly for the other four coefficients in Eqs. (15). □

The new relations between the coefficients of the first and second fundamental forms of a surface patch are contained in the following result.

Proposition 10.4 (Codazzi–Mainardi Equations)

Define the Christoffel symbols of a surface patch $\boldsymbol{\sigma}(u, v)$ as above. Then,

$$\begin{aligned} L_v - M_u &= L\Gamma_{12}^1 + M(\Gamma_{12}^2 - \Gamma_{11}^1) - N\Gamma_{11}^2, \\ M_v - N_u &= L\Gamma_{22}^1 + M(\Gamma_{22}^2 - \Gamma_{12}^1) - N\Gamma_{12}^2. \end{aligned}$$

Proof 10.4

We write down the equation $(\boldsymbol{\sigma}_{uu})_v = (\boldsymbol{\sigma}_{uv})_u$, using the Gauss equations for $\boldsymbol{\sigma}_{uu}$ and $\boldsymbol{\sigma}_{uv}$:

$$\begin{aligned} &(\Gamma_{11}^1 \boldsymbol{\sigma}_u + \Gamma_{11}^2 \boldsymbol{\sigma}_v + L\mathbf{N})_v = (\Gamma_{12}^1 \boldsymbol{\sigma}_u + \Gamma_{12}^2 \boldsymbol{\sigma}_v + M\mathbf{N})_u, \\ \therefore \quad &\left(\frac{\partial \Gamma_{11}^1}{\partial v} - \frac{\partial \Gamma_{12}^1}{\partial u}\right)\boldsymbol{\sigma}_u + \left(\frac{\partial \Gamma_{11}^2}{\partial v} - \frac{\partial \Gamma_{12}^2}{\partial u}\right)\boldsymbol{\sigma}_v + (L_v - M_u)\mathbf{N} \\ &= \Gamma_{12}^1 \boldsymbol{\sigma}_{uu} + (\Gamma_{12}^2 - \Gamma_{11}^1)\boldsymbol{\sigma}_{uv} - \Gamma_{11}^2 \boldsymbol{\sigma}_{vv} - L\mathbf{N}_v + M\mathbf{N}_u \\ &= \Gamma_{12}^1(\Gamma_{11}^1 \boldsymbol{\sigma}_u + \Gamma_{11}^2 \boldsymbol{\sigma}_v + L\mathbf{N}) + (\Gamma_{12}^2 - \Gamma_{11}^1)(\Gamma_{12}^1 \boldsymbol{\sigma}_u + \Gamma_{12}^2 \boldsymbol{\sigma}_v + M\mathbf{N}) \\ &\quad - \Gamma_{11}^2(\Gamma_{22}^1 \boldsymbol{\sigma}_u + \Gamma_{22}^2 \boldsymbol{\sigma}_v + N\mathbf{N}) - L\mathbf{N}_v + M\mathbf{N}_u, \end{aligned} \tag{16}$$

using the Gauss equations again. Now, $\mathbf{N}_u$ and $\mathbf{N}_v$ are perpendicular to $\mathbf{N}$, and so are linear combinations of $\boldsymbol{\sigma}_u$ and $\boldsymbol{\sigma}_v$. Hence, equating $\mathbf{N}$ components on both sides of the last equation gives

$$L_v - M_u = L\Gamma_{12}^1 + M(\Gamma_{12}^2 - \Gamma_{11}^1) - N\Gamma_{11}^2,$$

which is the first of the Codazzi–Mainardi equations. The other equation follows in a similar way from $(\boldsymbol{\sigma}_{uv})_v = (\boldsymbol{\sigma}_{vv})_u$. □

At first sight, it seems that we could get four other identities like those in Theorem 10.2 by equating the coefficients of $\boldsymbol{\sigma}_u$ and $\boldsymbol{\sigma}_v$ in Eq. (16) and in its analogue coming from $(\boldsymbol{\sigma}_{uv})_v = (\boldsymbol{\sigma}_{vv})_u$. It turns out, however, that these identites are all equivalent to the formula in Corollary 10.1 (and so, in particular, they give another proof of the Theorema Egregium). In fact, there are no further identites to be discovered, as the following theorem shows.

Theorem 10.3

Let $\boldsymbol{\sigma} : U \to \mathbf{R}^3$ and $\tilde{\boldsymbol{\sigma}} : U \to \mathbf{R}^3$ be surface patches with the same first and second fundamental forms. Then, there is a rigid motion M of $\mathbf{R}^3$ such that $\tilde{\boldsymbol{\sigma}} = M \circ \boldsymbol{\sigma}$.

Moreover, let V be an open subset of $\mathbf{R}^3$ and let E, F, G, L, M and N be smooth functions on V. Assume that $E > 0, G > 0, EG - F^2 > 0$ and that the equations in Corollary 10.1 and Proposition 10.4 hold, with $K = \frac{LN-M^2}{EG-F^2}$ and the Christoffel symbols defined as in Proposition 10.3. Then, if $(u_0, v_0) \in V$, there is an open set U contained in V and containing (u_0, v_0), and a surface patch $\boldsymbol{\sigma} : U \to \mathbf{R}^3$, such that $Edu^2+2Fdudv+Gdv^2$ and $Ldu^2+2Mdudv+Ndv^2$ are the first and second fundamental forms of $\boldsymbol{\sigma}$, respectively.

This theorem is the analogue for surfaces of Theorem 2.3, which shows that unit-speed plane curves are determined up to a rigid motion by their signed curvature. We shall not prove Theorem 10.3 here. The first part depends on uniqueness theorems for the solution of systems of ordinary differential equations, and is not particularly difficult. The second part is more sophisticated and depends on existence theorems for the solution of certain *partial* differential equations. The following example will illustrate what is involved.

Example 10.2

Consider the first and second fundamental forms $du^2 + dv^2$ and $-du^2$, respectively. Let us first see whether a surface patch with these first and second fundamental forms exists. Since all the coefficients of these forms are constant, all the Christoffel symbols are zero and the Codazzi–Mainardi equations are obviously satisfied. The formula in Corollary 10.1 gives $K = 0$, so the only other condition to be checked is $LN - M^2 = 0$, and this clearly holds since $M = N = 0$. Theorem 10.3 therefore tells us that a surface patch with the given first and second fundamental forms exists.

To find it, we note that the Gauss equations give

$$\boldsymbol{\sigma}_{uu} = -\mathbf{N}, \quad \boldsymbol{\sigma}_{uv} = \mathbf{0}, \quad \boldsymbol{\sigma}_{vv} = \mathbf{0}.$$

The last two equations tell us that $\boldsymbol{\sigma}_v$ is a constant vector, say $\mathbf{a}$, so

$$\boldsymbol{\sigma}(u, v) = \mathbf{b}(u) + \mathbf{a}v, \tag{17}$$

where $\mathbf{b}$ is a function of u only. The first equation then gives $\mathbf{N} = -\mathbf{b}''$ (a dash denoting d/du). We now need to use the expressions for $\mathbf{N}_u$ and $\mathbf{N}_v$ in terms of $\boldsymbol{\sigma}_u$ and $\boldsymbol{\sigma}_v$ in Proposition 6.4. The Weingarten matrix is

$$\mathcal{W} = \mathcal{F}_I^{-1}\mathcal{F}_{II} = \begin{pmatrix} 1 & 0 \\ 0 & 1 \end{pmatrix}^{-1} \begin{pmatrix} -1 & 0 \\ 0 & 0 \end{pmatrix} = \begin{pmatrix} -1 & 0 \\ 0 & 0 \end{pmatrix},$$

so Proposition 6.4 gives

$$\mathbf{N}_u = \boldsymbol{\sigma}_u, \quad \mathbf{N}_v = \mathbf{0}.$$

The second equation tells us nothing new, since we already knew that $\mathbf{N} = -\mathbf{b}''$ depends only on u. The first equation gives

$$\mathbf{b}''' + \mathbf{b}' = \mathbf{0}.$$

Hence, $\mathbf{b}'' + \mathbf{b}$ is a constant vector, which we can take to be zero by applying a translation to $\boldsymbol{\sigma}$ (see Eq. (17)). Then,

$$\mathbf{b}(u) = \mathbf{c}\cos u + \mathbf{d}\sin u,$$

where $\mathbf{c}$ and $\mathbf{d}$ are constant vectors, and $\mathbf{N} = -\mathbf{b}'' = \mathbf{b}$. This must be a unit vector for all values of u. It is easy to see that this is possible only if $\mathbf{c}$ and $\mathbf{d}$ are perpendicular unit vectors, in which case we can arrange that $\mathbf{c} = (1, 0, 0)$ and $\mathbf{d} = (0, 1, 0)$ by applying a rigid motion, giving $\mathbf{b}(u) = (\cos u, \sin u, 0)$. Finally, $\boldsymbol{\sigma}_u \times \boldsymbol{\sigma}_v = \lambda\mathbf{N}$ for some non-zero scalar λ, so $\mathbf{b}' \times \mathbf{a} = \lambda\mathbf{b}$. This forces $\mathbf{a} = (0, 0, \lambda)$, and the patch is given by

$$\boldsymbol{\sigma}(u, v) = (\cos u, \sin u, \lambda v),$$

a parametrisation of a circular cylinder of radius 1 (which the reader had probably guessed some time ago).

EXERCISES

10.9 A surface patch has first and second fundamental forms

$$\cos^2 v\, du^2 + dv^2 \quad \text{and} \quad -\cos^2 v\, du^2 - dv^2,$$

respectively. Show that the surface is part of a sphere of radius one. (Compute the Weingarten matrix.) Write down a parametrisation of the unit sphere with these first and second fundamental forms.

10.10 Show that there is no surface patch whose first and second fundamental forms are

$$du^2 + \cos^2 u\, dv^2 \quad \text{and} \quad \cos^2 u\, du^2 + dv^2,$$

respectively.

10.11 Suppose that the first and second fundamental forms of a surface patch are $Edu^2 + Gdv^2$ and $Ldu^2 + Ndv^2$, respectively (cf. Proposition 7.2). Show that the Codazzi–Mainardi equations reduce to

$$L_v = \frac{1}{2}E_v\left(\frac{L}{E} + \frac{N}{G}\right), \quad N_u = \frac{1}{2}G_u\left(\frac{L}{E} + \frac{N}{G}\right).$$

Deduce that the principal curvatures $\kappa_1 = L/E$ and $\kappa_2 = N/G$ satisfy the equations

$$(\kappa_1)_v = \frac{E_v}{2E}(\kappa_2 - \kappa_1), \quad (\kappa_2)_u = \frac{G_u}{2G}(\kappa_1 - \kappa_2).$$

10.4. Compact Surfaces of Constant Gaussian Curvature

We conclude this chapter with a beautiful theorem that is the analogue for surfaces of the characterisation given in Example 2.2 of circles as the plane curves with constant curvature.

Theorem 10.4

Every compact surface whose gaussian curvature is constant is a sphere.

Note that, by Proposition 7.6, the value of the constant gaussian curvature in this theorem must be > 0.

The proof of this theorem depends on the following lemma.

Lemma 10.2

Let $\boldsymbol{\sigma} : U \to \mathbf{R}^3$ be a surface patch containing a point $P = \boldsymbol{\sigma}(u_0, v_0)$ that is not an umbilic. Let $\kappa_1 \geq \kappa_2$ be the principal curvatures of $\boldsymbol{\sigma}$ and suppose that κ_1 has a local maximum at P and κ_2 has a local minimum there. Then, the gaussian curvature of $\boldsymbol{\sigma}$ at P is ≤ 0.

Proof 10.2

Since P is not an umbilic, $\kappa_1 > \kappa_2$ at P, so by shrinking U if necessary, we may assume that $\kappa_1 > \kappa_2$ everywhere.

By Proposition 7.2, we can assume that the first and second fundamental forms of $\boldsymbol{\sigma}$ are

$$Edu^2 + Gdv^2 \quad \text{and} \quad Ldu^2 + Ndv^2,$$

respectively. By Exercise 10.11,

$$E_v = -\frac{2E}{\kappa_1 - \kappa_2}(\kappa_1)_v, \quad G_u = \frac{2G}{\kappa_1 - \kappa_2}(\kappa_2)_u,$$

and by Corollary 10.2(ii), the gaussian curvature

$$K = -\frac{1}{2\sqrt{EG}}\left(\frac{\partial}{\partial u}\left(\frac{G_u}{\sqrt{EG}}\right) + \frac{\partial}{\partial v}\left(\frac{E_v}{\sqrt{EG}}\right)\right).$$

Since P is a stationary point of κ_1 and κ_2, we have $(\kappa_1)_v = (\kappa_2)_u = 0$, and hence $E_v = G_u = 0$, at P. Hence, at P,

$$K = -\frac{1}{2EG}(G_{uu} + E_{vv}) = -\frac{1}{2EG}\left(\frac{2G}{\kappa_1 - \kappa_2}(\kappa_2)_{uu} - \frac{2E}{\kappa_1 - \kappa_2}(\kappa_1)_{vv}\right)$$

(again dropping terms involving E_v, G_u and the first derivatives of κ_1 and κ_2). Since κ_1 has a local maximum at P, $(\kappa_1)_{vv} \leq 0$ there, and since κ_2 has a local minimum at P, $(\kappa_2)_{uu} \geq 0$ there. Hence, the last equation shows that $K \leq 0$ at P. □

Proof 10.4

The proof of Theorem 10.4 will use a little point set topology. We consider the continuous function on the surface $\mathcal{S}$ given by $J = (\kappa_1 - \kappa_2)^2$, where κ_1 and κ_2 are the principal curvatures. Note that this function is well defined even though κ_1 and κ_2 are not, partly because we do not know which principal curvature is to be called κ_1 and which κ_2, and partly because the sign of the principal curvatures depends on the choice of parametrisation of $\mathcal{S}$. We shall prove that this function is identically zero on $\mathcal{S}$, so that every point of $\mathcal{S}$ is an umbilic. Since the gaussian curvature $K > 0$, it follows from Proposition 6.5 that $\mathcal{S}$ is part of a sphere, say $\mathbf{S}$. In fact, $\mathcal{S}$ must be the whole of $\mathbf{S}$. For, any point P of

$\mathcal{S}$ is contained in a patch $\boldsymbol{\sigma} : U \to \mathbf{R}^3$ of $\mathcal{S}$, and $\boldsymbol{\sigma}(U) = \mathcal{S} \cap W$, where W is an open subset of $\mathbf{R}^3$; it follows that $\mathcal{S}$ is an *open* subset of $\mathbf{S}$. On the other hand, since $\mathcal{S}$ is compact, it is necessarily a closed subset of $\mathbf{R}^3$, and hence a *closed* subset of $\mathbf{S}$. But since $\mathbf{S}$ is connected, the only non-empty subset of $\mathbf{S}$ that is both open and closed is $\mathbf{S}$ itself.

Suppose then, to get a contradiction, that J is not identically zero on $\mathcal{S}$. Since $\mathcal{S}$ is compact, J must attain its maximum value at some point P of $\mathcal{S}$, and this maximum value is > 0. Choose a patch $\boldsymbol{\sigma} : U \to \mathbf{R}^3$ of $\mathcal{S}$ containing P, and let κ_1 and κ_2 be its principal curvatures. Since $\kappa_1\kappa_2 > 0$, by reparametrising if necessary, we can assume that κ_1 and κ_2 are both > 0 (see Exercise 6.17). Suppose that $\kappa_1 > \kappa_2$ at P; then by shrinking U if necessary, we can assume that $\kappa_1 > \kappa_2$ everywhere on U. Since K is a constant > 0, the function $\left(x - \frac{K}{x}\right)^2$ increases with x provided that $x > K/x > 0$. Since $\kappa_1 > K/\kappa_1 = \kappa_2 > 0$, this function is increasing at $x = \kappa_1$, so κ_1 must have a local maximum at P, and then $\kappa_2 = K/\kappa_1$ must have a local minimum there. By Lemma 10.2, $K \leq 0$ at P. This contradicts the assumption that $K > 0$. □

EXERCISES

10.12 Show that a compact surface with gaussian curvature > 0 everywhere and constant *mean* curvature is a sphere. (As in the proof of Theorem 10.4, if κ_1 has a local maximum at a point P of the surface, then $\kappa_2 = 2H - \kappa_1$ has a local minimum there.)

11
The Gauss–Bonnet Theorem

The Gauss–Bonnet theorem is the most beautiful and profound result in the theory of surfaces. Its most important version relates the average over a surface of its gaussian curvature to a property of the surface called its 'Euler number' which is 'topological', i.e. it is unchanged by any continuous deformation of the surface. Such deformations will in general change the value of the gaussian curvature, but the theorem says that its average over the surface does *not* change. The real importance of the Gauss–Bonnet theorem is as a prototype of analogous results which apply in higher dimensional situations, and which relate *geometrical* properties to *topological* ones. The study of such relations is one of the most important themes of 20th century Mathematics.

11.1. Gauss–Bonnet for Simple Closed Curves

The simplest version of the Gauss–Bonnet Theorem involves *simple closed curves* on a surface. In the special case when the surface is a plane, these curves have been discussed in Section 3.1. For a general surface, we make

Definition 11.1

A curve $\boldsymbol{\gamma}(t) = \boldsymbol{\sigma}(u(t), v(t))$ on a surface patch $\boldsymbol{\sigma} : U \to \mathbf{R}^3$ is called a *simple closed curve* with *period* a if $\boldsymbol{\pi}(t) = (u(t), v(t))$ is a simple closed curve in $\mathbf{R}^2$ with period a such that the region $\mathrm{int}(\boldsymbol{\pi})$ of $\mathbf{R}^2$ enclosed by $\boldsymbol{\pi}$ is entirely contained in U (see the diagrams below). The curve $\boldsymbol{\gamma}$ is said to be *positively-oriented* if $\boldsymbol{\pi}$ is positively-oriented. Finally, the image of $\mathrm{int}(\boldsymbol{\pi})$ under the map $\boldsymbol{\sigma}$ is defined to be the *interior* $\mathrm{int}(\boldsymbol{\gamma})$ of $\boldsymbol{\gamma}$.

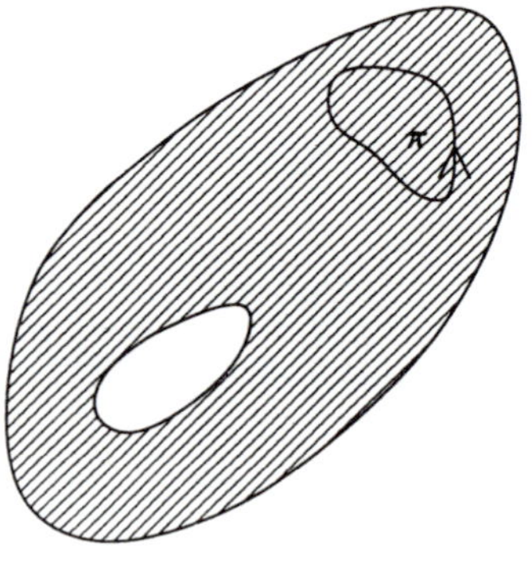

allowed

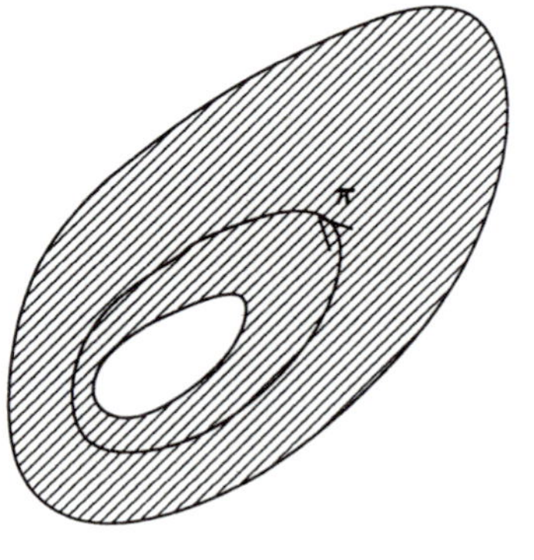

not allowed

We can now state the first version of the Gauss–Bonnet Theorem.

Theorem 11.1

Let $\boldsymbol{\gamma}(s)$ be a unit-speed simple closed curve on a surface $\boldsymbol{\sigma}$ of length $\ell(\boldsymbol{\gamma})$, and assume that $\boldsymbol{\gamma}$ is positively-oriented. Then,

$$\int_0^{\ell(\boldsymbol{\gamma})} \kappa_g ds = 2\pi - \iint_{\mathrm{int}(\boldsymbol{\gamma})} K d\mathcal{A}_{\boldsymbol{\sigma}},$$

where κ_g is the geodesic curvature of $\boldsymbol{\gamma}$, K is the gaussian curvature of $\boldsymbol{\sigma}$ and $d\mathcal{A}_{\boldsymbol{\sigma}} = (EG - F^2)^{1/2} dudv$ is the area element on $\boldsymbol{\sigma}$ (see Section 5.4).

We use s to denote the parameter of $\boldsymbol{\gamma}$ to emphasize that $\boldsymbol{\gamma}$ is unit-speed.

Proof 11.1

As in the proof of Theorem 10.1, choose a smooth orthonormal basis $\{\mathbf{e}', \mathbf{e}''\}$ of the tangent plane of $\boldsymbol{\sigma}$ at each point such that $\{\mathbf{e}', \mathbf{e}'', \mathbf{N}\}$ is a right-handed orthonormal basis of $\mathbf{R}^3$, where $\mathbf{N}$ is the unit normal to $\boldsymbol{\sigma}$. Consider the following integral:

$$\begin{aligned}\mathcal{I} &= \int_0^{\ell(\boldsymbol{\gamma})} \mathbf{e}'.\dot{\mathbf{e}}''ds \\ &= \int_0^{\ell(\boldsymbol{\gamma})} \mathbf{e}'.(\mathbf{e}''_u\dot{u} + \mathbf{e}''_v\dot{v})ds \\ &= \int_{\boldsymbol{\pi}} (\mathbf{e}'.\mathbf{e}''_u)du + (\mathbf{e}'.\mathbf{e}''_v)dv.\end{aligned}$$

By Green's theorem (see Section 3.1), this can be rewritten as a double integral:

$$\begin{aligned}\mathcal{I} &= \iint_{\text{int}(\boldsymbol{\pi})} \{(\mathbf{e}'.\mathbf{e}''_v)_u - (\mathbf{e}'.\mathbf{e}''_u)_v\}dudv \\ &= \iint_{\text{int}(\boldsymbol{\pi})} \{(\mathbf{e}'_u.\mathbf{e}''_v) - (\mathbf{e}'_v.\mathbf{e}''_u)\}dudv \\ &= \iint_{\text{int}(\boldsymbol{\pi})} \frac{LN - M^2}{(EG - F^2)^{1/2}}dudv \quad \text{(by Lemma 10.1)} \\ &= \iint_{\text{int}(\boldsymbol{\pi})} \frac{LN - M^2}{EG - F^2}(EG - F^2)^{1/2}dudv \\ &= \iint_{\text{int}(\boldsymbol{\pi})} K d\mathcal{A}_{\boldsymbol{\sigma}}. \qquad (1)\end{aligned}$$

Now let $\theta(s)$ be the angle between the unit tangent vector $\dot{\boldsymbol{\gamma}}$ of $\boldsymbol{\gamma}$ at $\boldsymbol{\gamma}(s)$ and the unit vector $\mathbf{e}'$ at the same point. More precisely, θ is the angle, uniquely determined up to a multiple of 2π, such that

$$\dot{\boldsymbol{\gamma}} = \cos\theta\mathbf{e}' + \sin\theta\mathbf{e}''. \qquad (2)$$

Then,

$$\mathbf{N} \times \dot{\boldsymbol{\gamma}} = -\sin\theta\mathbf{e}' + \cos\theta\mathbf{e}''. \qquad (3)$$

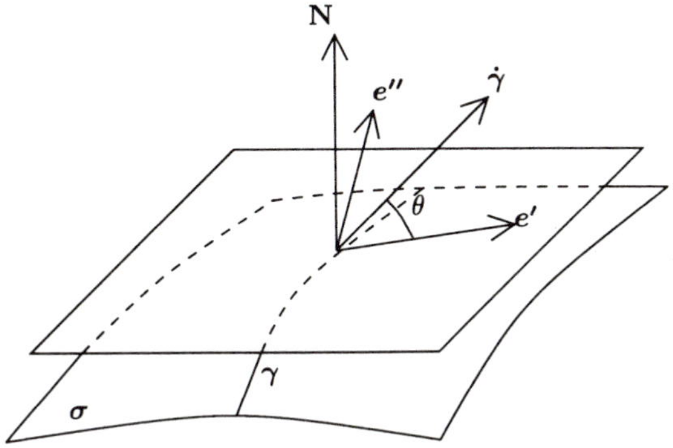

Now, by Eq. (2),

$$\ddot{\boldsymbol{\gamma}} = \cos\theta\dot{\mathbf{e}}' + \sin\theta\dot{\mathbf{e}}'' + \dot{\theta}(-\sin\theta\mathbf{e}' + \cos\theta\mathbf{e}''), \tag{4}$$

so by Eqs. (3) and (4) the geodesic curvature of $\boldsymbol{\gamma}$ is

$$\begin{aligned}\kappa_g &= (\mathbf{N} \times \dot{\boldsymbol{\gamma}}).\ddot{\boldsymbol{\gamma}} \quad \text{(see Section 6.2)}\\ &= \dot{\theta}(-\sin\theta\mathbf{e}' + \cos\theta\mathbf{e}'').(-\sin\theta\mathbf{e}' + \cos\theta\mathbf{e}'')\\ &\qquad + (-\sin\theta\mathbf{e}' + \cos\theta\mathbf{e}'').(\cos\theta\dot{\mathbf{e}}' + \sin\theta\dot{\mathbf{e}}'')\\ &= \dot{\theta} + \cos^2\theta(\dot{\mathbf{e}}'.\mathbf{e}'') - \sin^2\theta(\dot{\mathbf{e}}''.\mathbf{e}')\\ &\qquad + \sin\theta\cos\theta(\dot{\mathbf{e}}''.\mathbf{e}'' - \dot{\mathbf{e}}'.\mathbf{e}') \quad \text{(by Eqs. (2) and (3))}.\end{aligned}$$

Since $\mathbf{e}'$ and $\mathbf{e}''$ are perpendicular unit vectors,

$$\mathbf{e}'.\dot{\mathbf{e}}' = \mathbf{e}''.\dot{\mathbf{e}}'' = 0, \quad \dot{\mathbf{e}}'.\mathbf{e}'' = -\mathbf{e}'.\dot{\mathbf{e}}''.$$

Hence,

$$\kappa_g = \dot{\theta} - \mathbf{e}'.\dot{\mathbf{e}}'',$$

and by the definition of $\mathcal{I}$,

$$\mathcal{I} = \int_0^{\ell(\boldsymbol{\gamma})} (\dot{\theta} - \kappa_g)ds.$$

Thus, to complete the proof of Theorem 11.1, we must show that

$$\int_0^{\ell(\boldsymbol{\gamma})} \dot{\theta}\, ds = 2\pi. \tag{5}$$

Equation (5) is called 'Hopf's Umlaufsatz' – literally 'rotation theorem' in German. We cannot give a fully satisfactory proof of it here because the proof would take us too far into the realm of topology. Instead, we shall justify Eq. (5) by means of the following heuristic argument.

The main observation is that, if $\tilde{\boldsymbol{\gamma}}$ is any other simple closed curve contained in the interior of $\boldsymbol{\gamma}$, there is a smooth family of simple closed curves $\boldsymbol{\gamma}^\tau$, defined for $0 \leq \tau \leq 1$, say, with $\boldsymbol{\gamma}^0 = \boldsymbol{\gamma}$ and $\boldsymbol{\gamma}^1 = \tilde{\boldsymbol{\gamma}}$ (see Section 8.4 for the notion of a smooth family of curves). The existence of such a family is supposed to be 'intuitively obvious'.

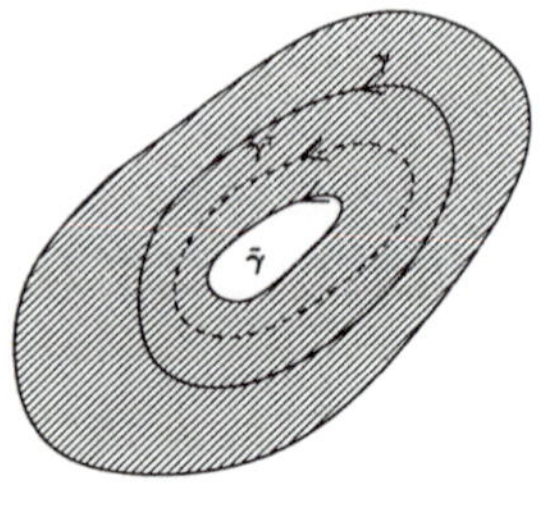

Note, however, that it is crucial that the interior of $\boldsymbol{\pi}$ is entirely contained in U, otherwise such a family will not exist, in general:

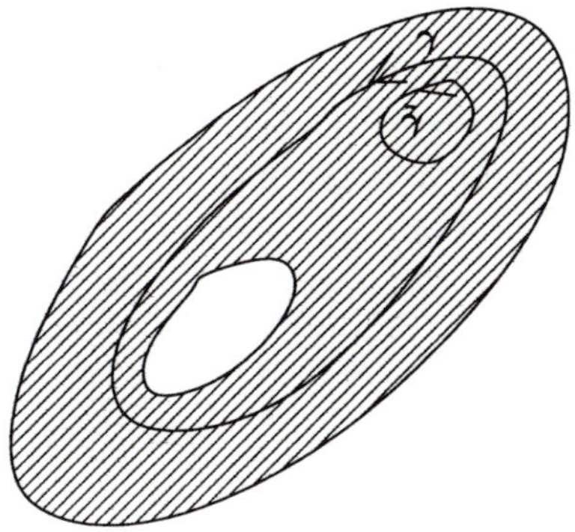

Observe next that the integral $\int_0^{\ell(\boldsymbol{\gamma}^\tau)} \dot{\theta}\, ds$ should depend continuously on τ. Further, since $\boldsymbol{\gamma}^\tau$ and $\mathbf{e}'$ return to their original values as one goes once round $\boldsymbol{\gamma}^\tau$, the integral is always an integer multiple of 2π. These two facts imply that the integral must be independent of τ – for by the Intermediate Value Theorem a continuous variable cannot change from one integer to a different integer without passing through some non-integer value. To compute $\int_0^{\ell(\boldsymbol{\gamma})} \dot{\theta}\, ds$, we can therefore replace $\boldsymbol{\gamma}$ by any other simple closed curve $\tilde{\boldsymbol{\gamma}}$ in the interior of $\boldsymbol{\gamma}$, since this will not change the value of the integral. We take $\tilde{\boldsymbol{\gamma}}$ to be the image under $\boldsymbol{\sigma}$ of a small circle in the interior of $\boldsymbol{\pi}$. It is 'intutively clear' that

$$\int_0^{\ell(\tilde{\boldsymbol{\gamma}})} \dot{\theta}\, ds = 2\pi,$$

because

(i) $\mathbf{e}'$ is essentially constant at all points of $\tilde{\boldsymbol{\gamma}}$ (because the circle is very small), and

(ii) the tangent vector to $\tilde{\boldsymbol{\gamma}}$ rotates by 2π on going once round $\tilde{\boldsymbol{\gamma}}$ because the interior of $\tilde{\boldsymbol{\gamma}}$ can be considered to be essentially part of a plane, and it is 'intutively clear' that the tangent vector of a simple closed curve in the plane rotates by 2π on going once round the curve.

This completes the 'proof' of Hopf's Umlaufsatz, and hence that of Theorem 11.1.

EXERCISES

11.1 A surface patch $\boldsymbol{\sigma}$ has gaussian curvature ≤ 0 everywhere. Prove that there are no simple closed geodesics in $\boldsymbol{\sigma}$. How do you reconcile this with the fact that the parallels of a circular cylinder are geodesics?

11.2 Let $\boldsymbol{\gamma}(s)$ be a simple closed curve in $\mathbf{R}^2$, parametrised by arc-length and of total length $\ell(\boldsymbol{\gamma})$. Deduce from Hopf's Umlaufsatz that, if $\kappa_s(s)$ is the signed curvature of $\boldsymbol{\gamma}$, then

$$\int_0^{\ell(\boldsymbol{\gamma})} \kappa_s(s)\, ds = 2\pi.$$

(Use Proposition 2.2.)

11.2. Gauss–Bonnet for Curvilinear Polygons

For the next version of Gauss–Bonnet, we shall have to generalise our notion of a curve by allowing the possibility of 'corners'. More precisely, we make the following definition.

Definition 11.2

A *curvilinear polygon* in $\mathbf{R}^2$ is a continuous map $\boldsymbol{\pi} : \mathbf{R} \to \mathbf{R}^2$ such that, for some real number a and some points $0 = t_0 < t_1 < \cdots < t_n = a$,

(i) $\boldsymbol{\pi}(t) = \boldsymbol{\pi}(t')$ if and only if $t' - t$ is an integer multiple of a;

(ii) $\boldsymbol{\pi}$ is smooth on each of the open intervals $(t_0, t_1), (t_1, t_2), \ldots, (t_{n-1}, t_n)$;

(iii) the one-sided derivatives

$$\dot{\boldsymbol{\pi}}^-(t_i) = \lim_{t \uparrow t_i} \frac{\boldsymbol{\pi}(t) - \boldsymbol{\pi}(t_i)}{t - t_i}, \quad \dot{\boldsymbol{\pi}}^+(t_i) = \lim_{t \downarrow t_i} \frac{\boldsymbol{\pi}(t) - \boldsymbol{\pi}(t_i)}{t - t_i} \tag{6}$$

exist for $i = 1, \ldots, n$ and are non-zero and not parallel.

The points $\boldsymbol{\gamma}(t_i)$ for $i = 1, \ldots, n$ are called the *vertices* of the curvilinear polygon $\boldsymbol{\pi}$, and the segments of it corresponding to the open intervals (t_{i-1}, t_i) are called its *edges*.

It makes sense to say that a curvilinear polygon $\boldsymbol{\pi}$ is positively-oriented: for all t such that $\boldsymbol{\pi}(t)$ is not a vertex, the vector $\mathbf{n}_s$ obtained by rotating $\dot{\boldsymbol{\pi}}$ anticlockwise by $\pi/2$ should point into $\mathrm{int}(\boldsymbol{\pi})$. (The region $\mathrm{int}(\boldsymbol{\pi})$ enclosed by $\boldsymbol{\pi}$ makes sense because the Jordan Curve Theorem applies to curvilinear polygons in the plane.)

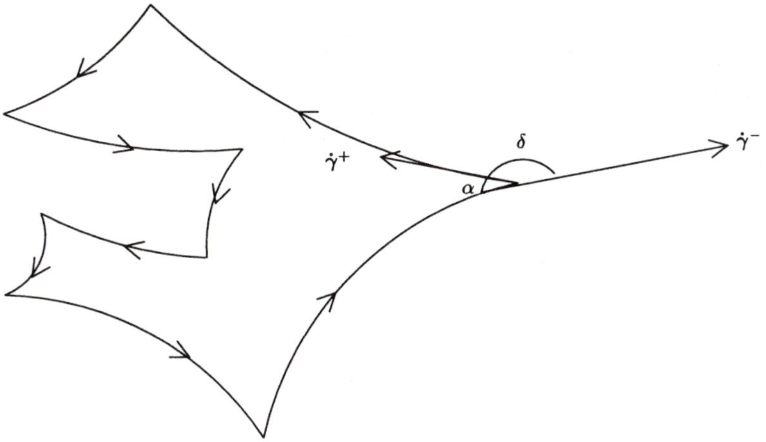

Now let $\boldsymbol{\sigma} : U \to \mathbf{R}^3$ be a surface patch and let $\boldsymbol{\pi} : \mathbf{R} \to U$ be a curvilinear polygon in U, as in Definition 11.2. Then, $\boldsymbol{\gamma} = \boldsymbol{\sigma} \circ \boldsymbol{\pi}$ is called a curvilinear polygon on the surface patch $\boldsymbol{\sigma}$, int($\boldsymbol{\gamma}$) is the image under $\boldsymbol{\sigma}$ of int($\boldsymbol{\pi}$), the vertices of $\boldsymbol{\gamma}$ are the points $\boldsymbol{\gamma}(t_i)$ for $i = 1, \ldots, n$, and the edges of $\boldsymbol{\sigma}$ are the segments of it corresponding to the open intervals (t_{i-1}, t_i). Since $\boldsymbol{\sigma}$ is allowable, the one-sided derivatives

$$\dot{\boldsymbol{\gamma}}^-(t_i) = \lim_{t \uparrow t_i} \frac{\boldsymbol{\gamma}(t) - \boldsymbol{\gamma}(t_i)}{t - t_i}, \quad \dot{\boldsymbol{\gamma}}^+(t_i) = \lim_{t \downarrow t_i} \frac{\boldsymbol{\gamma}(t) - \boldsymbol{\gamma}(t_i)}{t - t_i}$$

exist and are not parallel.

Let $\theta_i^{\pm}$ be the angles between $\dot{\boldsymbol{\gamma}}^{\pm}(t_i)$ and $\mathbf{e}'$, defined as in Eq. (2), let $\delta_i = \theta_i^+ - \theta_i^-$ be the external angle at the vertex $\boldsymbol{\gamma}(t_i)$, and let $\alpha_i = \pi - \delta_i$ be the internal angle. Since the tangent vectors $\dot{\boldsymbol{\gamma}}^+(t_i)$ and $\dot{\boldsymbol{\gamma}}^-(t_i)$ are not parallel, the angle δ_i is not a multiple of π. Note that all of these angles are well defined only up to multiples of 2π. We assume from now on that $0 < \alpha_i < 2\pi$ for $i = 1, \ldots, n$.

A curvilinear polygon $\boldsymbol{\gamma}$ is said to be unit-speed if $\| \dot{\boldsymbol{\gamma}} \| = 1$ whenever $\dot{\boldsymbol{\gamma}}$ is defined, i.e. for all t such that $\boldsymbol{\gamma}(t)$ is not a vertex of $\boldsymbol{\gamma}$. We denote the parameter of $\boldsymbol{\gamma}$ by s if $\boldsymbol{\gamma}$ is unit-speed. The period of $\boldsymbol{\gamma}$ is then equal to its length $\ell(\boldsymbol{\gamma})$, which is the sum of the lengths of the edges of $\boldsymbol{\gamma}$.

Theorem 11.2

Let $\boldsymbol{\gamma}$ be a positively-oriented unit-speed curvilinear polygon with n edges on a surface $\boldsymbol{\sigma}$, and let $\alpha_1, \alpha_2, \ldots, \alpha_n$ be the interior angles at its vertices. Then,

$$\int_0^{\ell(\boldsymbol{\gamma})} \kappa_g ds = \sum_{i=1}^{n} \alpha_i - (n-2)\pi - \iint_{\text{int}(\boldsymbol{\gamma})} K dA_{\boldsymbol{\sigma}}.$$

Proof 11.2

Exactly the same argument as in the proof of Theorem 11.1 shows that

$$\int_0^{\ell(\boldsymbol{\gamma})} \kappa_g ds = \int_0^{\ell(\boldsymbol{\gamma})} \dot{\theta} ds - \iint_{\text{int}(\boldsymbol{\gamma})} K d\mathcal{A}_{\boldsymbol{\sigma}}.$$

We shall prove that

$$\int_0^{\ell(\boldsymbol{\gamma})} \dot{\theta} ds = 2\pi - \sum_{i=1}^{n} \delta_i. \tag{7}$$

Assuming this, we get

$$\begin{aligned}\int_0^{\ell(\boldsymbol{\gamma})} \kappa_g ds &= 2\pi - \sum_{i=1}^{n} \delta_i - \iint_{\text{int}(\boldsymbol{\gamma})} K d\mathcal{A}_{\boldsymbol{\sigma}} \\ &= 2\pi - \sum_{i=1}^{n} (\pi - \alpha_i) - \iint_{\text{int}(\boldsymbol{\gamma})} K d\mathcal{A}_{\boldsymbol{\sigma}} \\ &= \sum_{i=1}^{n} \alpha_i - (n-2)\pi - \iint_{\text{int}(\boldsymbol{\gamma})} K d\mathcal{A}_{\boldsymbol{\sigma}}.\end{aligned}$$

To establish Eq. (7), we imagine 'smoothing' each vertex of $\boldsymbol{\gamma}$ as shown in the following diagram.

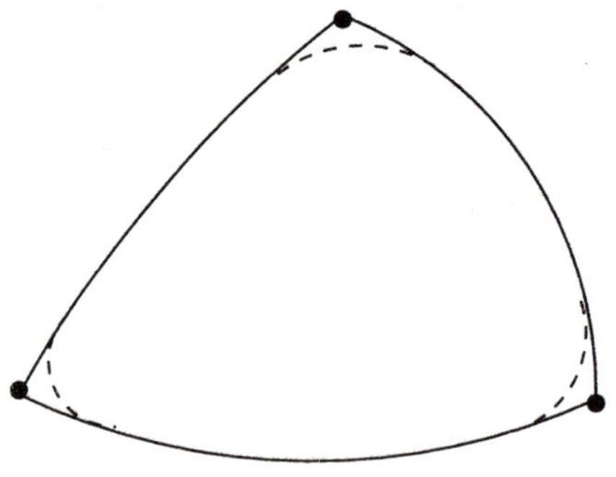

If the 'smoothed' curve $\tilde{\boldsymbol{\gamma}}$ is smooth (!), then, in an obvious notation,

$$\int_0^{\ell(\tilde{\boldsymbol{\gamma}})} \dot{\tilde{\theta}} ds = 2\pi. \tag{8}$$

Since $\boldsymbol{\gamma}$ and $\tilde{\boldsymbol{\gamma}}$ are the same except near the vertices of $\boldsymbol{\gamma}$, the difference

$$\int_0^{\ell(\tilde{\boldsymbol{\gamma}})} \dot{\tilde{\theta}} ds - \int_0^{\ell(\boldsymbol{\gamma})} \dot{\theta} ds \tag{9}$$

is a sum of n contributions, one from near each vertex. Near $\boldsymbol{\gamma}(s_i)$, the picture is

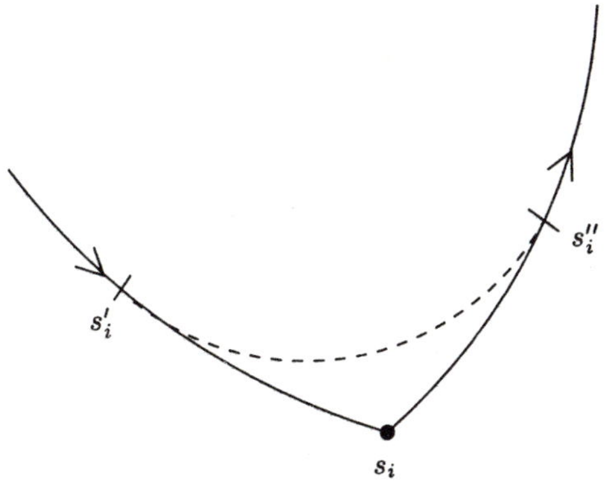

i.e. $\boldsymbol{\gamma}$ and $\tilde{\boldsymbol{\gamma}}$ agree except when s belongs to a small interval (s_i', s_i''), say, containing s_i, so the contribution from the ith vertex is

$$\int_{s_i'}^{s_i''} \dot{\tilde{\theta}} ds - \int_{s_i'}^{s_i} \dot{\theta} ds - \int_{s_i}^{s_i''} \dot{\theta} ds.$$

The first integral is the angle between $\dot{\tilde{\boldsymbol{\gamma}}}(s_i'')$ and $\dot{\tilde{\boldsymbol{\gamma}}}(s_i')$, which as s_i' and s_i'' tend to s_i becomes the angle between $\dot{\boldsymbol{\gamma}}^+(s_i)$ and $\dot{\boldsymbol{\gamma}}^-(s_i)$, i.e. δ_i. On the other hand, since $\boldsymbol{\gamma}(s)$ is smooth on each of the intervals (s_i', s_i) and (s_i, s_i''), the last two integrals go to zero as s_i' and s_i'' tend to s_i. Thus, the contribution to the expression (9) from the ith vertex tends to δ_i as s_i' and s_i'' tend to s_i . Summing over all the vertices, we get

$$\int_0^{\ell(\tilde{\boldsymbol{\gamma}})} \dot{\tilde{\theta}} ds - \int_0^{\ell(\boldsymbol{\gamma})} \dot{\theta} ds = \sum_{i=1}^{n} \delta_i.$$

Equation (7) now follows from this and Eq. (8). □

Corollary 11.1

If $\boldsymbol{\gamma}$ is a curvilinear polygon with n edges each of which is an arc of a geodesic, then the internal angles $\alpha_1, \alpha_2, \ldots, \alpha_n$ of the polygon satisfy the equation

$$\sum_{i=1}^{n} \alpha_i = (n-2)\pi + \iint_{\text{int}(\boldsymbol{\gamma})} K dA_{\boldsymbol{\sigma}}.$$

Proof 11.1

This is immediate from Theorem 11.2, since $\kappa_g = 0$ along a geodesic. □

As a special case of Corollary 11.1, consider an n-gon in the plane with straight edges. Since $K = 0$ for the plane, Corollary 11.1 gives

$$\sum_{i=1}^{n} \alpha_i = (n-2)\pi,$$

a well known result of elementary geometry.

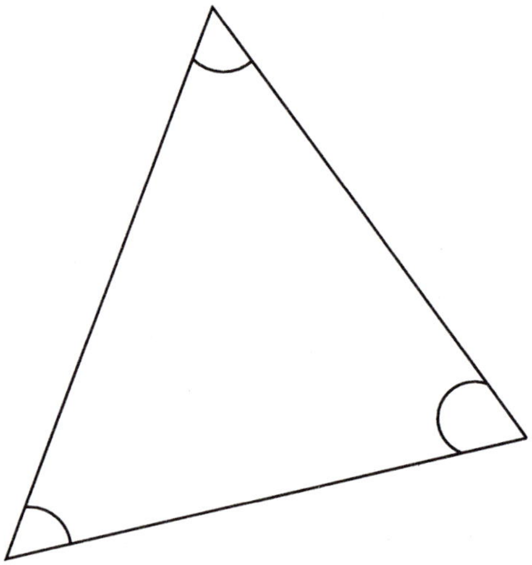

For a curvilinear n-gon on the unit sphere whose sides are arcs of great circles, we have $K = 1$ so $\sum \alpha_i$ exceeds the plane value $(n-2)\pi$ by the area $\iint d\mathcal{A}_{\boldsymbol{\sigma}}$ of the polygon. Taking $n = 3$, we get for a spherical triangle ABC whose edges are arcs of great circles,

$$\mathcal{A}(ABC) = \angle A + \angle B + \angle C - \pi.$$

This is just Theorem 5.5, which is therefore a special case of Gauss–Bonnet.

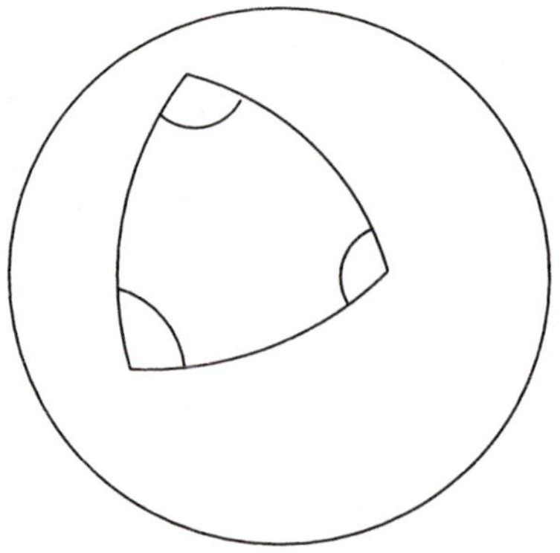

Finally, for a geodesic n-gon on the pseudosphere (see Section 7.2), for which $K = -1$, we see that $\sum \alpha_i$ is *less* than $(n-2)\pi$ by the area of the polygon. In particular, for a geodesic triangle ABC on the pseudosphere,

$$\mathcal{A}(ABC) = \pi - \angle A - \angle B - \angle C.$$

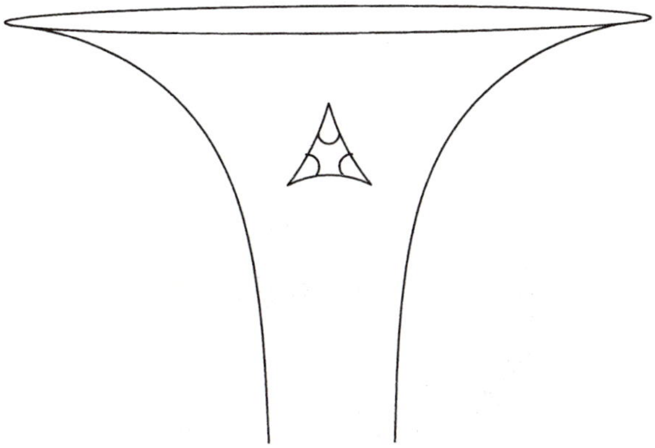

EXERCISES

11.3 Suppose that the gaussian curvature K of a surface patch $\boldsymbol{\sigma}$ satisfies $K \leq -1$ everywhere and that $\boldsymbol{\gamma}$ is a curvilinear n-gon on $\boldsymbol{\sigma}$ whose sides are geodesics. Show that $n \geq 3$, and that, if $n = 3$, the area enclosed by $\boldsymbol{\gamma}$ cannot exceed π.

11.4 Consider the surface of revolution

$$\boldsymbol{\sigma}(u, v) = (f(u)\cos v, f(u)\sin v, g(u)),$$

where $\boldsymbol{\gamma}(u) = (f(u), 0, g(u))$ is a unit-speed curve in the xz-plane. Let $u_1 < u_2$ be constants, let $\boldsymbol{\gamma}_1$ and $\boldsymbol{\gamma}_2$ be the two parallels $u = u_1$ and $u = u_2$ on $\boldsymbol{\sigma}$, and let R be the region of the uv-plane given by

$$u_1 \leq u \leq u_2, \quad 0 < v < 2\pi.$$

Compute

$$\int_0^{\ell(\boldsymbol{\gamma}_1)} \kappa_g ds, \quad \int_0^{\ell(\boldsymbol{\gamma}_2)} \kappa_g ds \quad \text{and} \quad \iint_R K d\mathcal{A}_{\boldsymbol{\sigma}},$$

and explain your result on the basis of the Gauss–Bonnet theorem.

11.3. Gauss–Bonnet for Compact Surfaces

The most important version of the Gauss–Bonnet Theorem applies to a *compact* surface $\mathcal{S}$. It is a surprising result that there are very few compact surfaces in $\mathbf{R}^3$ up to diffeomorphism, and they can all be described explicitly. The simplest example is, of course, the sphere. The next simplest is the *torus*, which can be obtained by rotating around the z-axis a circle in the xz-plane which does not intersect the z-axis:

One can also join such tori together:

This surface is denoted by T_g, where g is the number of holes, called the *genus* of the surface (we take $g = 0$ for the sphere). We accept the following theorem without proof:

Theorem 11.3

For any integer $g \geq 0$, T_g can be given an atlas making it a smooth surface. Moreover, every compact surface is diffeomorphic to one of the T_g.

The version of the Gauss–Bonnet theorem we are aiming for is obtained by covering a compact surface $\mathcal{S}$ with curvilinear polygons that fit together nicely, applying Theorem 11.2 to each one, and adding up the results. We begin to make this precise with

Definition 11.3

Let $\mathcal{S}$ be a surface, with atlas consisting of the patches $\boldsymbol{\sigma}_i : U_i \to \mathbf{R}^3$. A *triangulation* of $\mathcal{S}$ is a collection of curvilinear polygons, (the interior of) each of which is contained in one of the $\boldsymbol{\sigma}_i(U_i)$, such that

(i) every point of $\mathcal{S}$ is in at least one of the curvilinear polygons;
(ii) two curvilinear polygons are either disjoint, or their intersection is a common edge or a common vertex;
(iii) each edge is an edge of exactly two polygons.

Thus, situations like

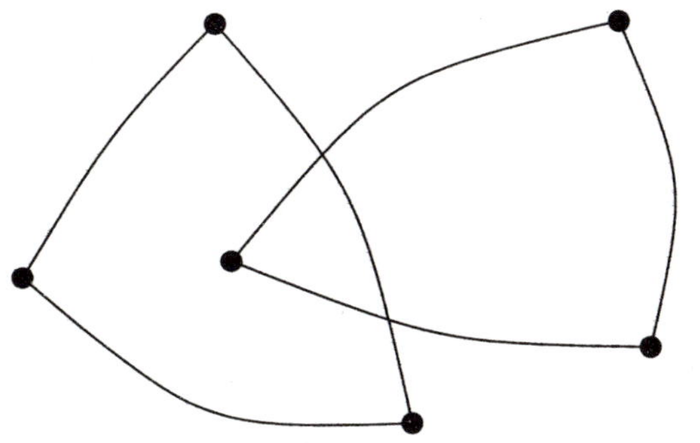

are not allowed. A triangulation of the unit sphere with eight polygons is obtained by intersecting the sphere with the three coordinate planes:

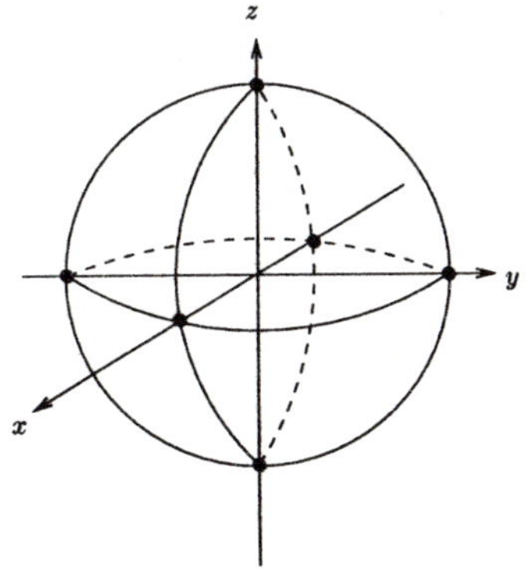

We state without proof:

Theorem 11.4

Every compact surface has a triangulation with finitely many polygons.

We introduce the following number associated to any triangulation:

Definition 11.4

The *Euler number* χ of a triangulation of a compact surface S is

$$\chi = V - E + F,$$

where

$$\begin{aligned} V &= \text{ the total number of vertices of the triangulation,} \\ E &= \text{ the total number of edges of the triangulation,} \\ F &= \text{ the total number of polygons of the triangulation.} \end{aligned}$$

For the triangulation of the sphere given above, $V = 6$, $E = 12$ and $F = 8$, so $\chi = 6 - 12 + 8 = 2$. The importance of the Euler number is that, although different triangulations of a given surface will in general have different numbers of vertices, edges and polygons, χ is actually *independent* of the triangulation and depends only on the surface. For example, we can get another triangulation of the sphere by 'inflating' a regular tetrahedron:

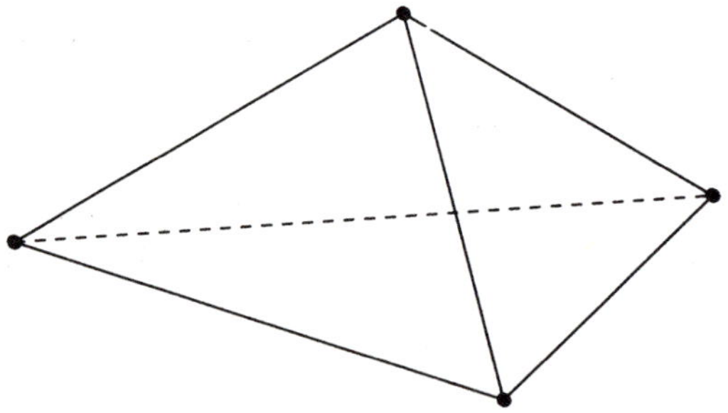

This time, $V = 4$, $E = 6$ and $F = 4$, so $\chi = 4 - 6 + 4 = 2$, the same as before. This property of χ is a consequence of the following theorem, which is the desired extension of Theorem 11.2 to compact surfaces:

Theorem 11.5

Let S be a compact surface. Then, for any triangulation of S,

$$\int\int_S K dA = 2\pi\chi,$$

where χ is the Euler number of the triangulation.

We need to explain what is meant by the left-hand side of the equation in Theorem 11.5. Fix a triangulation of $\mathcal{S}$ with polygons P_i, say. Each P_i is contained in the image of some patch $\boldsymbol{\sigma}_i : U_i \to \mathbf{R}^3$ in the atlas of $\mathcal{S}$, say $P_i = \boldsymbol{\sigma}_i(R_i)$, where $R_i \subseteq U_i$. Then, by definition,

$$\int\!\!\int_{\mathcal{S}} K dA = \sum_i \int\!\!\int_{R_i} K dA_{\boldsymbol{\sigma}_i},$$

where K is the gaussian curvature of $\boldsymbol{\sigma}_i$. Unfortunately, we have to show that this is a good definition, i.e. that it does not depend on our choice of patches $\boldsymbol{\sigma}_i$ (since, even for a given triangulation, P_i may be contained in more than one patch), nor does it depend on the triangulation itself.

To see this, we note first that, if $\tilde{\boldsymbol{\sigma}}_i : \tilde{U}_i \to \mathbf{R}^3$ is a reparametrisation of $\boldsymbol{\sigma}_i$ and if $P_i = \tilde{\boldsymbol{\sigma}}_i(\tilde{R}_i)$, where $\tilde{R}_i \subseteq \tilde{U}_i$, then

$$\int\!\!\int_{R_i} K dA_{\boldsymbol{\sigma}_i} = \int\!\!\int_{\tilde{R}_i} K dA_{\tilde{\boldsymbol{\sigma}}_i}$$

because both the area element $dA_{\boldsymbol{\sigma}}$ and the gaussian curvature K are unchanged by reparametrisation (see Proposition 5.3 and Exercise 6.17).

Next, if $\{P_i\}$ and $\{P'_j\}$ are two triangulations of $\mathcal{S}$, it is intuitively clear that we can find a third triangulation $\{P''_k\}$ of $\mathcal{S}$ such that each P_i is the union of some of the P''_k, as is each P'_j. For example, if some P_i and some P'_j overlap as follows,

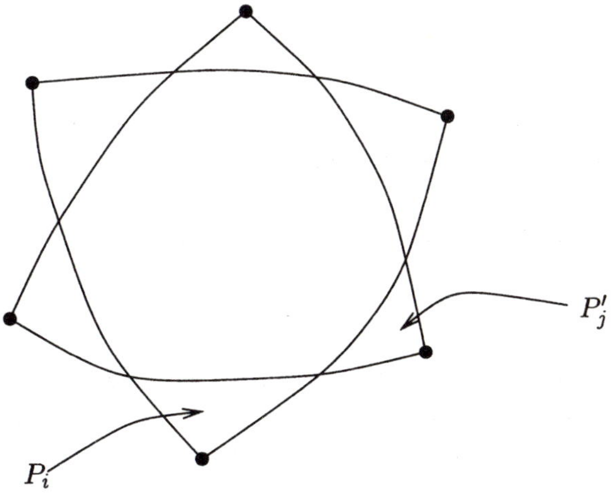

then by inserting additional vertices we can create the appropriate polygons P''_k:

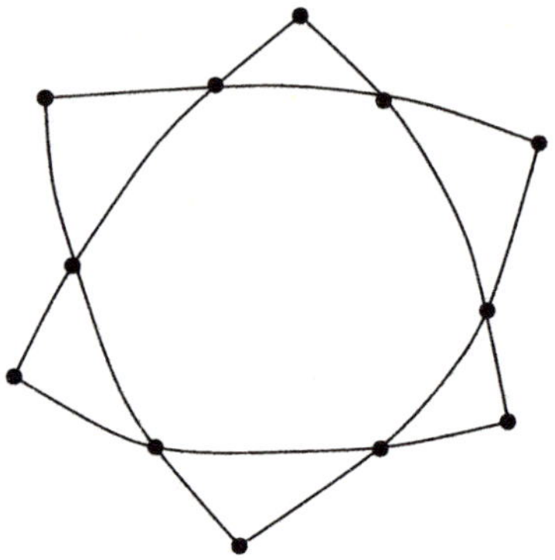

It is then clear that (in an obvious notation)

$$\sum_i \iint_{R_i} K dA_{\boldsymbol{\sigma}_i} = \sum_j \iint_{R'_j} K dA_{\boldsymbol{\sigma}'_j},$$

since both sides are equal to

$$\sum_k \iint_{R''_k} K dA_{\boldsymbol{\sigma}''_k}.$$

This is just because the integral of K over the union of finitely many polygons, all contained in a single surface patch and any two of which are either disjoint or intersect only in a common edge or vertex, is simply the sum of the integrals of K over each polygon.

Together with the fact that $\iint_{\mathcal{S}} K dA$ is independent of the triangulation, Theorem 11.5 thus implies

Corollary 11.2

The Euler number χ of a triangulation of a compact surface $\mathcal{S}$ depends only on $\mathcal{S}$ and not on the choice of triangulation.

We now give the proof of Theorem 11.5.

Proof 11.5

As above, we fix a triangulation of $\mathcal{S}$ with polygons P_i, say, each of which is contained in the image of some patch $\boldsymbol{\sigma}_i : U_i \to \mathbf{R}^3$ in the atlas of $\mathcal{S}$, say $P_i = \boldsymbol{\sigma}_i(R_i)$, where $R_i \subseteq U_i$. By Theorem 11.2,

$$\iint_{R_i} K dA_{\boldsymbol{\sigma}_i} = \angle_i - (n_i - 2)\pi + \int_0^{\ell(\boldsymbol{\gamma}_i)} \kappa_g \, ds, \qquad (10)$$

where n_i is the number of vertices of P_i, $\boldsymbol{\gamma}_i$ is the curvilinear polygon that forms the boundary of P_i, $\ell(\boldsymbol{\gamma}_i)$ is its length, and $\angle_i$ is the sum of its interior angles. We must therefore sum the contributions of each of the three terms on the right-hand side of Eq. (10) over all the polygons P_i in the triangulation.

First, $\sum_i \angle_i$ is the sum of all the internal angles of all the polygons. At each vertex, several polygons meet, but the sum of the angles at the vertex is obviously 2π, so

$$\sum_i \angle_i = 2\pi V, \tag{11}$$

where V is the total number of vertices.

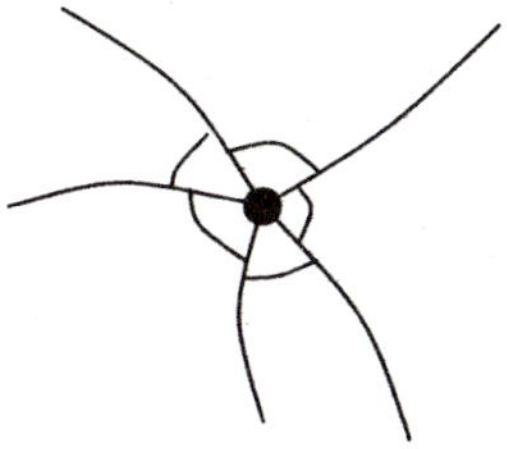

Next,

$$\sum_i (n_i - 2)\pi = \left(\sum_i n_i\right)\pi - 2\pi F = 2\pi E - 2\pi F, \tag{12}$$

where F is the total number of polygons and E the total number of edges, since in the sum $\sum_i n_i$ each edge is counted twice (as each edge is an edge of exactly two polygons).

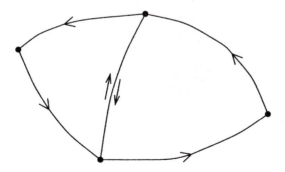

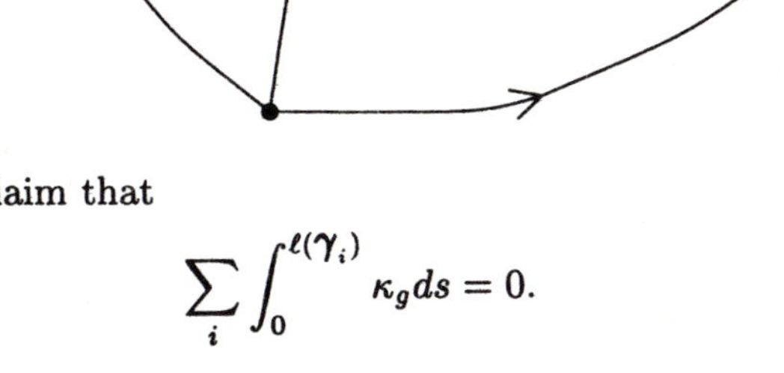

Finally, we claim that

$$\sum_i \int_0^{\ell(\boldsymbol{\gamma}_i)} \kappa_g ds = 0. \tag{13}$$

Indeed, note that in the sum in Eq. (13), we integrate twice along each edge, once in each direction. By Eq. (5) in Chapter 6, κ_g changes sign when we traverse a given curve in the opposite direction, so these two integrals cancel out. The various contributions to the sum in Eq. (13) therefore cancel out in pairs, thus proving Eq. (13).

Putting Eqs. (10), (11), (12) and (13) together gives

$$\begin{aligned}\int\int_{S} K d\mathcal{A} &= \sum_i \int\int_{R_i} K d\mathcal{A}_{\boldsymbol{\sigma}_i} \\ &= \sum_i \angle_i - \sum_i (n_i - 2)\pi + \sum_i \int_0^{\ell(\boldsymbol{\gamma}_i)} \kappa_g \, ds \\ &= 2\pi V - (2\pi E - 2\pi F) + 0 \\ &= 2\pi\chi,\end{aligned}$$

proving Theorem 11.5. □

To see why Theorem 11.5 is so remarkable, let us apply it to the unit sphere S^2. Then, $\chi = 2$ so we get

$$\int\int_{S^2} K d\mathcal{A} = 4\pi. \tag{14}$$

Of course, this result is not remarkable at all because $K = 1$ so the left-hand side of Eq. (14) is just the area of the sphere. But now suppose that we deform the sphere, i.e. we think of the sphere as being a rubber sheet and we pull and stretch it in any way we like, but without tearing:

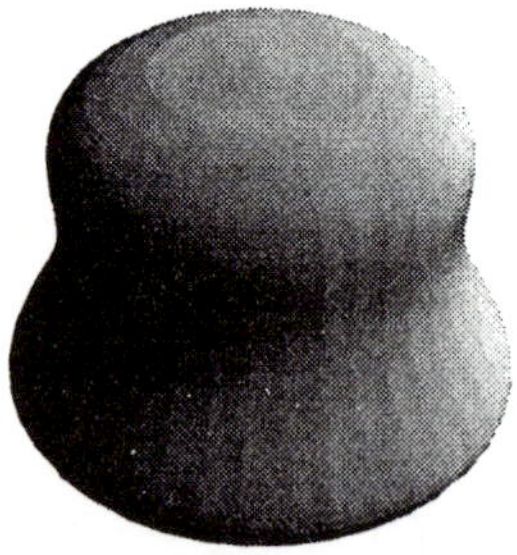

For such a deformed sphere $\mathcal{S}$, K will not be constant and the direct computation of the integral $\int\int_{\mathcal{S}} K d\mathcal{A}$ will be difficult. But if we start with a triangulation of the undeformed sphere, then after deformation we shall have a triangulation of the deformed sphere *with the same number of vertices, edges and polygons as the original triangulation.* It follows that the Euler number of the deformed sphere is *the same* as that of the undeformed sphere, i.e. 2, so by Theorem 11.5,

$\iint_S K dA = 4\pi$. (More generally, this discussion shows that the Euler number of *any* compact surface is unchanged when the surface is deformed without tearing.)

We complete the picture by determining the Euler numbers of all the compact surfaces.

Theorem 11.6

The Euler number of the compact surface T_g of genus g is $2 - 2g$.

Proof 11.6

The formula is correct when $g = 0$, since we know that $\chi = 2$ for a sphere. We now prove it for the torus T_1. To find a triangulation of the torus, we use the fact that it can be obtained from a square in the plane by gluing opposite edges:

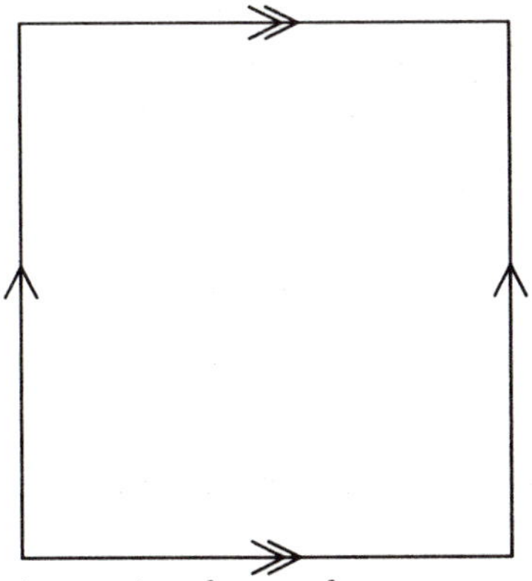

We subdivide the square into triangles as shown:

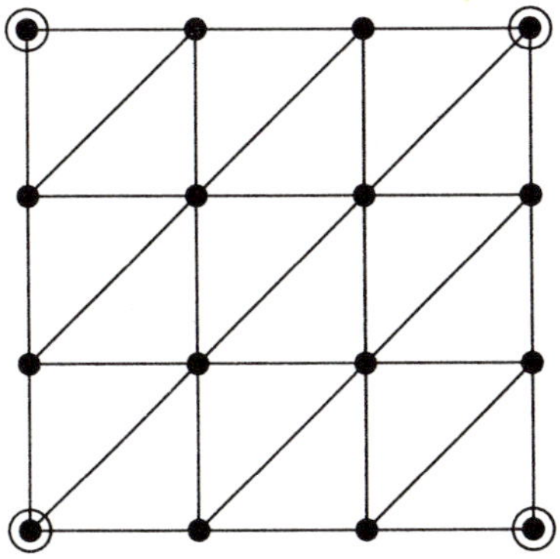

This leads to a triangulation of T_1 with $V = 9$, $E = 27$ and $F = 18$. One must count carefully: for example, the four circled vertices of the square correspond to a single vertex on the torus. Note also that not just any subdivision of the square into triangles is acceptable. For example, the subdivision

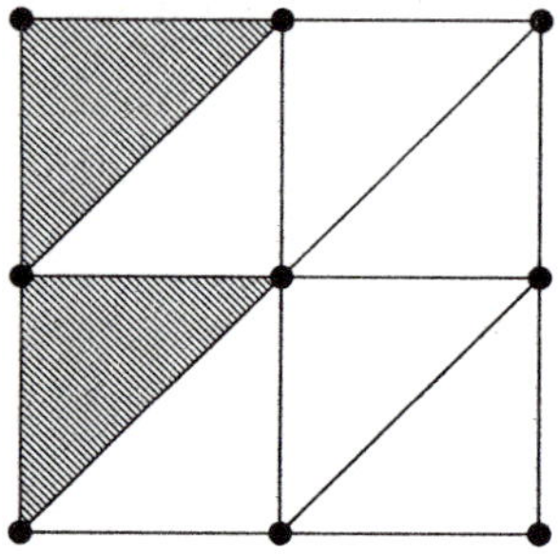

is *not* acceptable, since after gluing, the two shaded triangles intersect in *two* vertices, which is not allowed:

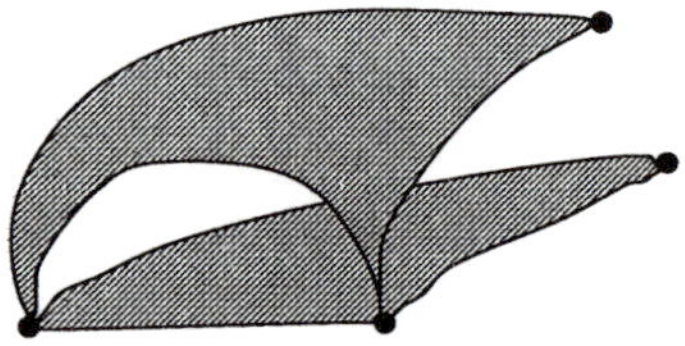

But the finer subdivision above does work, and gives

$$\chi = 9 - 27 + 18 = 0 = 2 - 2 \times 1,$$

proving the theorem when $g = 1$.

We now complete the proof by induction on g, using the fact that T_{g+1} can be obtained from T_g by gluing on a copy of T_1:

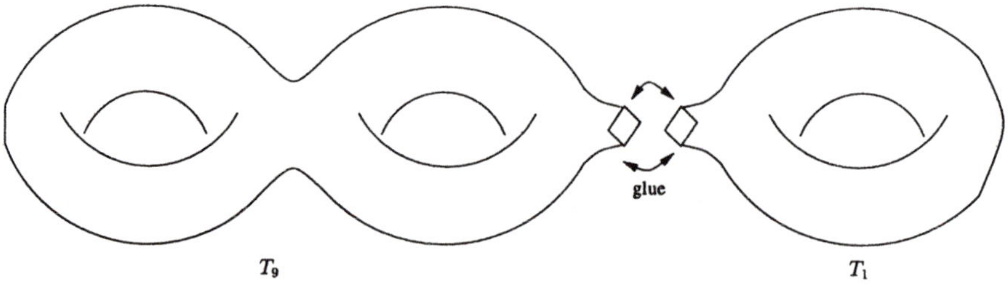

Suppose we carry out the gluing by removing a curvilinear n-gon from T_g and T_1 and gluing corresponding edges (having fixed suitable triangulations of T_g and T_1). If V', E' and F' are the numbers of vertices, edges and polygons in the triangulation of T_g, and V'', E'' and F'' those for T_1, the numbers V, E and F for T_{g+1} are given by

$$\begin{aligned} V &= V' - n + V'' - n + n = V' + V'' - n, \\ E &= E' - n + E'' - n + n = E' + E'' - n, \\ F &= F' - 1 + F'' - 1 = F' + F'' - 2. \end{aligned}$$

Indeed, V is the number V' of vertices in T_g plus the number V'' in T_1, except that the n vertices of the polygon along which T_1 and T_g are glued have been counted twice, so $V = V' + V'' - n$; a similar argument applies to the edges; and F is as stated because the polygon along which T_1 and T_g are glued is not part of the triangulation of T_{g+1}. Hence,

$$\begin{aligned} \chi(T_{g+1}) &= V - E + F \\ &= (V' + V'' - n) - (E' + E'' - n) + (F' + F'' - 2) \\ &= V' - E' + F' + V'' - E'' + F'' - 2 \\ &= \chi(T_g) + \chi(T_1) - 2 \\ &= 2 - 2g + 0 - 2 \quad \text{(by the induction hypothesis)} \\ &= 2 - 2(g+1), \end{aligned}$$

proving the result for genus $g + 1$. □

Corollary 11.3

We have

$$\iint_{T_g} K dA = 4\pi(1 - g).$$

Proof 11.3

Just combine Theorems 11.5 and 11.6. □

EXERCISES

11.5 Show that, if a triangulation of a compact surface with Euler number χ by curvilinear triangles has V vertices, E edges and F triangles, then

$$3F = 2E, \quad E = 3(V - \chi), \quad V \geq \frac{1}{2}(7 + \sqrt{49 - 24\chi}).$$

11.6 A triangulation of the sphere has n curvilinear triangles, and r triangles meet at each vertex. Explain why there are $3n/r$ vertices altogether in the triangulation, and write down the total number of edges. Show that

$$\frac{6}{r} - \frac{4}{n} = 1.$$

Deduce that $r \leq 5$, and sketch triangulations of the sphere corresponding to $r = 3$, 4 and 5.

11.7 Show that, given 5 points on a sphere, it is impossible to connect each pair by curves on the sphere that intersect only at the given points. (Such a collection of curves would give a triangulation of the sphere for which $2E \geq 3F$, since each face would have at least 3 edges.) Deduce that the same result holds if 'sphere' is replaced by 'plane'. (Use stereographic projection.)

11.8 Let P_1, P_2, P_3 and Q_1, Q_2, Q_3 be points on a sphere. Show that it is impossible to join each P_i to each Q_j by 9 curves on the sphere that intersect only at the given points. (This is sometimes called the 'Utilities Problem', thinking of P_1, P_2 and P_3 as the gas, water and electricity supplies to three homes Q_1, Q_2 and Q_3.)

11.9 Show that, if a compact surface $\mathcal{S}$ is diffeomorphic to the torus T_1, then

$$\iint_{\mathcal{S}} K\,d\mathcal{A} = 0.$$

Can such a surface have $K = 0$ everywhere?

11.10 Show that, if $\mathcal{S}$ is the ellipsoid

$$\frac{x^2 + y^2}{a^2} + \frac{z^2}{b^2} = 1,$$

where a and b are positive constants,

$$\iint_{\mathcal{S}} K\,d\mathcal{A} = 4\pi.$$

By computing the above integral directly, deduce that

$$\int_{-\pi/2}^{\pi/2} \frac{ab^2 \cos\theta}{(a^2 \sin^2\theta + b^2 \cos^2\theta)^{3/2}}\,d\theta = 2.$$

11.11 Suppose that $\mathcal{S}$ is a compact surface whose gaussian curvature K is > 0 everywhere. Show that $\mathcal{S}$ is diffeomorphic to a sphere. Is the converse of this statement true?

11.12 One of the following surfaces is compact and one is not:

(i) $x^2 - y^2 + z^4 = 1$;

(ii) $x^2 + y^2 + z^4 = 1$.
Which is which, and why? Sketch the compact surface and write down its Euler number.

11.4. Singularities of Vector Fields

Suppose that $\mathcal{S}$ is a surface and that $\mathbf{V}$ is a *smooth tangent vector field* on $\mathcal{S}$. This means that, if $\boldsymbol{\sigma} : U \to \mathbf{R}^3$ is a patch of $\mathcal{S}$ and (u, v) are coordinates on U, then

$$\mathbf{V} = \alpha(u, v)\boldsymbol{\sigma}_u + \beta(u, v)\boldsymbol{\sigma}_v,$$

where α and β are smooth functions on U. It is easy to see that this smoothness condition is independent of the choice of patch $\boldsymbol{\sigma}$ (see Exercise 11.14).

Definition 11.5

If $\mathbf{V}$ is a smooth tangent vector field on a surface $\mathcal{S}$, a point P of $\mathcal{S}$ at which $\mathbf{V} = \mathbf{0}$ is called a *stationary point* of $\mathbf{V}$.

The reason for this terminology is as follows. We saw in the proof of Proposition 7.4 that, if P is any point of $\mathcal{S}$, there is a unique curve $\boldsymbol{\gamma}(t)$ on $\mathcal{S}$ such that $\dot{\boldsymbol{\gamma}} = \mathbf{V}$ and $\boldsymbol{\gamma}(0) = P$; $\boldsymbol{\gamma}$ is called an *integral curve* of $\mathbf{V}$. We can think of $\boldsymbol{\gamma}$ as the path followed by a particle of some fluid that is flowing over the surface. If $\mathbf{V} = \mathbf{0}$ at P, the velocity $\dot{\boldsymbol{\gamma}}$ of the flow is zero at P, so the fluid is stationary there.

We are going to prove a theorem which says that the number of stationary points of any smooth tangent vector field on a compact surface $\mathcal{S}$, counted with the appropriate multiplicity, is equal to the Euler number of $\mathcal{S}$. To define this multiplicity, let P be a stationary point of $\mathbf{V}$ contained in a surface patch $\boldsymbol{\sigma} : U \to \mathbf{R}^3$ of $\mathcal{S}$, say, with $\boldsymbol{\sigma}(u_0, v_0) = P$. Assume that P is the only stationary point of $\mathbf{V}$ in the region $\boldsymbol{\sigma}(U)$ of $\mathcal{S}$. Let $\boldsymbol{\xi}$ be a nowhere vanishing smooth tangent vector field on $\boldsymbol{\sigma}(U)$ (for example, we may choose $\boldsymbol{\xi} = \boldsymbol{\sigma}_u$ or $\boldsymbol{\sigma}_v$), and let ψ be the angle between $\mathbf{V}$ and $\boldsymbol{\xi}$.

Definition 11.6

With the above notation and assumption, the *multiplicity* of the stationary point P of the tangent vector field $\mathbf{V}$ is

$$\mu(P) = \frac{1}{2\pi}\int_0^{\ell(\boldsymbol{\gamma})} \frac{d\psi}{ds}\, ds,$$

where $\boldsymbol{\gamma}(s)$ is any positively-oriented unit-speed simple closed curve of length $\ell(\boldsymbol{\gamma})$ in $\boldsymbol{\sigma}(U)$ with P in its interior.

It is clear that $\mu(P)$ is an integer, and an argument similar to our heuristic proof of Hopf's Umlaufsatz in Section 11.1 shows that $\mu(P)$ does not depend on the choice of simple closed curve $\boldsymbol{\gamma}$. It is also easy to see that it is independent of the choice of 'reference' vector field $\boldsymbol{\xi}$ (see Exercise 11.15).

Example 11.1

The following smooth tangent vector fields in the plane have stationary points of the indicated multiplicity at the origin (we have shown the integral curves of the vector fields for the sake of clarity):

(i) $\mathbf{V}(x,y) = (x,y)$; $\mu = +1$.

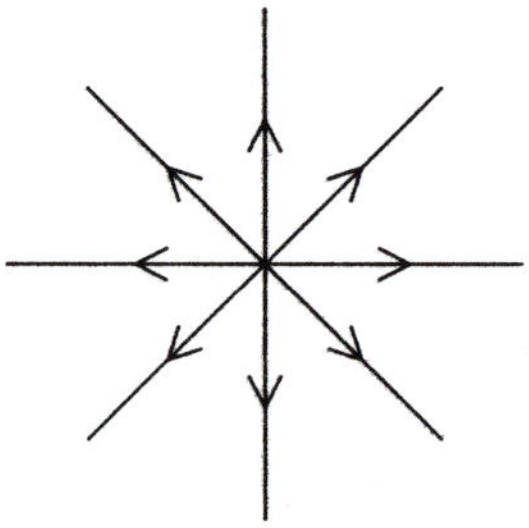

(ii) $\mathbf{V}(x,y) = (-x,-y)$; $\mu = +1$

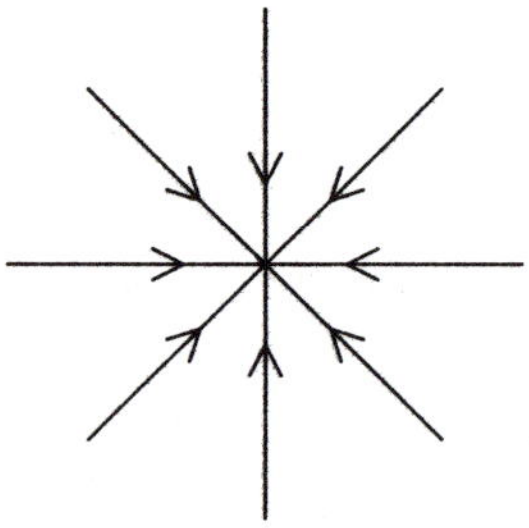

(iii) $\mathbf{V}(x,y) = (y,-x)$; $\mu = +1$

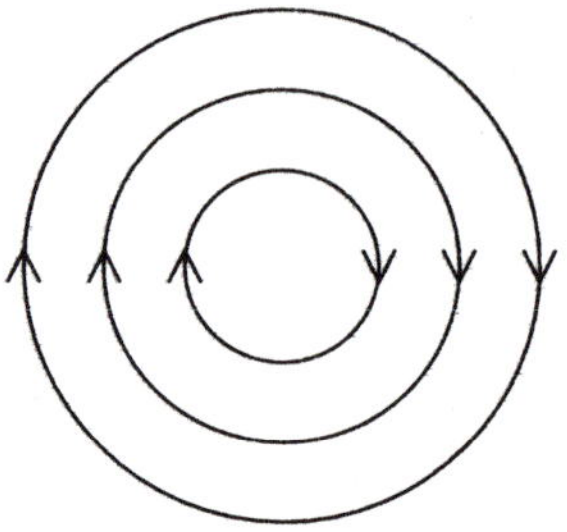

(iv) $\mathbf{V}(x,y) = (x,-y)$; $\mu = -1$

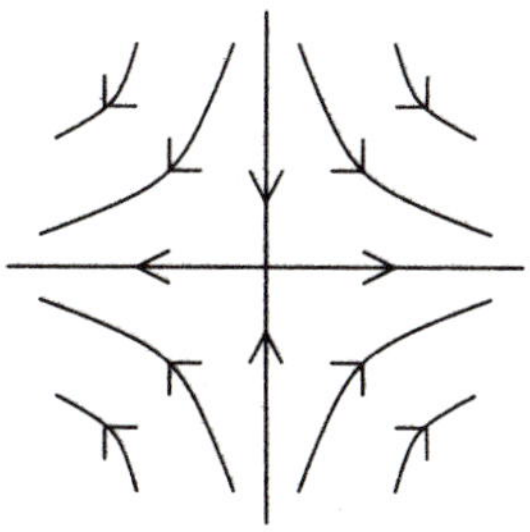

The stationary point in examples (i), (ii), (iii) and (iv) is called a *source*, *sink*, *vortex* and *bifurcation*, respectively.

Let us verify the multiplicity in case (iv), for example. Take the 'reference' tangent vector field to be the constant vector field $\boldsymbol{\xi} = (1,0)$. Then, the angle ψ is given by

$$(\cos\psi, \sin\psi) = \frac{\mathbf{V}}{\parallel \mathbf{V} \parallel} = \left(\frac{x}{\sqrt{x^2+y^2}}, -\frac{y}{\sqrt{x^2+y^2}}\right).$$

Taking $\boldsymbol{\gamma}(s) = (\cos s, \sin s)$ to be the unit circle, at $\boldsymbol{\gamma}(s)$ the angle ψ satisfies

$$(\cos\psi, \sin\psi) = (\cos s, -\sin s),$$

so $\psi = 2\pi - s$. Hence,

$$\mu(0,0) = \frac{1}{2\pi}\int_0^{2\pi} \frac{d}{ds}(2\pi - s)\, ds = -1.$$

Theorem 11.7

Let $\mathbf{V}$ be a smooth tangent vector field on a compact surface $\mathcal{S}$ which has only finitely many stationary points, say $P_1, P_2, \ldots, P_n$. Then,

$$\sum_{i=1}^{n} \mu(P_i) = \chi,$$

the Euler number of $\mathcal{S}$.

Proof 11.7

Let $\boldsymbol{\gamma}_i$ be a positively-oriented unit-speed simple closed curve contained in a patch $\boldsymbol{\sigma}_i$ of $\mathcal{S}$ with P_i in the interior of $\boldsymbol{\gamma}_i$. Assume that the $\boldsymbol{\gamma}_i$ are chosen so small that their interiors are disjoint. Choose a triangulation of the part $\mathcal{S}'$ of $\mathcal{S}$ outside $\boldsymbol{\gamma}_1, \boldsymbol{\gamma}_2, \ldots, \boldsymbol{\gamma}_n$ by curvilinear polygons $\boldsymbol{\Gamma}_j$. Note that the edges of some of these curvilinear polygons will be segments of the curves $\boldsymbol{\gamma}_i$:

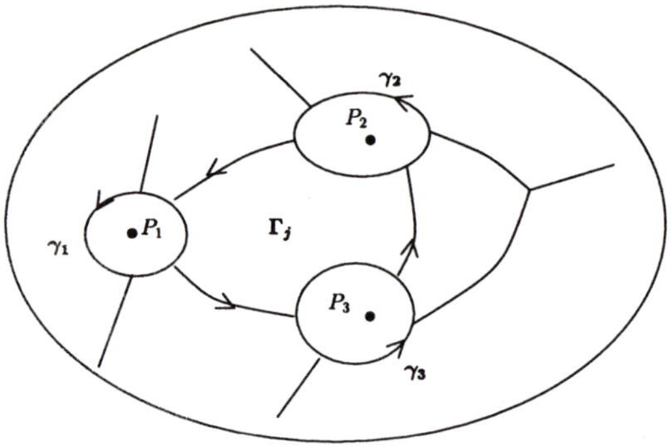

Note also that, when these polygons are positively-oriented, the induced orientation of the $\boldsymbol{\gamma}_i$ is *opposite* to their positive orientation (see the diagram above, in which the arrows indicate the sense of positive orientation).

We can regard the curvilinear polygons in $\mathcal{S}'$, together with the simple closed curves $\boldsymbol{\gamma}_i$ and their interiors, as a triangulation of $\mathcal{S}$, so by Theorem 11.3,

$$\int_{\mathcal{S}'} K\,d\mathcal{A} + \sum_{i=1}^{n} \int_{\text{int}(\boldsymbol{\gamma}_i)} K\,d\mathcal{A} = 2\pi\chi, \tag{15}$$

where χ is the Euler number of $\mathcal{S}$. On $\mathcal{S}'$, we choose an orthonormal basis $\{\mathbf{e}', \mathbf{e}''\}$ of the tangent plane of $\mathcal{S}$ at each point so that $\mathbf{e}'$ is parallel to the tangent vector field $\mathbf{V}$. Arguing as in the proof of Theorem 11.1, we see that

$$\int_{\mathcal{S}'} K\,d\mathcal{A} = \sum_j \int_0^{\ell(\boldsymbol{\Gamma}_j)} \mathbf{e}'.\dot{\mathbf{e}}''\,ds, \tag{16}$$

where s is arc-length on $\boldsymbol{\Gamma}_j$ and $\ell(\boldsymbol{\Gamma}_j)$ is its length. Any common edge of two of the curvilinear polygons $\boldsymbol{\Gamma}_j$ is traversed once in each direction so their contributions to the sum in Eq. (16) cancel out. What remains is the integral along the segments of the curves $\boldsymbol{\gamma}_i$ that are part of the polygons $\boldsymbol{\Gamma}_j$. In view of the remark about orientations above, we get

$$\int_{\mathcal{S}'} K\,d\mathcal{A} = -\sum_{i=1}^{n} \int_0^{\ell(\boldsymbol{\gamma}_i)} \mathbf{e}'.\dot{\mathbf{e}}''\,ds, \tag{17}$$

where s is arc-length along $\boldsymbol{\gamma}_i$ and $\ell(\boldsymbol{\gamma}_i)$ is its length.

Now choose an orthonormal basis $\{\mathbf{f}', \mathbf{f}''\}$ of the tangent plane of $\mathcal{S}$ on each patch $\boldsymbol{\sigma}_i$. By the proof of Theorem 11.1,

$$\int_{\text{int}(\boldsymbol{\gamma}_i)} K\,d\mathcal{A} = \int_0^{\ell(\boldsymbol{\gamma}_i)} \mathbf{f}'.\dot{\mathbf{f}}''\,ds. \tag{18}$$

Combining Eqs. (15), (17) and (18), we get

$$\sum_{i=1}^{n} \int_0^{\ell(\boldsymbol{\gamma}_i)} (\mathbf{f}'.\dot{\mathbf{f}}'' - \mathbf{e}'.\dot{\mathbf{e}}'')\, ds = 2\pi\chi. \tag{19}$$

But, from the proof of Theorem 11.2,

$$\mathbf{e}'.\dot{\mathbf{e}}'' = \dot{\theta} - \kappa_g, \quad \mathbf{f}'.\dot{\mathbf{f}}'' = \dot{\varphi} - \kappa_g,$$

where κ_g is the geodesic curvature of $\boldsymbol{\gamma}_i$ and θ and φ are the angles between $\dot{\boldsymbol{\gamma}}_i$ and $\mathbf{e}'$ and $\mathbf{f}'$, respectively. Then, $\psi = \varphi - \theta$ is the angle between $\mathbf{e}'$ and $\mathbf{f}'$, i.e. between $\mathbf{V}$ and the 'reference' tangent vector field $\mathbf{f}'$ on $\boldsymbol{\sigma}_i$. So the left-hand side of Eq. (19) is

$$\sum_{i=1}^{n} \int_0^{\ell(\boldsymbol{\gamma}_i)} \frac{d\psi}{ds}\, ds = 2\pi \sum_{i=1}^{n} \mu(P_i),$$

as we want. □

We now give some simple examples of vector fields on surfaces (we show their integral curves for clarity).

Example 11.2

A vector field on the sphere with 1 source and 1 sink: $\chi = 2$

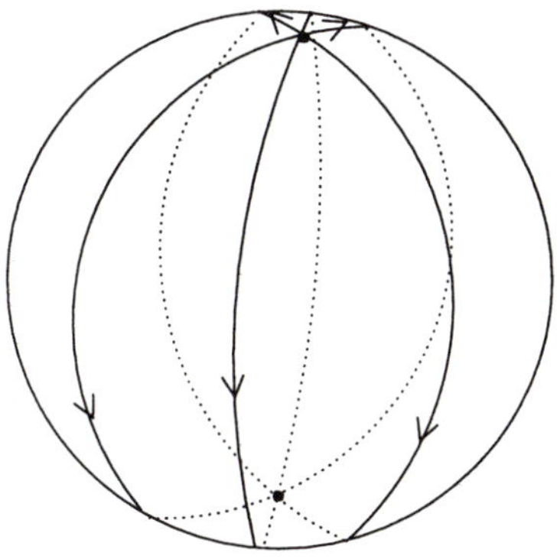

Example 11.3

A vector field on the torus with no stationary points: $\chi = 0$

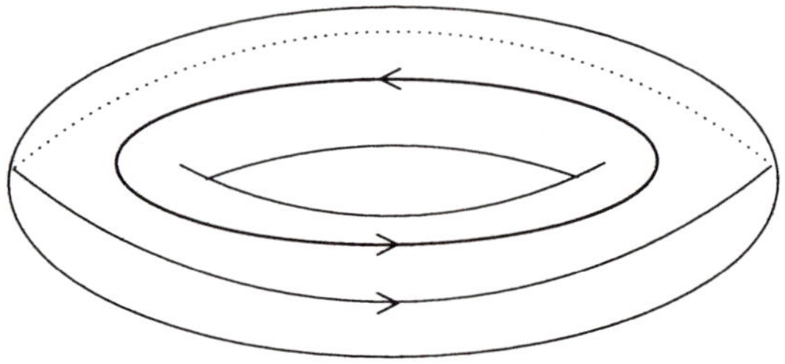

Example 11.4

A vector field on the double torus T_2 with 2 bifurcations: $\chi = -2$

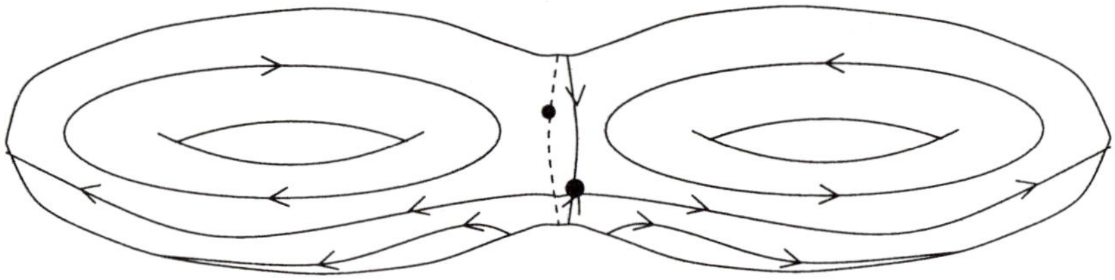

EXERCISES

11.13 Let k be a non-zero integer and let $\mathbf{V}(x, y) = (\alpha, \beta)$ be the vector field on the plane given by

$$\alpha + i\beta = \begin{cases} (x + iy)^k & \text{if } k > 0, \\ (x - iy)^{-k} & \text{if } k < 0. \end{cases}$$

Show that the origin is a stationary point of $\mathbf{V}$ of multiplicity k. (Take $\boldsymbol{\gamma}$ in Definition 11.6 to be the unit circle and use de Moivre's theorem.)

11.14 Show that the definition of a smooth tangent vector field is independent of the choice of surface patch. Show also that a tangent vector field $\mathbf{V}$ on S is smooth if and only if, for any surface patch

$\boldsymbol{\sigma}$ of $\mathcal{S}$, the three components of $\mathbf{V}$ at the point $\boldsymbol{\sigma}(u, v)$ are smooth functions of (u, v).

11.15 Show that the Definition 11.6 of the multiplicity of a stationary point of a tangent vector field $\mathbf{V}$ is independent of the 'reference' vector field $\boldsymbol{\xi}$. (If $\tilde{\boldsymbol{\xi}}$ is another reference tangent vector field, and θ is the angle between $\tilde{\boldsymbol{\xi}}$ and $\boldsymbol{\xi}$, then $d\theta/ds = -(1-\rho^2)^{-1/2}\dot{\rho}$, where $\rho = \cos\theta$. Now use Green's theorem to show that $\int_0^{\ell(\boldsymbol{\gamma})} (d\theta/ds)ds = 0$.)

11.5. Critical Points

If $f(u, v)$ is a smooth function defined on an open subset U of $\mathbf{R}^2$, we say that a point (u_0, v_0) is a *critical point* of f if $\partial f/\partial u$ and $\partial f/\partial v$ both vanish at (u_0, v_0). Equivalently, the *gradient* of f,

$$\nabla f = \left(\frac{\partial f}{\partial u}, \frac{\partial f}{\partial v}\right),$$

should vanish at (u_0, v_0).

If now $F : \mathcal{S} \to \mathbf{R}$ is a smooth function on a surface $\mathcal{S}$ (see Exercise 4.11), and if $\boldsymbol{\sigma} : U \to \mathbf{R}^3$ is a surface patch of $\mathcal{S}$, then $f = F \circ \boldsymbol{\sigma}$ is a smooth function on the open subset U of $\mathbf{R}^2$. This suggests

Definition 11.7

Let $\mathcal{S}$ be a surface and $F : \mathcal{S} \to \mathbf{R}$ a smooth function on $\mathcal{S}$. A point P of $\mathcal{S}$ is a *critical point* of F if there is a surface patch $\boldsymbol{\sigma}$ of $\mathcal{S}$, with $P = \boldsymbol{\sigma}(u_0, v_0)$, say, such that $f = F \circ \boldsymbol{\sigma}$ has a critical point at (u_0, v_0).

It is easy to check directly that the definition of a critical point is independent of the choice of patch $\boldsymbol{\sigma}$ (see Exercise 11.18), but this will follow immediately from another characterisation of critical points that we shall now give and which is independent of any arbitrary choices.

If $F : \mathcal{S} \to \mathbf{R}$ is a smooth function, we want to define an analogue of the gradient of f that is a tangent vector field on $\mathcal{S}$. This is accomplished in

Proposition 11.1

If F is a smooth function on a surface $\mathcal{S}$, there is a unique smooth tangent vector field $\nabla_{\mathcal{S}} F$ on $\mathcal{S}$ such that, if P is a point of $\mathcal{S}$ and $\boldsymbol{\gamma}(t)$ is a curve in $\mathcal{S}$

which passes through P when $t = 0$, we have

$$(\nabla_{\mathcal{S}} F).\dot{\boldsymbol{\gamma}}(0) = \left.\frac{d}{dt}\right|_{t=0} F(\boldsymbol{\gamma}(t)). \tag{20}$$

Moreover, P is a critical point of F if and only if $\nabla_{\mathcal{S}} F = \mathbf{0}$ at P.

Proof 11.1

The tangent vector field $\nabla_{\mathcal{S}} F$ is obviously unique, if it exists. Indeed, every tangent vector to $\mathcal{S}$ at P is of the form $\dot{\boldsymbol{\gamma}}(0)$ for some curve $\boldsymbol{\gamma}$ on $\mathcal{S}$ with $\boldsymbol{\gamma}(0) = P$, so any two choices of $\nabla_{\mathcal{S}} F$ at P would differ by a vector perpendicular to every tangent vector to $\mathcal{S}$ at P, which must be zero.

To see that $\nabla_{\mathcal{S}} F$ exists, choose a surface patch $\boldsymbol{\sigma}(u, v)$ for $\mathcal{S}$ with $\boldsymbol{\sigma}(u_0, v_0) = P$, say, and let $f = F \circ \boldsymbol{\sigma}$. Let $\{\mathbf{e}', \mathbf{e}''\}$ be the basis of the tangent plane of $\mathcal{S}$ at P such that

$$\mathbf{e}'.\boldsymbol{\sigma}_u = \mathbf{e}''.\boldsymbol{\sigma}_v = 1, \quad \mathbf{e}'.\boldsymbol{\sigma}_v = \mathbf{e}''.\boldsymbol{\sigma}_u = 0. \tag{21}$$

Explicitly,

$$\mathbf{e}' = \frac{G\boldsymbol{\sigma}_u - F\boldsymbol{\sigma}_v}{EG - F^2}, \quad \mathbf{e}'' = \frac{E\boldsymbol{\sigma}_v - F\boldsymbol{\sigma}_u}{EG - F^2}, \tag{22}$$

where $Edu^2 + 2Fdudv + Gdv^2$ is the first fundamental form of $\boldsymbol{\sigma}$. We take

$$\nabla_{\mathcal{S}} F = f_u \mathbf{e}' + f_v \mathbf{e}'', \tag{23}$$

where the derivatives are evaluated at (u_0, v_0). If $\boldsymbol{\gamma}$ is as in the statement of the proposition, say $\boldsymbol{\gamma}(t) = \boldsymbol{\sigma}(u(t), v(t))$, then (with d/dt denoted by a dot)

$$\begin{aligned}(\nabla_{\mathcal{S}} F).\dot{\boldsymbol{\gamma}}(0) &= (f_u \mathbf{e}' + f_v \mathbf{e}'').(\boldsymbol{\sigma}_u \dot{u} + \boldsymbol{\sigma}_v \dot{v}) \\ &= f_u \dot{u} + f_v \dot{v} \quad \text{(by Eqs. (21))} \\ &= \dot{f},\end{aligned}$$

the derivatives with respect to t being evaluated at $t = 0$. It is clear from Eqs. (22) and (23) that $\nabla_{\mathcal{S}} F$ is smooth and that P is a critical point of F if and only if $\nabla_{\mathcal{S}} F = \mathbf{0}$ at P. □

Since $\nabla_{\mathcal{S}} F$ is a smooth tangent vector field on $\mathcal{S}$, we can apply Theorem 11.7 to it. To do so, we must compute the multiplicity of the stationary points of $\nabla_{\mathcal{S}} F$. For this, we shall make an additional assumption about F, contained in

Definition 11.8

A critical point P of a smooth function F on a surface $\mathcal{S}$ is said to be *non-degenerate* if, whenever $\boldsymbol{\sigma}(u, v)$ is a patch of $\mathcal{S}$ with $P = \boldsymbol{\sigma}(u_0, v_0)$, say, the matrix

$$\mathcal{H} = \begin{pmatrix} \frac{\partial^2 f}{\partial u^2} & \frac{\partial^2 f}{\partial u \partial v} \\ \frac{\partial^2 f}{\partial u \partial v} & \frac{\partial^2 f}{\partial v^2} \end{pmatrix}$$

is invertible, where $f = F \circ \boldsymbol{\sigma}$ and the derivatives are evaluated at (u_0, v_0). In this case, the point P is called a *local maximum*, a *saddle point* or a *local minimum* if $\mathcal{H}$ has $2, 1$ or 0 negative eigenvalues, respectively.

It is not difficult to show that this definition is sensible, i.e. independent of the choice of patch $\boldsymbol{\sigma}$ (see Exercise 11.16). Note that the matrix $\mathcal{H}$ is real and symmetric, so it always has two real eigenvalues (not necessarily distinct).

Proposition 11.2

Let P be a critical point of a smooth function F on a surface $\mathcal{S}$. Then, the multiplicity of P as a stationary point of $\nabla_{\mathcal{S}} F$ is

$$\mu(P) = \begin{cases} 1 & \textit{if } P \textit{ is a local maximum or a local minimum,} \\ -1 & \textit{if } P \textit{ is a saddle point.} \end{cases}$$

Example 11.5

The function on the plane given by $F(u, v) = -u^2 - v^2$ (resp. $u^2 - v^2$, $u^2 + v^2$) has a local maximum (resp. saddle point, local minimum) at the origin.

We shall not give a complete proof of Proposition 11.2 here. But the following argument should convince the reader of its truth. Let us assume that $(u_0, v_0) = (0, 0)$ for simplicity, and write the matrix in Definition 11.8 at $u = v = 0$ as

$$\mathcal{H} = \begin{pmatrix} \lambda & \mu \\ \mu & \nu \end{pmatrix}.$$

Then, Taylor's theorem tells us that

$$f(u, v) = \frac{1}{2}(\lambda u^2 + 2\mu uv + \nu v^2) + r(u, v),$$

where $r(u, v)/(u^2 + v^2)$ tends to zero as u and v tend to zero. It is plausible, then, that the behaviour of $\nabla_{\mathcal{S}} F$ near P is the same as that of $\nabla_{\mathcal{S}} \tilde{F}$, where

$$\tilde{F}(\boldsymbol{\sigma}(u, v)) = \frac{1}{2}(\lambda u^2 + 2\mu uv + \nu v^2).$$

In particular, F and $\tilde{F}$ should have the same type of critical point at P.

But the multiplicity of P as a critical point of $\tilde{F}$ is easy to compute. To do so, note first that there is a rotation matrix P such that

$$P^t \begin{pmatrix} \lambda & \mu \\ \mu & \nu \end{pmatrix} P = \begin{pmatrix} \epsilon_1 & 0 \\ 0 & \epsilon_2 \end{pmatrix}$$

is a diagonal matrix. This means that, by applying a rotation in the uv-plane (i.e. a reparametrisation of $\boldsymbol{\sigma}$), we can assume that

$$\tilde{F}(\boldsymbol{\sigma}(u,v)) = \frac{1}{2}(\epsilon_1 u^2 + \epsilon_2 v^2).$$

We can also assume that the patch $\boldsymbol{\sigma}$ is *conformal*, so that angles measured in the uv-plane are the same as those measured on the surface (see Section 9.4).

Taking our 'reference' tangent vector field to be $\boldsymbol{\sigma}_u$, the angle ψ between $\nabla_S \tilde{F}$ and $\boldsymbol{\sigma}_u$ is given by

$$(\cos\psi, \sin\psi) = \frac{(\epsilon_1 u, \epsilon_2 v)}{\sqrt{\epsilon_1^2 u^2 + \epsilon_2^2 v^2}}.$$

We take the simple closed curve in the uv-plane given by the ellipse

$$\epsilon_1^2 u^2 + \epsilon_2^2 v^2 = r^2,$$

where r is a small positive number, which can be parametrised by

$$u = \frac{r}{|\epsilon_1|}\cos t, \quad v = \frac{r}{|\epsilon_2|}\sin t.$$

Hence,

$$\cos\psi = \frac{\epsilon_1}{|\epsilon_1|}\cos t, \quad \sin\psi = \frac{\epsilon_2}{|\epsilon_2|}\sin t,$$

and so

$$\psi = \begin{cases} t & \text{if } \epsilon_1 > 0 \text{ and } \epsilon_2 > 0, \\ 2\pi - t & \text{if } \epsilon_1 > 0 \text{ and } \epsilon_2 < 0, \\ \pi - t & \text{if } \epsilon_1 < 0 \text{ and } \epsilon_2 > 0, \\ \pi + t & \text{if } \epsilon_1 < 0 \text{ and } \epsilon_2 < 0. \end{cases}$$

This gives the multiplicity of P as a stationary point of $\nabla_S \tilde{F}$ as

$$\frac{1}{2\pi}\int_0^{2\pi} \frac{d\psi}{dt}\,dt = \begin{cases} 1 & \text{if } \epsilon_1 \text{ and } \epsilon_2 \text{ have the same sign}, \\ -1 & \text{otherwise}, \end{cases}$$

in accordance with Proposition 11.2.

If we accept this heuristic argument, we can combine Theorem 11.7 and Proposition 11.2 to give

Theorem 11.8

Let $F : S \to \mathbf{R}$ be a smooth function on a compact surface S with only finitely many critical points, all non-degenerate. Then,

$$\begin{pmatrix} \textit{number of local} \\ \textit{maxima of } F \end{pmatrix} - \begin{pmatrix} \textit{number of saddle} \\ \textit{points of } F \end{pmatrix} + \begin{pmatrix} \textit{number of local} \\ \textit{minima of } F \end{pmatrix} = \chi,$$

the Euler number of S.

Example 11.6

If we take the surface T_g of genus g described at the beginning of Section 11.2 and stand it upright on the xy-plane, the height above the plane is a smooth function F on T_g. The critical points of F are as shown in the following diagram, and they are all non-degenerate (cf. Exercise 11.18).

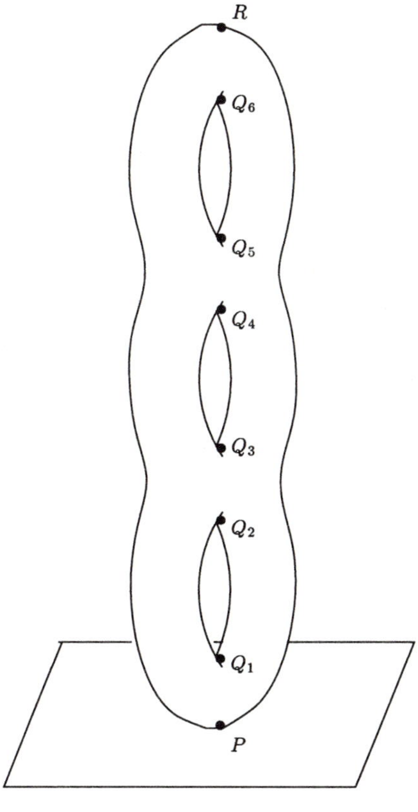

There is a unique local minimum P, $2g$ saddle points $Q_1, Q_2, \ldots, Q_{2g}$, and a unique local maximum R. Hence, Theorem 11.8 gives the Euler number of T_g as

$$1 - 2g + 1 = 2 - 2g,$$

in accordance with Theorem 11.6.

EXERCISES

11.16 Show directly that the definition of a critical point, and whether it is non-degenerate, is independent of the choice of surface patch in Definitions 11.7 and 11.8. Show that the classification of non-degenerate critical points into local maxima, local minima and saddle points is also independent of this choice. (Show that the critical point is a local maximum (resp. local minimum) if and only if $v^t\mathcal{H}v < 0$ (resp. > 0) for all non-zero 2×1 matrices v.)

11.17 For which of the following functions on the plane is the origin a non-degenerate critical point? In the non-degenerate case(s), classify the origin as a local maximum, local minimum or saddle point.
(i) $x^2 - 2xy + 4y^2$;
(ii) $x^2 + 4xy$;
(iii) $x^3 - 3xy^2$.

11.18 Let $\mathcal{S}$ be the torus obtained by rotating the circle $(x-2)^2+z^2 = 1$ in the xz-plane around the z-axis, and let $F : \mathcal{S} \to \mathbf{R}$ be the distance from the plane $x = -3$. Show that F has four critical points, all non-degenerate, and classify them as local maxima, saddle points, or local minima. (See Exercise 4.10 for a parametrisation of $\mathcal{S}$.)

Solutions

Chapter 1

1.1 It is a parametrisation of the part of the parabola with $x \geq 0$.

1.2 (i) $\boldsymbol{\gamma}(t) = (\sec t, \tan t)$ with $-\pi/2 < t < \pi/2$ and $\pi/2 < t < 3\pi/2$.
(ii) $\boldsymbol{\gamma}(t) = (2\cos t, 3\sin t)$.

1.3 (i) $x + y = 1$.
(ii) $y = (\ln x)^2$.

1.4 (i) $\dot{\boldsymbol{\gamma}}(t) = \sin 2t(-1, 1)$.
(ii) $\dot{\boldsymbol{\gamma}}(t) = (e^t, 2t)$.

1.5 $\dot{\boldsymbol{\gamma}}(t) = 3\sin t\cos t(-\cos t, \sin t)$ vanishes where $\sin t = 0$ or $\cos t = 0$, i.e. $t = n\pi/2$ where n is any integer. These points correspond to the four cusps of the astroid.

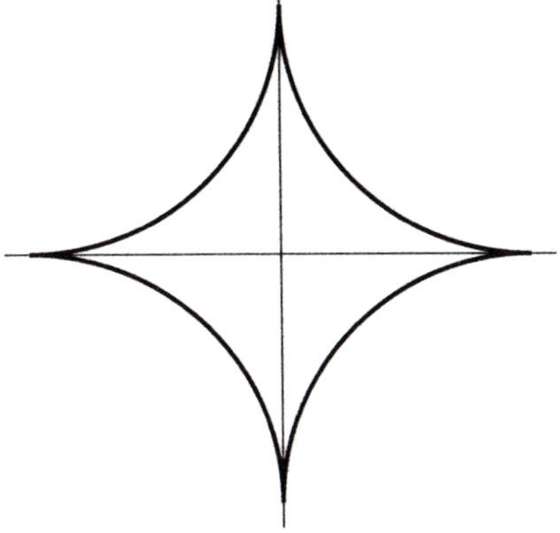

1.6 (i) Let OP make an angle θ with the positive x-axis. Then R has coordinates $\boldsymbol{\gamma}(\theta) = (2a\cot\theta, a(1-\cos 2\theta))$.
(ii) From $x = 2a\cot\theta$, $y = a(1-\cos 2\theta)$, we get $\sin^2\theta = y/2a$, $\cos^2\theta = \cot^2\theta\sin^2\theta = x^2y/8a^3$, so the cartesian equation is $y/2a + x^2y/8a^3 = 1$.

1.7 When the circle has rotated through an angle t, its centre has moved to (at, a), so the point on the circle initially at the origin is now at the point $(a(t-\sin t), a(1-\cos t))$.

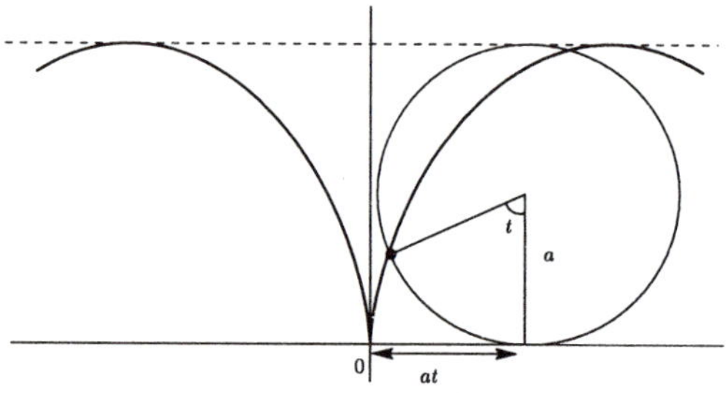

1.8 Let the fixed circle have radius a, and the moving circle radius b (so that $b < a$ in the case of the hypocycloid), and let the point P of the moving circle be initially in contact with the fixed circle at $(a, 0)$. When the moving circle has rotated through an angle φ, the line joining the origin to the point of contact of the circles makes an angle θ with the positive x-axis, where $a\theta = b\varphi$. The point P is then at the point

$$\begin{aligned}\boldsymbol{\gamma}(\theta) &= ((a+b)\cos\theta - b\cos(\theta+\varphi), (a+b)\sin\theta - b\sin(\theta+\varphi)) \\ &= ((a+b)\cos\theta - b\cos((a+b)\theta/b), (a+b)\sin\theta - b\sin((a+b)\theta/b))\end{aligned}$$

in the case of the epicycloid,

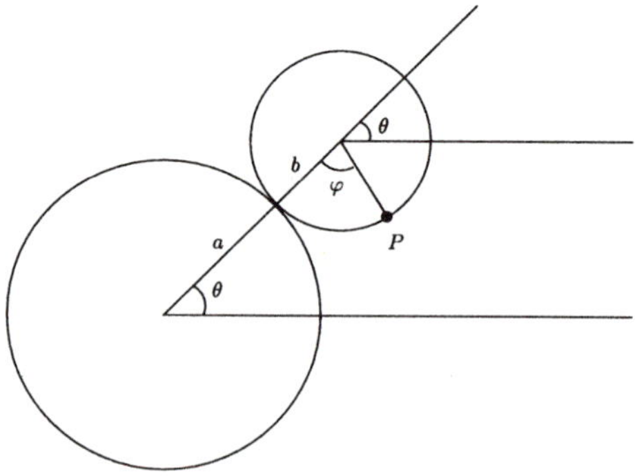

and

$$\begin{aligned}\boldsymbol{\gamma}(\theta) &= ((a-b)\cos\theta + b\cos(\varphi-\theta), (a-b)\sin\theta - b\sin(\varphi-\theta)) \\ &= ((a-b)\cos\theta + b\cos((a-b)\theta/b), (a-b)\sin\theta - b\sin((a-b)\theta/b))\end{aligned}$$

in the case of the hypocycloid.

1.9 A point (x, y, z) lies on the cylinder if $x^2 + y^2 = 1/4$ and on the sphere if $(x + \frac{1}{2})^2 + y^2 + z^2 = 1$. From the second equation, $-1 \leq z \leq 1$ so let $z = \sin t$. Subtracting the two equations gives $x + \frac{1}{4} + \sin^2 t = \frac{3}{4}$, so $x = \frac{1}{2} - \sin^2 t = \cos^2 t - \frac{1}{2}$. From either equation we then get $y = \sin t \cos t$ (or $y = -\sin t \cos t$, but the two solutions are interchanged by $t \mapsto \pi - t$).

1.10 $\dot{\boldsymbol{\gamma}}(t) = (e^t(\cos t - \sin t), e^t(\sin t + \cos t))$ so the angle θ between $\boldsymbol{\gamma}(t)$ and $\dot{\boldsymbol{\gamma}}(t)$ is given by

$$\cos\theta = \frac{\boldsymbol{\gamma}.\dot{\boldsymbol{\gamma}}}{\| \boldsymbol{\gamma} \| \| \dot{\boldsymbol{\gamma}} \|} = \frac{e^{2t}(\cos^2 t - \sin t\cos t + \sin^2 t + \sin t \cos t)}{e^{2t}((\cos t - \sin t)^2 + (\sin t + \cos t)^2)} = \frac{1}{2},$$

and so $\theta = \pi/3$.

1.11 $\dot{\boldsymbol{\gamma}}(t) = (1, \sinh t)$ so $\| \dot{\boldsymbol{\gamma}} \| = \cosh t$ and the arc-length is $s = \int_0^t \cosh u \, du = \sinh t$.

1.12 (i) $\| \dot{\boldsymbol{\gamma}} \|^2 = \frac{1}{4}(1+t) + \frac{1}{4}(1-t) + \frac{1}{2} = 1$.
(ii) $\| \dot{\boldsymbol{\gamma}} \|^2 = \frac{16}{25}\sin^2 t + \cos^2 t + \frac{9}{25}\sin^2 t = \cos^2 t + \sin^2 t = 1$.

1.13 The cycloid is parametrised by $\boldsymbol{\gamma}(t) = a(t - \sin t, 1 - \cos t)$, where t is the angle through which the circle has rotated. So $\dot{\boldsymbol{\gamma}} = a(1 - \cos t, \sin t)$, $\| \dot{\boldsymbol{\gamma}} \|^2 = a^2(2 - 2\cos t) = 4a^2 \sin^2 \frac{t}{2}$, and the arc-length is

$$\int_0^{2\pi} 2a \sin\frac{t}{2}\, dt = -4a\cos\frac{t}{2}\Big|_{t=0}^{t=2\pi} = 8a.$$

1.14 (i) $\dot{\boldsymbol{\gamma}} = \sin 2t(-1, 1)$ vanishes when t is an integer multiple of $\pi/2$, so $\boldsymbol{\gamma}$ is not regular.
(ii) Now $\boldsymbol{\gamma}$ is regular since $\dot{\boldsymbol{\gamma}} \neq \mathbf{0}$ for $0 < t < \pi/2$.
(iii) $\dot{\boldsymbol{\gamma}} = (1, \sinh t)$ is obviously never zero, so $\boldsymbol{\gamma}$ is regular.

1.15 $x = r\cos\theta = \sin^2\theta$, $y = r\sin\theta = \sin^2\theta\tan\theta$, so the parametrisation in terms of θ is $\theta \mapsto (\sin^2\theta, \sin^2\theta\tan\theta)$. Since $\theta \mapsto \sin\theta$ is a bijective smooth map $(-\pi/2, \pi/2) \to (-1, 1)$, with smooth inverse $t \mapsto \sin^{-1} t$, $t = \sin\theta$ is a reparametrisation map. Since $\sin^2\theta = t^2$, $\sin^2\theta\tan\theta = t^3/\sqrt{1-t^2}$, the reparametrised curve is as stated.

1.16 We have $s = \int_{t_0}^{t} \| d\boldsymbol{\gamma}/du \| \, du$, $\tilde{s} = \int_{\tilde{t}_0}^{\tilde{t}} \| d\tilde{\boldsymbol{\gamma}}/d\tilde{u} \| \, d\tilde{u}$. By the chain rule, $d\boldsymbol{\gamma}/du = (d\tilde{\boldsymbol{\gamma}}/d\tilde{u})(d\tilde{u}/du)$, so $s = \pm\int_{t_0}^{t} \| d\tilde{\boldsymbol{\gamma}}/d\tilde{u} \| \, (d\tilde{u}/du)\, du = \pm\tilde{s}$, the sign being that of $d\tilde{u}/du$.

1.17 If $\boldsymbol{\gamma}(t) = (x(t), y(t), z(t))$ is a curve in the surface $f(x, y, z) = 0$, differentiating $f(x(t), y(t), z(t)) = 0$ with respect to t gives $\dot{x}f_x + \dot{y}f_y + \dot{z}f_z = 0$, so $\dot{\boldsymbol{\gamma}}$ is perpendicular to (f_x, f_y, f_z). Since this holds for every curve in the surface, (f_x, f_y, f_z) is perpendicular to the surface. The surfaces $f = 0$ and $g = 0$ should intersect in a curve if the vectors (f_x, f_y, f_z) and (g_x, g_y, g_z) are not parallel.

1.18 Let $\boldsymbol{\gamma}(t) = (u(t), v(t), w(t))$ be a regular curve in $\mathbf{R}^3$. At least one of $\dot{u}, \dot{v}, \dot{w}$ is non-zero at each value of t. Suppose that $\dot{u}(t_0) \neq 0$ and $x_0 = u(t_0)$. As in the 'proof' of Theorem 1.2, there is a smooth function $h(x)$ defined for x near x_0 such that $t = h(x)$ is the unique solution of $x = u(t)$ for each t near t_0. Then, for t near t_0, $\boldsymbol{\gamma}(t)$ is contained in the level curve $f(x,y,z) = g(x,y,z) = 0$, where $f(x,y,z) = y - v(h(x))$ and $g(x,y,z) = z - w(h(x))$. The functions f and g satisfy the conditions in the previous exercise, since $(f_x, f_y, f_z) = (-\dot{v}h', 1, 0)$, $(g_x, g_y, g_z) = (-\dot{w}h', 0, 1)$, a dash denoting d/dx.

1.19 Define $\Theta(t) = \tan \pi\theta(t)/2$, where θ is the function defined in Exercise 8.20. Then Θ is smooth, $\Theta(t) = 0$ if $t \leq 0$, and $\Theta : (0,\infty) \to (0,\infty)$ is a bijection. The curve

$$\boldsymbol{\gamma}(t) = \begin{cases} (\Theta(t), \Theta(t)) & \text{if } t \geq 0, \\ (-\Theta(-t), \Theta(-t)) & \text{if } t \leq 0, \end{cases}$$

is a smooth parametrisation of $y = |x|$.

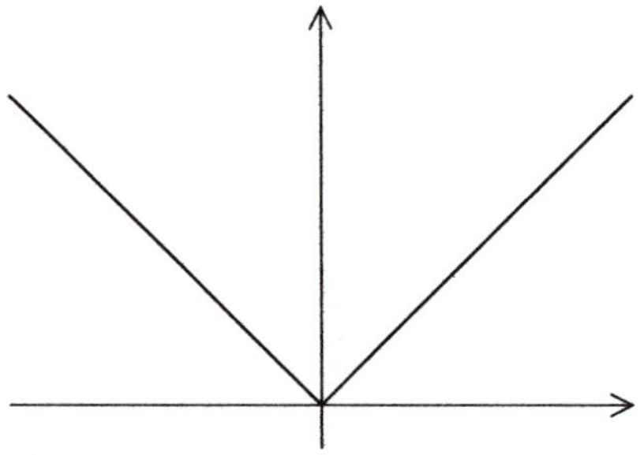

There is no regular parametrisation of $y = |x|$. For if there were, there would be a unit-speed parametrisation $\tilde{\boldsymbol{\gamma}}(t)$, say, and we can assume that $\tilde{\boldsymbol{\gamma}}(0) = (0,0)$. The unit tangent vector $\dot{\tilde{\boldsymbol{\gamma}}}$ would have to be either $\frac{1}{\sqrt{2}}(1,1)$ or $-\frac{1}{\sqrt{2}}(1,1)$ when $x > 0$, so by continuity we would have $\dot{\tilde{\boldsymbol{\gamma}}}(0) = \pm\frac{1}{\sqrt{2}}(1,1)$. But, by considering the part $x < 0$ in the same way, we see that $\dot{\tilde{\boldsymbol{\gamma}}}(0) = \pm\frac{1}{\sqrt{2}}(1,-1)$. These statements are contradictory.

Chapter 2

2.1 (i) $\boldsymbol{\gamma}$ is unit-speed (Exercise 1.12(i)) so

$$\kappa = \parallel \ddot{\boldsymbol{\gamma}} \parallel = \parallel (\frac{1}{4}(1+t)^{-1/2}, \frac{1}{4}(1-t)^{-1/2}, 0) \parallel = \frac{1}{\sqrt{8(1-t^2)}}.$$

(ii) $\boldsymbol{\gamma}$ is unit-speed (Exercise 1.12(ii)) so

$$\kappa = \parallel \ddot{\boldsymbol{\gamma}} \parallel = \parallel (-\frac{4}{5}\cos t, \sin t, \frac{3}{5}\cos t) \parallel = 1.$$

(iii) Using Proposition 2.1,

$$\kappa = \frac{\| (1, \sinh t, 0) \times (0, \cosh t, 0) \|}{\| (1, \sinh t, 0) \|^3} = \frac{\cosh t}{\cosh^3 t} = \text{sech}^2 t.$$

(iv) $(-3\cos^2 t \sin t, 3\sin^2 t \cos t, 0) \times (-3\cos^3 t + 6\cos t \sin^2 t, 6\sin t \cos^2 t - 3\sin^3 t, 0) = (0, 0, -9\sin^2 t \cos^2 t)$, so

$$\kappa = \frac{\| (0, 0, -9\sin^2 t \cos^2 t) \|}{\| (-3\cos^2 t \sin t, 3\sin^2 t \cos t, 0) \|^3} = \frac{1}{3|\sin t \cos t|}.$$

This becomes infinite when t is an integer multiple of $\pi/2$, i.e. at the four cusps $(\pm 1, 0)$ and $(0, \pm 1)$ of the astroid.

2.2 The proof of Proposition 1.4 shows that, if $\mathbf{v}(t)$ is a smooth (vector) function of t, then $\| \mathbf{v}(t) \|$ is a smooth (scalar) function of t provided $\mathbf{v}(t)$ is never zero. The result now follows from the formula in Proposition 2.1. The curvature of the regular curve $\boldsymbol{\gamma}(t) = (t, t^3)$ is $\kappa(t) = 6|t|/(1+9t^4)^{3/2}$, which is not differentiable at $t = 0$.

2.3 Differentiate $\mathbf{t}.\mathbf{n}_s = 0$ and use $\dot{\mathbf{t}} = \kappa_s \mathbf{n}_s$.

2.4 If $\boldsymbol{\gamma}$ is smooth, $\mathbf{t} = \dot{\boldsymbol{\gamma}}$ is smooth and hence so is $\mathbf{n}_s$ (since it is obtained by applying a rotation to $\mathbf{t}$) and $\dot{\mathbf{t}}$. So $\kappa_s = \dot{\mathbf{t}}.\mathbf{n}_s$ is smooth.

2.5 The arc-length $s = \int \| \dot{\boldsymbol{\gamma}} \| \, dt = \int \| e^{kt}(k\cos t - \sin t, k\sin t + \cos t) \| \, dt$ $= \int \sqrt{k^2+1} e^{kt} \, dt = \frac{\sqrt{k^2+1}}{k} e^{kt} + c$, where c is a constant. Taking $c = 0$ makes $s \to 0$ as $t \to \mp\infty$ if $\pm k > 0$.
Since $\mathbf{t} = \dot{\boldsymbol{\gamma}} / \| \dot{\boldsymbol{\gamma}} \| = \frac{1}{\sqrt{k^2+1}}(k\cos t - \sin t, k\sin t + \cos t)$, we have $\mathbf{n}_s = \frac{1}{\sqrt{k^2+1}}(-k\sin t - \cos t, k\cos t - \sin t)$. So $d\mathbf{t}/ds = (d\mathbf{t}/dt)/(ds/dt) = \frac{e^{-kt}}{k^2+1}(-k\sin t - \cos t, k\cos t - \sin t) = \frac{e^{-kt}}{\sqrt{k^2+1}}\mathbf{n}_s$ so $\kappa_s = 1/ks$. By Theorem 2.1, any other curve with the same signed curvature is obtained from the logarithmic spiral by applying a rigid motion.

2.6 (i) Differentiating $\boldsymbol{\gamma} = r\mathbf{t}$ gives $\mathbf{t} = \dot{r}\mathbf{t} + \kappa_s r \mathbf{n}_s$. Since $\mathbf{t}$ and $\mathbf{n}_s$ are perpendicular unit vectors, it follows that $\kappa_s = 0$ and $\boldsymbol{\gamma}$ is part of a straight line.
(ii) Differentiating $\boldsymbol{\gamma} = r\mathbf{n}_s$ gives $\mathbf{t} = \dot{r}\mathbf{n}_s + r\dot{\mathbf{n}}_s = \dot{r}\mathbf{n}_s - \kappa_s r\mathbf{t}$ (Exercise 2.3). Hence, $\dot{r} = 0$, so r is constant, and $\kappa_s = -1/r$, hence κ_s is constant. So $\boldsymbol{\gamma}$ is part of a circle.
(iii) Write $\boldsymbol{\gamma} = r(\mathbf{t}\cos\theta + \mathbf{n}_s \sin\theta)$. Differentiating and equating coefficients of $\mathbf{t}$ and $\mathbf{n}_s$ gives $\dot{r}\cos\theta - \kappa_s r \sin\theta = 1$, $\dot{r}\sin\theta + \kappa_s r\cos\theta = 0$, from which $\dot{r} = \cos\theta$ and $\kappa_s r = -\sin\theta$. From the first equation, $r = s\cos\theta$ (we can assume the arbitrary constant is zero by adding a suitable constant to s) so $\kappa_s = -1/s \cot\theta$. By Exercise 2.5, $\boldsymbol{\gamma}$ is obtained by applying a rigid motion to the logarithmic spiral defined there with $k = -\cot\theta$.

2.7 We can assume that $\boldsymbol{\gamma}$ is unit-speed. Then $\dot{\boldsymbol{\gamma}}^\lambda = (1 - \lambda\kappa_s)\mathbf{t}$, and this is non-zero since $1 - \lambda\kappa_s > 0$. The unit tangent vector of $\boldsymbol{\gamma}^\lambda$ is $\mathbf{t}$, and the arc-length s of $\boldsymbol{\gamma}^\lambda$ satisfies $ds/dt = 1 - \lambda\kappa_s$. Hence, the curvature of $\boldsymbol{\gamma}^\lambda$ is $\| d\mathbf{t}/ds \| = \| \dot{\mathbf{t}} \| /(1-\lambda\kappa_s) = \kappa_s/(1-\lambda\kappa_s)$.

2.8 The circle passes through $\boldsymbol{\gamma}(s)$ because $\| \boldsymbol{\epsilon} - \boldsymbol{\gamma} \| = \| \frac{1}{\kappa_s \mathbf{n}_s} \| = 1/|\kappa_s|$, which is the radius of the circle. It is tangent to $\boldsymbol{\gamma}$ at this point because $\boldsymbol{\epsilon} - \boldsymbol{\gamma} = \frac{1}{\kappa_s}\mathbf{n}_s$ is perpendicular to the tangent $\mathbf{t}$ of $\boldsymbol{\gamma}$. The curvature of the circle is the reciprocal of its radius, i.e. $|\kappa_s|$, which is also the curvature of $\boldsymbol{\gamma}$.

2.9 The tangent vector of $\boldsymbol{\epsilon}$ is $\mathbf{t} + \frac{1}{\kappa_s}(-\kappa_s\mathbf{t}) - \frac{\dot{\kappa}_s}{\kappa_s^2}\mathbf{n}_s = -\frac{\dot{\kappa}_s}{\kappa_s^2}\mathbf{n}_s$ so its arc-length is $u = \int \| \dot{\boldsymbol{\epsilon}} \| \, ds = \int \frac{\dot{\kappa}_s}{\kappa_s^2} ds = u_0 - \frac{1}{\kappa_s}$, where u_0 is a constant. Hence, the unit tangent vector of $\boldsymbol{\epsilon}$ is $-\mathbf{n}_s$ and its signed unit normal is $\mathbf{t}$. Since $-d\mathbf{n}_s/du = \kappa_s\mathbf{t}/(du/ds) = \frac{\kappa_s^3}{\dot{\kappa}_s}\mathbf{t}$, the signed curvature of $\boldsymbol{\epsilon}$ is $\kappa_s^3/\dot{\kappa}_s$.
Denoting d/dt by a dash, $\boldsymbol{\gamma}' = a(1-\cos t, \sin t)$ so the arc-length s of $\boldsymbol{\gamma}$ is given by $ds/dt = 2a\sin(t/2)$ and its unit tangent vector is $\mathbf{t} = \dot{\boldsymbol{\gamma}} = (\sin(t/2), \cos(t/2))$. So $\mathbf{n}_s = (-\cos(t/2), \sin(t/2))$ and $\dot{\mathbf{t}} = (d\mathbf{t}/dt)/(ds/dt) = \frac{1}{4a\sin(t/2)}(\cos(t/2), -\sin(t/2)) = -1/4a\sin(t/2)\mathbf{n}_s$, so the signed curvature of $\boldsymbol{\gamma}$ is $-1/4a\sin(t/2)$. Its evolute is therefore $\boldsymbol{\epsilon} = a(t-\sin t, 1-\cos t) - 4a\sin(t/2)(-\cos(t/2), \sin(t/2)) = a(t+\sin t, -1+\cos t)$. Reparametrising by $\tilde{t} = \pi + t$, we get $a(\tilde{t}-\sin\tilde{t}, 1-\cos\tilde{t}) + a(-\pi, -2)$, which is obtained from a reparametrisation of $\boldsymbol{\gamma}$ by translating by the vector $a(-\pi, -2)$.

2.10 The free part of the string is tangent to $\boldsymbol{\gamma}$ at $\boldsymbol{\gamma}(s)$ and has length $\ell - s$, hence the stated formula for $\boldsymbol{\iota}(s)$. The tangent vector of $\boldsymbol{\iota}$ is $\dot{\boldsymbol{\gamma}} - \dot{\boldsymbol{\gamma}} + (\ell - s)\ddot{\boldsymbol{\gamma}} = \kappa_s(\ell - s)\mathbf{n}_s$ (a dot denotes d/ds). The arc-length v of $\boldsymbol{\iota}$ is given by $dv/ds = \kappa_s(\ell - s)$ so its unit tangent vector is $\mathbf{n}_s$ and its signed unit normal is $-\mathbf{t}$. Now $d\mathbf{n}_s/dv = \frac{1}{\kappa_s(\ell-s)}\dot{\mathbf{n}}_s = \frac{-1}{\ell-s}\mathbf{t}$, so the signed curvature of $\boldsymbol{\iota}$ is $1/(\ell - s)$.

2.11 (i) With the notation in Exercise 2.9, the involute of $\boldsymbol{\epsilon}$ is

$$\boldsymbol{\iota}(u) = \boldsymbol{\epsilon} + (\ell - u)\frac{d\boldsymbol{\epsilon}}{du} = \boldsymbol{\gamma} + \frac{1}{\kappa_s}\mathbf{n}_s - (\ell - u)\mathbf{n}_s = \boldsymbol{\gamma} - (\ell - u_0)\mathbf{n}_s,$$

since $u = u_0 - \frac{1}{\kappa_s}$, so $\boldsymbol{\iota}$ is the parallel curve $\boldsymbol{\gamma}^{-(\ell - u_0)}$.
(ii) Using the results of Exercise 2.10, the evolute of $\boldsymbol{\iota}$ is

$$\boldsymbol{\iota} + (\ell - s)(-\mathbf{t}) = \boldsymbol{\gamma} + (\ell - s)\mathbf{t} - (\ell - s)\mathbf{t} = \boldsymbol{\gamma}.$$

2.12 If we take the fixed vector in Proposition 2.2 to be parallel to the line of reflection, the effect of the reflection is to change φ to $-\varphi$, and hence to change κ_s to $-\kappa_s$.

2.13 If two unit-speed curves have the same non-zero curvature, their signed curvatures are either the same or differ in sign. In the first case the curves differ by a rigid motion by Theorem 2.1; in the latter case, applying a reflection to one curve gives two curves with the same signed curvature by Exercise 2.12, and these curves then differ by a rigid motion.

2.14 (i) $\mathbf{t} = (\frac{1}{2}(1+t)^{1/2}, -\frac{1}{2}(1-t)^{1/2}, \frac{1}{\sqrt{2}})$ is a unit vector so $\boldsymbol{\gamma}$ is unit-speed; $\dot{\mathbf{t}} = (\frac{1}{4}(1+t)^{-1/2}, \frac{1}{4}(1-t)^{-1/2}, 0)$, so $\kappa = \| \dot{\mathbf{t}} \| = 1/\sqrt{8(1-t^2)}$; $\mathbf{n} = \frac{1}{\kappa}\dot{\mathbf{t}} = \frac{1}{\sqrt{2}}((1-t)^{1/2}, (1+t)^{1/2}, 0)$; $\mathbf{b} = \mathbf{t} \times \mathbf{n} = (-\frac{1}{2}(1+t)^{1/2}, \frac{1}{2}(1-t)^{1/2}, \frac{1}{\sqrt{2}})$; $\dot{\mathbf{b}} = (-\frac{1}{4}(1+t)^{-1/2}, -\frac{1}{4}(1-t)^{-1/2}, 0)$ so the torsion $\tau = 1/\sqrt{8(1-t^2)}$. The equation $\dot{\mathbf{n}} = -\kappa\mathbf{t} + \tau\mathbf{n}$ is easily checked.
(ii) $\mathbf{t} = (-\frac{4}{5}\sin t, -\cos t, \frac{3}{5}\sin t)$ is a unit vector so $\boldsymbol{\gamma}$ is unit-speed; $\dot{\mathbf{t}} = (-\frac{4}{5}\cos t, \sin t, \frac{3}{5}\cos t)$, so $\kappa = \| \dot{\mathbf{t}} \| = 1$; $\mathbf{n} = \frac{1}{\kappa}\dot{\mathbf{t}} = (-\frac{4}{5}\cos t, \sin t, \frac{3}{5}\cos t)$; $\mathbf{b} = \mathbf{t} \times \mathbf{n} = (-\frac{3}{5}, 0, -\frac{4}{5})$, so $\dot{\mathbf{b}} = \mathbf{0}$ and $\tau = 0$.

2.15 Show that $\tau = 0$ or observe that $x = \frac{1+t^2}{t}, y = t+1, z = \frac{1-t}{t}$ satisfy $x - y - z = 0$.

2.16 By Proposition 2.5, $\boldsymbol{\gamma}$ is a circle of radius $1/\kappa = 1$ with centre $\boldsymbol{\gamma} + \frac{1}{\kappa}\mathbf{n} = (0,1,0)$ in the plane passing through $(0,1,0)$ perpendicular to $\mathbf{b} = (-\frac{3}{5}, 0, -\frac{4}{5})$, i.e. the plane $3x + 4z = 0$.

2.17 Let $a = \kappa/(\kappa^2+\tau^2)$, $b = \tau/(\kappa^2+\tau^2)$. By Examples 2.1 and 2.4, the circular helix with parameters a and b has curvature $a/(a^2+b^2) = \kappa$ and torsion $b/(a^2+b^2) = \tau$. By Theorem 2.3, every curve with curvature κ and torsion τ is obtained by applying a rigid motion to this helix.

2.18 This follows from Proposition 2.3 since the numerator and denominator of the expression in (11) are smooth functions of t.

2.19 Let a dot denote d/dt. Then, $\dot{\boldsymbol{\delta}} = \ddot{\boldsymbol{\gamma}} = \kappa\mathbf{n}$, so the unit tangent vector of $\boldsymbol{\delta}$ is $\mathbf{T} = \mathbf{n}$ and its arc-length s satisfies $ds/dt = \kappa$. Now $d\mathbf{T}/ds = \dot{\mathbf{n}}/(ds/dt) = \kappa^{-1}(-\kappa\mathbf{t} + \tau\mathbf{b}) = -\mathbf{t} + \frac{\tau}{\kappa}\mathbf{b}$. Hence, the curvature of $\boldsymbol{\delta}$ is $\| -\mathbf{t} + \frac{\tau}{\kappa}\mathbf{b} \| = (1 + \frac{\tau^2}{\kappa^2})^{1/2} = \mu$, say. The principal normal of $\boldsymbol{\delta}$ is $\mathbf{N} = \mu^{-1}(-\mathbf{t} + \frac{\tau}{\kappa}\mathbf{b})$ and its binormal is $\mathbf{B} = \mathbf{T} \times \mathbf{N} == \mu^{-1}(\mathbf{b} + \frac{\tau}{\kappa}\mathbf{t})$. The torsion T of $\boldsymbol{\delta}$ is given by $d\mathbf{B}/ds = -T\mathbf{N}$, i.e. $\kappa^{-1}\dot{\mathbf{B}} = -T\mathbf{N}$. Computing the derivatives and equating coefficients of $\mathbf{b}$ gives $T = (\kappa\dot{\tau} - \tau\dot{\kappa})/\kappa(\kappa^2+\tau^2)$.

2.20 Differentiating $\mathbf{t.a}$ (= constant) gives $\mathbf{n.a} = 0$; since $\mathbf{t}, \mathbf{n}, \mathbf{b}$ are an orthonormal basis of $\mathbf{R}^3$, $\mathbf{a} = \mathbf{t}\cos\theta + \mu\mathbf{b}$ for some scalar μ; since $\mathbf{a}$ is a unit vector, $\mu = \pm\sin\theta$; differentiating $\mathbf{a} = \mathbf{t}\cos\theta \pm \mathbf{b}\sin\theta$ gives $\tau = \kappa\cot\theta$. Conversely, if $\tau = \lambda\kappa$, there exists θ with $\lambda = \cot\theta$; differentiating shows that $\mathbf{a} = \mathbf{t}\cos\theta + \mathbf{b}\sin\theta$ is a constant vector and $\mathbf{t.a} = \cos\theta$ shows that θ is the angle between $\mathbf{t}$ and $\mathbf{a}$.

2.21 Differentiating $(\boldsymbol{\gamma} - \mathbf{a}).(\boldsymbol{\gamma} - \mathbf{a}) = r^2$ repeatedly gives $\mathbf{t}.(\boldsymbol{\gamma} - \mathbf{a}) = 0$; $\mathbf{t.t} + \kappa\mathbf{n}.(\boldsymbol{\gamma} - \mathbf{a}) = 0$, and so $\mathbf{n}.(\boldsymbol{\gamma} - \mathbf{a}) = -1/\kappa$; $\mathbf{n.t} + (-\kappa\mathbf{t} + \tau\mathbf{b}).(\boldsymbol{\gamma} - \mathbf{a}) = \dot{\kappa}/\kappa^2$, and so $\mathbf{b}.(\boldsymbol{\gamma} - \mathbf{a}) = \dot{\kappa}/\tau\kappa^2$; and finally $\mathbf{b.t} - \tau\mathbf{n}.(\boldsymbol{\gamma} - \mathbf{a}) = (\dot{\kappa}/\tau\kappa^2)\dot{}$, and so $\tau/\kappa = (\dot{\kappa}/\tau\kappa^2)\dot{}$.
Conversely, if Eq. (20) holds, then $\rho = -\sigma(\dot{\rho}\sigma)\dot{}$, so $(\rho^2 + (\dot{\rho}\sigma)^2)\dot{} = 2\rho\dot{\rho} + 2(\dot{\rho}\sigma)(\dot{\rho}\sigma)\dot{} = 0$, hence $\rho^2 + (\dot{\rho}\sigma)^2$ is a constant, say r^2 (where $r > 0$). Let $\mathbf{a} = \boldsymbol{\gamma} + \rho\mathbf{n} + \dot{\rho}\sigma\mathbf{b}$; then $\dot{\mathbf{a}} = \mathbf{t} + \dot{\rho}\mathbf{n} + \rho(-\kappa\mathbf{t} + \tau\mathbf{b}) + (\dot{\rho}\sigma)\dot{}\mathbf{b} + (\dot{\rho}\sigma)(-\tau\mathbf{n}) = \mathbf{0}$ using Eq. (20); so $\mathbf{a}$ is a constant vector and $\| \boldsymbol{\gamma} - \mathbf{a} \|^2 = \rho^2 + (\dot{\rho}\sigma)^2 = r^2$,

hence $\boldsymbol{\gamma}$ is contained in the sphere with centre $\mathbf{a}$ and radius r.

2.22 Let $\lambda_{ij} = \mathbf{v}_i.\mathbf{v}_j$. The vectors $\mathbf{v}_1, \mathbf{v}_2, \mathbf{v}_3$ are orthonormal if and only if $\lambda_{ij} = \delta_{ij}$ ($= 1$ if $i = j$ and $= 0$ if $i \neq j$). So it is enough to prove that $\lambda_{ij} = \delta_{ij}$ for all values of s given that it holds for $s = s_0$. Differentiating $\mathbf{v}_i.\mathbf{v}_j$ gives

$$\dot{\lambda}_{ij} = \sum_{k=1}^{3}(a_{ik}\lambda_{kj} + a_{jk}\lambda_{ik}).$$

Now $\lambda_{ij} = \delta_{ij}$ is a solution of this system of differential equations because $a_{ij} + a_{ji} = 0$. But the theory of ordinary differential equations tells us that there is a unique solution with given values when $s = s_0$.

Chapter 3

3.1 If $\tilde{\boldsymbol{\gamma}}$ is obtained from $\boldsymbol{\gamma}$ by a translation, then $\dot{\tilde{\boldsymbol{\gamma}}} = \dot{\boldsymbol{\gamma}}$ so by Eqs. (1) and (3) the length and area of $\tilde{\boldsymbol{\gamma}}$ are the same as those of $\boldsymbol{\gamma}$. If $\tilde{\boldsymbol{\gamma}}$ is obtained from $\boldsymbol{\gamma}$ by a rotation by an angle θ about the origin, then $\tilde{x} = x\cos\theta - y\sin\theta$, $\tilde{y} = x\sin\theta + y\cos\theta$. This implies that $\dot{\tilde{x}}^2 + \dot{\tilde{y}}^2 = \dot{x}^2 + \dot{y}^2$ and $\tilde{x}\dot{\tilde{y}} - \tilde{y}\dot{\tilde{x}} = x\dot{y} - y\dot{x}$, so Eqs. (1) and (3) again show that the length and area are unchanged.

3.2 $\boldsymbol{\gamma}(t') = \boldsymbol{\gamma}(t) \iff \cos t' = \cos t$ and $\sin t' = \sin t \iff t' - t$ is a multiple of 2π, so $\boldsymbol{\gamma}$ is simple closed with period 2π. Taking $x = a\cos t, y = b\sin t$ in Eq. (3) gives the area as $\frac{1}{2}\int_0^{2\pi} ab\, dt = \pi ab$.

3.3 $\dot{\boldsymbol{\gamma}} = (-\sin t - 2\sin 2t, \cos t + 2\cos 2t)$ so $\parallel \dot{\boldsymbol{\gamma}} \parallel = \sqrt{5 + 4\cos t}$, which is never zero, so $\boldsymbol{\gamma}$ is regular. $\boldsymbol{\gamma}(t + 2\pi) = \boldsymbol{\gamma}(t)$ is obvious. If $\boldsymbol{\gamma}$ is simple closed with period a, then $\boldsymbol{\gamma}(0) = \boldsymbol{\gamma}(a)$ implies $(1 + 2\cos a)\cos a = 3$ so $\cos a = 1$ and $a = 2\pi$. But $\boldsymbol{\gamma}(2\pi/3) = \boldsymbol{\gamma}(4\pi/3)$ and $4\pi/3 - 2\pi/3$ is not a multiple of 2π.

3.4 Differentiating $\boldsymbol{\gamma}(t + a) = \boldsymbol{\gamma}(t)$ gives $\mathbf{t}(t + a) = \mathbf{t}(a)$; rotating anti-clockwise by $\pi/2$ gives $\mathbf{n}_s(t+a) = \mathbf{n}_s(t)$; differentiating again gives $\kappa_s(t+a)\mathbf{n}_s(t+a) = \kappa_s(t)\mathbf{n}_s(t)$, so $\kappa_s(t + a) = \kappa_s(t)$.

3.5 By Exercise 3.2 and the isoperimetric inequality, the length ℓ of the ellipse satisfies $\ell \geq \sqrt{4\pi \times \pi ab} = 2\pi\sqrt{ab}$, with equality if and only if the ellipse is a circle, i.e. $a = b$. But parametrising the ellipse as in Exercise 3.2, its length is

$$\int_0^{2\pi} \parallel \dot{\boldsymbol{\gamma}} \parallel\, dt = \int_0^{2\pi} \sqrt{a^2\sin^2 t + b^2\cos^2 t}\, dt.$$

3.6 Let (x_1, y_1) and (x_2, y_2) be points in the interior of the ellipse, so that $\frac{x_i^2}{a^2} + \frac{y_i^2}{b^2} < 1$ for $i = 1, 2$. A point of the line segment joining the two points is $(tx_1 + (1-t)x_2, ty_1 + (1-t)y_2)$ for some $0 \leq t \leq 1$. This is in the interior of the ellipse because

$$\frac{(tx_1+(1-t)x_2)^2}{a^2}+\frac{(ty_1+(1-t)y_2)^2}{b^2}$$
$$=t^2\left(\frac{x_1^2}{a^2}+\frac{y_1^2}{b^2}\right)+(1-t)^2\left(\frac{x_2^2}{a^2}+\frac{y_2^2}{b^2}\right)+2t(1-t)\left(\frac{x_1x_2}{a^2}+\frac{y_1y_2}{b^2}\right)$$
$$<t^2+(1-t)^2+t(1-t)\left(\frac{x_1^2}{a^2}+\frac{y_1^2}{b^2}+\frac{x_2^2}{a^2}+\frac{y_2^2}{b^2}\right)$$
$$<t^2+(1-t)^2+2t(1-t)=1.$$

3.7 If $\tilde{\boldsymbol{\gamma}}(\tilde{t})$ is a reparametrisation of $\boldsymbol{\gamma}(t)$, the vertices of $\boldsymbol{\gamma}$ (resp. $\tilde{\boldsymbol{\gamma}}$) are given by $d\kappa_s/dt = 0$ (resp. $d\kappa_s/d\tilde{t} = 0$). Since $d\kappa_s/dt = (d\kappa_s/d\tilde{t})(d\tilde{t}/dt)$ and since $d\tilde{t}/dt$ is never zero, $\boldsymbol{\gamma}$ and $\tilde{\boldsymbol{\gamma}}$ have the same vertices.

3.8 $\dot{\boldsymbol{\gamma}} = (-\sin t - 2\sin 2t, \cos t + 2\cos 2t)$ and $\parallel \dot{\boldsymbol{\gamma}} \parallel = \sqrt{5+4\cos t}$, so the angle φ between $\dot{\boldsymbol{\gamma}}$ and the x-axis is given by

$$\cos\varphi = \frac{-\sin t - 2\sin 2t}{\sqrt{5+4\cos t}}, \quad \sin\varphi = \frac{\cos t + 2\cos 2t}{\sqrt{5+4\cos t}}.$$

Differentiating the second equation gives $\dot{\varphi}\cos\varphi = -\sin t(24\cos^2 t + 42\cos t + 9)/(5+4\cos t)^{3/2}$, so $\dot{\varphi} = \sin t(24\cos^2 t + 42\cos t + 9)/(5 + 4\cos t)(\sin t + 2\sin 2t) = (9+6\cos t)/(5+4\cos t)$. Hence, if s is the arc-length of $\boldsymbol{\gamma}$, $\kappa_s = d\varphi/ds = (d\varphi/dt)/(ds/dt) = (9+6\cos t)/(5+4\cos t)^{3/2}$, so $\dot{\kappa}_s = 12\sin t(2+\cos t)/(5+4\cos t)^{5/2}$. This vanishes at only two points of the curve, where $t = 0$ and $t = \pi$.

Chapter 4

4.1 Let U be an open disc in $\mathbf{R}^2$ and $\mathcal{S} = \{(x,y,z) \in \mathbf{R}^3 \mid (x,y) \in U,\ z = 0\}$. If $W = \{(x,y,z) \in \mathbf{R}^3 \mid (x,y) \in U\}$, then W is an open subset of $\mathbf{R}^3$, and $\mathcal{S} \cap W$ is homeomorphic to U by $(x,y,0) \mapsto (x,y)$. So $\mathcal{S}$ is a surface.

4.2 Let $U = \{(u,v) \in \mathbf{R}^2 \mid 0 < u^2 + v^2 < \pi^2\}$, let $r = \sqrt{u^2+v^2}$, and define $\boldsymbol{\sigma} : U \to \mathbf{R}^3$ by $\boldsymbol{\sigma}(u,v) = (\frac{u}{r}, \frac{v}{r}, \tan(r - \frac{\pi}{2}))$.

4.3 The image of $\boldsymbol{\sigma}_\pm^x$ is the intersection of the sphere with the open set $\pm x > 0$ in $\mathbf{R}^3$, and its inverse is the projection $(x,y,z) \mapsto (y,z)$. Similarly for $\boldsymbol{\sigma}_\pm^y$ and $\boldsymbol{\sigma}_\pm^z$. A point of the sphere not in the image of any of the six patches would have to have x, y and z all zero, which is impossible.

4.4 Multiplying the two equations gives $(x^2-z^2)\sin\theta\cos\theta = (1-y^2)\sin\theta\cos\theta$, so $x^2+y^2-z^2 = 1$ unless $\cos\theta = 0$ or $\sin\theta = 0$; if $\cos\theta = 0$, then $x = -z$ and $y = 1$ and if $\sin\theta = 0$ then $x = z$ and $y = -1$, and both of these lines are also contained in the surface.

The given line L_θ passes through $(\sin 2\theta, -\cos 2\theta, 0)$ and is parallel to the vector $(\cos 2\theta, \sin 2\theta, 1)$; it follows that we get all of the lines by taking $0 \leq \theta < \pi$. Let (x, y, z) be a point of the surface; if $x \neq z$, let θ be such that $\cot\theta = (1-y)/(x-z)$; then (x, y, z) is on L_θ; similarly if $x \neq -z$. The only remaining cases are the points $(0, 0, \pm 1)$, which lie on the lines $L_{\pi/2}$ and L_0.

With the notation of Exercise 4.2, define $\boldsymbol{\sigma} : U \to \mathbf{R}^3$ by $\boldsymbol{\sigma}(u, v) = (\sin 2\theta, -\cos 2\theta, 0) + t(\cos 2\theta, \sin 2\theta, 1)$, where $t = \tan(r - \frac{\pi}{2})$, $\cos\theta = u/r$ and $\sin\theta = v/r$. This is a surface patch which, by the preceding paragraph, covers the whole surface.

Let M_φ be the line

$$(x - z)\cos\varphi = (1 + y)\sin\varphi, \quad (x + z)\sin\varphi = (1 - y)\cos\varphi.$$

By the same argument as above, M_φ is contained in the surface and every point of the surface lies on some M_φ with $0 \leq \varphi < \pi$. If $\theta + \varphi$ is not a multiple of π, the lines L_θ and M_φ intersect in the point $\left(\frac{\cos(\theta-\varphi)}{\sin(\theta+\varphi)}, \frac{\sin(\theta-\varphi)}{\sin(\theta+\varphi)}, \frac{\cos(\theta+\varphi)}{\sin(\theta+\varphi)}\right)$; for each θ with $0 \leq \theta < \pi$, there is exactly one φ with $0 \leq \varphi < \pi$ such that $\theta + \varphi$ is a multiple of π, and the lines L_θ and M_φ do not intersect. If (x, y, z) lies on both L_θ and L_φ, with $\theta \neq \varphi$, then $(1 - y)\tan\theta = (1 - y)\tan\varphi$ and $(1 + y)\cot\theta = (1 - y)\cot\varphi$, which gives both $y = 1$ and $y = -1$ (the case in which $\theta = 0$ and $\varphi = \pi/2$, or vice versa, has to be treated separately, but the conclusion is the same). This shows that L_θ and L_φ do not intersect; similarly, M_θ and M_φ do not intersect.

4.5 If the sphere $\mathcal{S}$ could be covered by a single surface patch $\boldsymbol{\sigma} : U \to \mathbf{R}^3$, then $\mathcal{S}$ would be homeomorphic to the open subset U of $\mathbf{R}^2$. As $\mathcal{S}$ is a closed and bounded subset of $\mathbf{R}^3$, it is compact. Hence, U would be compact, and hence closed. But, since $\mathbf{R}^2$ is connected, the only non-empty subset of $\mathbf{R}^2$ that is both open and closed is $\mathbf{R}^2$ itself, and this is not compact as it is not bounded.

4.6 $\boldsymbol{\sigma}$ is obviously smooth and $\boldsymbol{\sigma}_u \times \boldsymbol{\sigma}_v = (-f_u, -f_v, 1)$ is nowhere zero, so $\boldsymbol{\sigma}$ is regular.

4.7 $\boldsymbol{\sigma}^z_\pm$ is a special case of Exercise 4.6, with $f = \pm\sqrt{1 - u^2 - v^2}$ ($\sqrt{1 - u^2 - v^2}$ is smooth because $1 - u^2 - v^2 > 0$ if $(u, v) \in U$); similarly for the other patches. The transition map from $\boldsymbol{\sigma}^x_+$ to $\boldsymbol{\sigma}^y_+$, for example, is $\Phi(\tilde{u}, \tilde{v}) = (u, v)$, where $\boldsymbol{\sigma}^y_+(\tilde{u}, \tilde{v}) = \boldsymbol{\sigma}^x_+(u, v)$; so $u = \sqrt{1 - \tilde{u}^2 - \tilde{v}^2}$, $v = \tilde{v}$, and this is smooth since $1 - \tilde{u}^2 - \tilde{v}^2 > 0$ if $(\tilde{u}, \tilde{v}) \in U$.

4.8 $\boldsymbol{\sigma}$ is a smooth map on the open set $R = \{(r, \theta) \in \mathbf{R}^2 \mid r > 0\}$, and its image is contained in the surface because $\cosh^2\theta - \sinh^2\theta = 1$; and $\boldsymbol{\sigma}_r \times \boldsymbol{\sigma}_\theta = (-2r^2\cosh\theta, 2r^2\sinh\theta, r)$, which is never zero on R.

Exercise 4.6 gives the parametrisation $\tilde{\boldsymbol{\sigma}}(u, v) = (u, v, u^2 - v^2)$, defined on

the open set $U = \{(u,v) \in \mathbf{R}^2 \mid u^2 - v^2 > 0\}$. This is a reparametrisation of $\boldsymbol{\sigma}$ via the reparametrisation map $(u,v) \mapsto (r,\theta)$, where $r = \sqrt{u^2 - v^2}$, $\theta = \cosh^{-1}(u/\sqrt{u^2 - v^2})$ (this is smooth because $u^2 - v^2 > 0$).
For the part with $z < 0$, we can take $\boldsymbol{\sigma}(r,\theta) = (r\sinh\theta, r\cosh\theta, -r^2)$ defined on the open set R; and $\tilde{\boldsymbol{\sigma}}(u,v) = (u, v, u^2 - v^2)$ defined on the open set $V = \{(u,v) \in \mathbf{R}^2 \mid u^2 - v^2 < 0\}$.

4.9 This is similar to Example 4.5, but using the 'latitude longitude' patch $\boldsymbol{\sigma}(\theta,\varphi) = (a\cos\theta\cos\varphi, b\cos\theta\sin\varphi, c\sin\theta)$. Alternatively, one can use Theorem 4.1, noting that if $f(x,y,z) = \frac{x^2}{a^2} + \frac{y^2}{b^2} + \frac{z^2}{c^2} - 1$, then $(f_x, f_y, f_z) = (2x/a^2, 2y/b^2, 2z/c^2)$ vanishes only at the origin, and hence at no point of the ellipsoid.

4.10 A typical point on the circle $\mathcal{C}$ has coordinates $(a + b\cos\theta, 0, b\sin\theta)$; rotating this about the z-axis through an angle φ gives the point $\boldsymbol{\sigma}(\theta,\varphi)$; the whole of the torus is covered by the four patches obtained by taking (θ,φ) to lie in one of the following open sets: (i) $0 < \theta < 2\pi, 0 < \varphi < 2\pi$, (ii) $0 < \theta < 2\pi, -\pi < \varphi < \pi$, (iii) $-\pi < \theta < \pi, 0 < \varphi < 2\pi$, (iv) $-\pi < \theta < \pi, -\pi < \varphi < \pi$. Each patch is regular because $\boldsymbol{\sigma}_\theta \times \boldsymbol{\sigma}_\varphi = -b(a + b\cos\theta)(\cos\theta\cos\varphi, \cos\theta\sin\varphi, 1)$ is never zero (since $a + b\cos\theta \geq a - b > 0$).
Let $\boldsymbol{\sigma}(\theta,\varphi) = (x,y,z)$. Then, $x^2 + y^2 + z^2 + a^2 - b^2 = 2a(a + b\cos\theta)$, so $(x^2+y^2+z^2+a^2-b^2)^2 = 4a^2(a+b\cos\theta)^2 = 4a^2(x^2+y^2)$. Let $f(x,y,z)$ be the left-hand side minus the right-hand side; then, $f_x = 4x(x^2 + y^2 + z^2 - a^2 - b^2)$, $f_y = 4y(x^2 + y^2 + z^2 - a^2 - b^2)$, $f_z = 4z(x^2 + y^2 + z^2 + a^2 - b^2)$; if $f_z = 0$ then $z = 0$ since $x^2 + y^2 + z^2 + a^2 - b^2 > 0$ everywhere on the torus; if $f_x = f_y = 0$ too, then since the origin is not on the torus, we must have $x^2 + y^2 = a^2 + b^2$, but then substituting into the equation of the torus gives $(2a^2)^2 = 4a^2(a^2 + b^2)$, a contradiction. So (f_x, f_y, f_z) is nowhere zero on the torus, which is therefore a smooth surface by Theorem 4.1.

4.11 If $\mathcal{S}$ is covered by a single surface patch $\boldsymbol{\sigma} : U \to \mathbf{R}^3$, then $f : \mathcal{S} \to \mathbf{R}$ is smooth if and only if $f \circ \boldsymbol{\sigma} : U \to \mathbf{R}$ is smooth. We must check that, if $\tilde{\boldsymbol{\sigma}} : \tilde{U} \to \mathbf{R}^3$ is another patch covering $\mathcal{S}$, then $f \circ \tilde{\boldsymbol{\sigma}}$ is smooth if and only if $f \circ \boldsymbol{\sigma}$ is smooth. This is true because $f \circ \tilde{\boldsymbol{\sigma}} = (f \circ \boldsymbol{\sigma}) \circ \Phi$, where Φ is the transition map from $\boldsymbol{\sigma}$ to $\tilde{\boldsymbol{\sigma}}$, and both Φ and Φ^{-1} are smooth.

4.12 If $\tilde{\boldsymbol{\sigma}} = \boldsymbol{\sigma} + \mathbf{a}$, where $\mathbf{a}$ is a constant vector, then $\tilde{\boldsymbol{\sigma}}$ is smooth if $\boldsymbol{\sigma}$ is smooth, and $\tilde{\boldsymbol{\sigma}}_u = \boldsymbol{\sigma}_u, \tilde{\boldsymbol{\sigma}}_v = \boldsymbol{\sigma}_v$, so $\tilde{\boldsymbol{\sigma}}$ is regular if $\boldsymbol{\sigma}$ is regular.
If $\tilde{\boldsymbol{\sigma}} = A \circ \boldsymbol{\sigma}$, where A is a linear transformation of $\mathbf{R}^3$, then $\tilde{\boldsymbol{\sigma}}$ is smooth if $\boldsymbol{\sigma}$ is smooth, and $\tilde{\boldsymbol{\sigma}}_u = A(\boldsymbol{\sigma}_u), \tilde{\boldsymbol{\sigma}}_v = A(\boldsymbol{\sigma}_v)$, so, if A is invertible, $\tilde{\boldsymbol{\sigma}}_u$ and $\tilde{\boldsymbol{\sigma}}_v$ are linearly independent if $\boldsymbol{\sigma}_u$ and $\boldsymbol{\sigma}_v$ are linearly independent.

4.13 (i) At $(1,1,0)$, $\boldsymbol{\sigma}_u = (1,0,2)$, $\boldsymbol{\sigma}_v = (0,1,-2)$ so $\boldsymbol{\sigma}_u \times \boldsymbol{\sigma}_v = (-2,2,1)$ and the tangent plane is $-2x + 2y + z = 0$.
(ii) At $(1,0,1)$, where $r = 1, \theta = 0, \boldsymbol{\sigma}_r = (1,0,2), \boldsymbol{\sigma}_\theta = (0,1,2)$ so $\boldsymbol{\sigma}_r \times \boldsymbol{\sigma}_\theta =$

$(-2, -2, 1)$ and the equation of the tangent plane is $-2x - 2y + z = 0$.

4.14 Suppose the centre of the propeller is initially at the origin. At time t, the centre is at $(0, 0, \alpha t)$ where α is the speed of the aeroplane. If the propeller is initially along the x-axis, the point initially at $(v, 0, 0)$ is therefore at the point $(v \cos \omega t, v \sin \omega t, \alpha t)$ at time t, where ω is the angular velocity of the propeller. Let $u = \omega t$, $\lambda = \alpha/\omega$.
$\boldsymbol{\sigma}_u = (-v \sin u, v \cos u, \lambda)$, $\boldsymbol{\sigma}_v = (\cos u, \sin u, 0)$, so the standard unit normal is $\mathbf{N} = (\lambda^2 + v^2)^{-1/2}(-\lambda \sin u, \lambda \cos u, -v)$. If θ is the angle between $\mathbf{N}$ and the z-axis, $\cos\theta = -v/(\lambda^2 + v^2)^{1/2}$ and hence $\cot\theta = \pm v/\lambda$, while the distance from the z-axis is v.

4.15 Let $\tilde{\boldsymbol{\sigma}}(\tilde{u}, \tilde{v})$ be a reparametrisation of $\boldsymbol{\sigma}$. Then,

$$\boldsymbol{\sigma}_u = \frac{\partial \tilde{u}}{\partial u}\tilde{\boldsymbol{\sigma}}_{\tilde{u}} + \frac{\partial \tilde{v}}{\partial u}\tilde{\boldsymbol{\sigma}}_{\tilde{v}}, \quad \boldsymbol{\sigma}_v = \frac{\partial \tilde{u}}{\partial v}\tilde{\boldsymbol{\sigma}}_{\tilde{u}} + \frac{\partial \tilde{v}}{\partial v}\tilde{\boldsymbol{\sigma}}_{\tilde{v}},$$

so $\boldsymbol{\sigma}_u$ and $\boldsymbol{\sigma}_v$ are linear combinations of $\tilde{\boldsymbol{\sigma}}_{\tilde{u}}$ and $\tilde{\boldsymbol{\sigma}}_{\tilde{v}}$. Hence, any linear combination of $\boldsymbol{\sigma}_u$ and $\boldsymbol{\sigma}_v$ is a linear combination of $\tilde{\boldsymbol{\sigma}}_{\tilde{u}}$ and $\tilde{\boldsymbol{\sigma}}_{\tilde{v}}$. The converse is also true since $\boldsymbol{\sigma}$ is a reparametrisation of $\tilde{\boldsymbol{\sigma}}$.

4.16 If $\boldsymbol{\gamma}(t) = (x(t), y(t), z(t))$ is a curve in $\mathcal{S}$, differentiating $f(x(t), y(t), z(t)) = 0$ gives $f_x\dot{x} + f_y\dot{y} + f_z\dot{z} = 0$, i.e. $\boldsymbol{\nabla} f.\dot{\boldsymbol{\gamma}} = 0$, showing that $\boldsymbol{\nabla} f$ is perpendicular to the tangent vector of every curve in $\mathcal{S}$, and hence to the tangent plane of $\mathcal{S}$. Since $\mathcal{S}$ has a canonical (smooth) choice of unit normal $\boldsymbol{\nabla} f / \parallel \boldsymbol{\nabla} f \parallel$ at each point, it is orientable.

4.17 See the proof of Proposition 11.1 for the first part. By the argument in Exercise 4.16, $\boldsymbol{\nabla} F.\dot{\boldsymbol{\gamma}}(0) = \frac{d}{dt}F(\boldsymbol{\gamma}(t))|_{t=0}$, so $\boldsymbol{\nabla}_{\mathcal{S}} F - \boldsymbol{\nabla} F$ is perpendicular to the tangent plane of $\mathcal{S}$ at P. This proves that $\boldsymbol{\nabla}_{\mathcal{S}} F$ is the perpendicular projection of $\boldsymbol{\nabla} F$ onto the tangent plane. If F has a local maximum or minimum at P, then $t \mapsto F(\boldsymbol{\gamma}(t))$ has a local maximum or minimum at $t = 0$ for all curves $\boldsymbol{\gamma}$ on $\mathcal{S}$ with $\boldsymbol{\gamma}(0) = P$; hence, $\boldsymbol{\nabla}_{\mathcal{S}} F = \mathbf{0}$ at P, so $\boldsymbol{\nabla} F$ is perpendicular to the tangent plane of $\mathcal{S}$ at P, and hence is parallel to $\boldsymbol{\nabla} f$ by Exercise 4.16. So $\boldsymbol{\nabla} F = \lambda \boldsymbol{\nabla} f$ for some scalar λ.

4.18 From Example 4.13, the surface can be parametrised by $\boldsymbol{\sigma}(u, v) = (\cosh u \cos v, \cosh u \sin v, u)$, with $u \in \mathbf{R}$ and $-\pi < v < \pi$ or $0 < v < 2\pi$.

4.19 $\parallel \boldsymbol{\sigma}(u, v) \parallel^2 = \operatorname{sech}^2 u(\cos^2 v + \sin^2 v) + \tanh^2 u = \operatorname{sech}^2 u + \tanh^2 u = 1$, so $\boldsymbol{\sigma}$ parametrises part of the unit sphere; $\boldsymbol{\sigma}$ is clearly smooth; and $\boldsymbol{\sigma}_u \times \boldsymbol{\sigma}_v = -\operatorname{sech}^2 u\, \boldsymbol{\sigma}(u, v)$ is never zero, so $\boldsymbol{\sigma}$ is regular. Meridians correspond to the parameter curves $v =$ constant, and parallels to the curves $u =$ constant.

4.20 The vector $\boldsymbol{\sigma}_u$ is tangent to the meridians, so a unit-speed curve $\boldsymbol{\gamma}$ is a loxodrome if $\dot{\boldsymbol{\gamma}}.\boldsymbol{\sigma}_u / \parallel \boldsymbol{\sigma}_u \parallel = \cos\alpha$, which gives $\dot{u} = \cosh u \cos\alpha$; since $\boldsymbol{\gamma}$ is unit-speed, $\dot{\boldsymbol{\gamma}} = (-\dot{u} \operatorname{sech} u \tanh u \cos v - \dot{v} \operatorname{sech} u \sin v, -\dot{u} \operatorname{sech} u \tanh u \sin v + \dot{v} \operatorname{sech} u \cos v, \dot{u} \operatorname{sech}^2 u)$ is a unit vector; this gives $\dot{u}^2 + \dot{v}^2 = \cosh^2 u$, so $\dot{v} = \pm \cosh u \sin\alpha$. The correponding curve in the uv-plane is given by

$dv/du = \dot{v}/\dot{u} = \pm \tan\alpha$, and so is a straight line $v = \pm u \tan\alpha + c$, where c is a constant.

4.21 The point at a distance v from the z-axis on the ruling through $(0, 0, u)$ has position vector given by $\boldsymbol{\sigma}(u, v) = (0, 0, u) + v(\cos\theta(u), \sin\theta(u), 0) = (v\cos\theta(u), v\sin\theta(u), u)$; $\boldsymbol{\sigma}_u \times \boldsymbol{\sigma}_v = (-\sin\theta, -\cos\theta, -v d\theta/du)$ is clearly never zero, so $\boldsymbol{\sigma}$ is regular.

4.22 (i) $\tilde{\boldsymbol{\gamma}}.\mathbf{a} = 0$ so $\tilde{\boldsymbol{\gamma}}$ is contained in the plane perpendicular to $\mathbf{a}$ and passing through the origin; (ii) simple algebra; (iii) $\tilde{v}$ is clearly a smooth function of (u, v) and the jacobian matrix of the map $(u, v) \mapsto (u, \tilde{v})$ is $\begin{pmatrix} 1 & 0 \\ \dot{\boldsymbol{\gamma}}.\mathbf{a} & 1 \end{pmatrix}$, where a dot denotes d/du; this matrix is invertible so $\tilde{\boldsymbol{\gamma}}$ is a reparametrisation of $\boldsymbol{\gamma}$.

4.23 (i) $(a\cos u\cos v, b\cos u\sin v, c\sin u)$ (cf. Exercise 4.9); (ii) see Exercise 4.4; (iii) $(u, v, \pm\sqrt{1 + \frac{u^2}{p^2} + \frac{v^2}{q^2}})$; (iv), (v) see Exercise 4.6; (vi) see Example 4.3; (vii) $(p\cos u, q\cos u, v)$; (viii) $(\pm p\cosh u, q\sinh u, v)$; (ix) $(u, u^2/p^2, v)$; (x) $(0, u, v)$; (xi) $(\pm p, u, v)$.

4.24 (a) Types (vii)–(xi); (b) type (vi); (c) types (ii) (see Exercise 4.4), (v) (see Exercise 4.25) and (vi)–(x); (d) type (i) if p^2, q^2 and r^2 are not distinct, types (ii), (iii), (iv), (vi) and (vii) if $p^2 = q^2$, and type (x).

4.25 $z = \left(\frac{x}{p} - \frac{y}{q}\right)\left(\frac{x}{p} + \frac{y}{q}\right) = uv$, $x = \frac{1}{2}p(u + v)$, $y = \frac{1}{2}q(v - u)$, so a parametrisation is $\boldsymbol{\sigma}(u, v) = (\frac{1}{2}p(u + v), \frac{1}{2}q(v - u), uv)$; $\boldsymbol{\sigma}_u \times \boldsymbol{\sigma}_v = (-\frac{1}{2}q(u + v), \frac{1}{2}p(v - u), pq)$ is never zero so $\boldsymbol{\sigma}$ is regular. For a fixed value of u, $\boldsymbol{\sigma}(u, v) = (\frac{1}{2}pu, -\frac{1}{2}qu, 0) + v(\frac{1}{2}p, \frac{1}{2}q, u)$ is a straight line; similarly for a fixed value of v; hence the hyperbolic paraboloid is the union of each of two families of straight lines.

4.26 Substituting the components (x, y, z) of $\boldsymbol{\gamma}(t) = \mathbf{a} + \mathbf{b}t$ into the equation of the quadric gives a quadratic equation for t; if the quadric contains three points on the line, this quadratic equation has three roots, hence is identically zero, so the quadric contains the whole line.
Take three points on each of the given lines; substituting the coordinates of these nine points into the equation of the quadric gives a system of nine homogeneous linear equations for the ten coefficients $a_1, \ldots, c$ of the quadric; such a system always has a non-trivial solution. By the first part, the resulting quadric contains all three lines.

4.27 Let L_1, L_2, L_3 be three lines from the first family; by Exercise 4.26, there is a quadric $\mathcal{Q}$ containing all three lines; all but finitely many lines of the second family intersect each of the three lines; if L' is such a line, $\mathcal{Q}$ contains three points of L', and hence the whole of L' by Exercise 4.26; so $\mathcal{Q}$ contains all but finitely many lines of the second family; since any quadric is a closed subset of $\mathbf{R}^3$, $\mathcal{Q}$ must contain all the lines of the second family, and hence must contain $\mathcal{S}$.

4.28 Both parts are geometrically obvious.

4.29 Let (a,b,c) be a point of $\mathbf{R}^3$ with a and b non-zero. Then $F_t(a,b,c) \to \infty$ as $t \to \infty$ and as t approaches p^2 and q^2 from the left; and $F_t(a,b,c) \to -\infty$ as $t \to -\infty$ and as t approaches p^2 and q^2 from the right. From this and the fact that $F_t(a,b,c) = 0$ is equivalent to a cubic equation for t, it follows that there exist unique numbers u, v, w with $u < p^2$, $p^2 < v < q^2$ and $q^2 < w$ such that $F_t(a,b,c) = 0$ when $t = u, v$ or w. The surfaces $F_u(x,y,z) = 0$ and $F_w(x,y,z) = 0$ are elliptic paraboloids and $F_v(x,y,z) = 0$ is a hyperbolic paraboloid, and we have shown that there is one surface of each type passing through each point (a,b,c).
To parametrise these surfaces, write $F_t(x,y,z) = 0$ as the cubic equation $x^2(q^2-t)+y^2(p^2-t)-2z(p^2-t)(q^2-t)+t(p^2-t)(q^2-t) = 0$, and note that the left-hand side must be equal to $(t-u)(t-v)(t-w)$; putting $t = p^2, q^2$ and then equating coefficients of t^2 (say) gives $x = \pm\sqrt{\frac{(p^2-u)(p^2-v)(p^2-w)}{q^2-p^2}}$, $y = \pm\sqrt{\frac{(q^2-u)(q^2-v)(q^2-w)}{p^2-q^2}}$, $z = \frac{1}{2}(u+v+w-p^2-q^2)$.

4.30 Let $F : W \to V$ be the smooth bijective map with smooth inverse $F^{-1} : V \to W$ constructed in the proof of Proposition 4.1. Then, $(u(t), v(t)) = F^{-1}(\boldsymbol{\gamma}(t))$ is smooth.

4.31 Suppose, for example, that $f_y \neq 0$ at (x_0, y_0). Let $F(x,y) = (x, f(x,y))$; then F is smooth and its jacobian matrix $\begin{pmatrix} 1 & f_x \\ 0 & f_y \end{pmatrix}$ is invertible at (x_0, y_0). By the inverse function theorem, F has a smooth inverse G defined on an open subset of $\mathbf{R}^2$ containing $F(x_0, y_0) = (x_0, 0)$, and G must be of the form $G(x,z) = (x, g(x,z))$ for some smooth function g. Then $\boldsymbol{\gamma}(t) = (t, g(t,0))$ is a parametrisation of the level curve $f(x,y) = 0$ containing (x_0, y_0).
The matrix $\begin{pmatrix} f_x & f_y & f_z \\ g_x & g_y & g_z \end{pmatrix}$ has rank 2 everywhere; suppose that, at some point (x_0, y_0, z_0) on the level curve, the 2×2 submatrix $\begin{pmatrix} f_y & f_z \\ g_y & g_z \end{pmatrix}$ is invertible. Define $F(x,y,z) = (x, f(x,y,z), g(x,y,z))$; then F is smooth and its jacobian matrix $\begin{pmatrix} 1 & 0 & 0 \\ f_x & f_y & f_z \\ g_x & g_y & g_z \end{pmatrix}$ is invertible at (x_0, y_0, z_0). Let $G(x,u,v) = (x, \varphi(x,u,v), \psi(x,u,v))$ be the smooth inverse of F defined near $(x_0, 0, 0)$. Then $\boldsymbol{\gamma}(t) = (t, \varphi(t,0,0), \psi(t,0,0))$ is a parametrisation of the level curve $f(x,y,z) = g(x,y,z) = 0$ containing (x_0, y_0, z_0).

Chapter 5

5.1 (i) Quadric cone; we have $\boldsymbol{\sigma}_u = (\cosh u \sinh v, \cosh u \cosh v, \cosh u)$, $\boldsymbol{\sigma}_v = (\sinh u \cosh v, \sinh u \sinh v, 0)$, and so we get $\| \boldsymbol{\sigma}_u \|^2 = 2\cosh^2 u$, $\boldsymbol{\sigma}_u.\boldsymbol{\sigma}_v = 2\sinh u \cosh u \sinh v \cosh v$, $\| \boldsymbol{\sigma}_v \|^2 = \sinh^2 u$, and the first fundamental form is $2\cosh^2 u\, du^2 + 4\sinh u \cosh u \sinh v \cosh v\, dudv + \sinh^2 u\, dv^2$.
(ii) Paraboloid of revolution; $(2+4u^2)\,du^2 + 8uv\,dudv + (2+4v^2)\,dv^2$.
(iii) Hyperbolic cylinder; $(\cosh^2 u + \sinh^2 u)\,du^2 + dv^2$.
(iv) Paraboloid of revolution; $(1+4u^2)\,du^2 + 8uv\,dudv + (1+4v^2)\,dv^2$.

5.2 The first fundamental form is $2\,du^2 + u^2\,dv^2$, so the length of the curve is $\int_0^\pi (2\dot{u}^2 + u^2\dot{v}^2)^{1/2}\,dt = \int_0^\pi (2\lambda^2 e^{2\lambda t} + e^{2\lambda t})^{1/2}\,dt = \frac{(2\lambda^2+1)^{1/2}}{\lambda}(e^{\lambda t} - 1)$.

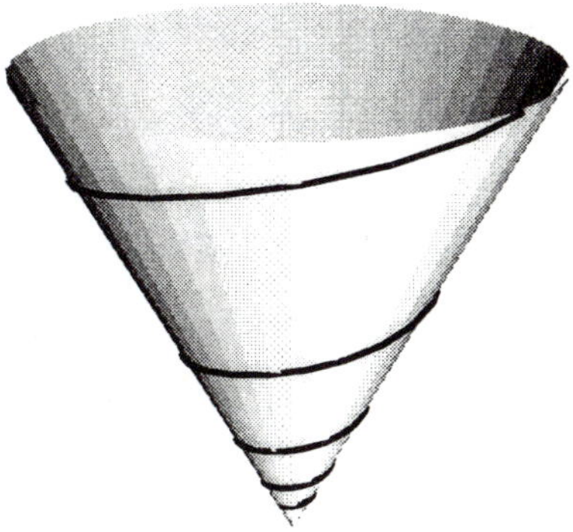

5.3 Applying a translation to a surface patch $\boldsymbol{\sigma}$ does not change $\boldsymbol{\sigma}_u$ or $\boldsymbol{\sigma}_v$. If A is a rotation about the origin, $(A\boldsymbol{\sigma})_u = A(\boldsymbol{\sigma}_u), (A\boldsymbol{\sigma})_v = A(\boldsymbol{\sigma}_v)$, and A preserves dot products ($A(\mathbf{p}).A(\mathbf{q}) = \mathbf{p}.\mathbf{q}$ for all vectors $\mathbf{p}, \mathbf{q} \in \mathbf{R}^3$).

5.4 By the chain rule, $\tilde{\boldsymbol{\sigma}}_{\tilde{u}} = \boldsymbol{\sigma}_u \frac{\partial u}{\partial \tilde{u}} + \boldsymbol{\sigma}_v \frac{\partial v}{\partial \tilde{u}}$, $\tilde{\boldsymbol{\sigma}}_{\tilde{v}} = \boldsymbol{\sigma}_u \frac{\partial u}{\partial \tilde{v}} + \boldsymbol{\sigma}_v \frac{\partial v}{\partial \tilde{v}}$, which gives $\tilde{E} = \tilde{\boldsymbol{\sigma}}_{\tilde{u}}.\tilde{\boldsymbol{\sigma}}_{\tilde{u}} = E\left(\frac{\partial u}{\partial \tilde{u}}\right)^2 + 2F\frac{\partial u}{\partial \tilde{u}}\frac{\partial v}{\partial \tilde{u}} + G\left(\frac{\partial v}{\partial \tilde{u}}\right)^2$. Similar expressions for $\tilde{F}$ and $\tilde{G}$ can be found; multiplying out the matrices shows that these formulas are equivalent to the matrix equation in the question.

5.5 The map is $\boldsymbol{\sigma}(u,v) \mapsto \left(u\sqrt{2}\cos\frac{v}{\sqrt{2}}, u\sqrt{2}\sin\frac{v}{\sqrt{2}}, 0\right) = \tilde{\boldsymbol{\sigma}}(u,v)$, say. The image of this map is the sector of the xy-plane whose polar coordinates (r,θ) satisfy $0 < \theta < \pi\sqrt{2}$. The first fundamental form of $\boldsymbol{\sigma}$ is $2\,du^2 + u^2\,dv^2$ (Exercise 5.2); $\tilde{\boldsymbol{\sigma}}_u = \left(\sqrt{2}\cos\frac{v}{\sqrt{2}}, \sqrt{2}\sin\frac{v}{\sqrt{2}}, 0\right)$, $\tilde{\boldsymbol{\sigma}}_v = \left(-u\sin\frac{v}{\sqrt{2}}, u\cos\frac{v}{\sqrt{2}}, 0\right)$, so $\| \tilde{\boldsymbol{\sigma}}_u \|^2 = 2$, $\boldsymbol{\sigma}_u.\boldsymbol{\sigma}_v = 0$, $\| \boldsymbol{\sigma}_v \|^2 = u^2$ and the first fundamental form of $\tilde{\boldsymbol{\sigma}}$ is also $2\,du^2 + u^2\,dv^2$.

5.6 No: the part of the ruling $(t,0,t)$ with $1 \le t \le 2$ (say) has length $\sqrt{2}$ and is mapped to the straight line segment $(t,0,0)$ with $1 \le t \le 2$, which has length 1.

5.7 For the generalised cylinder, Example 5.3 shows that the first fundamental form is $du^2 + dv^2$, so $\boldsymbol{\sigma}(u, v) \mapsto (u, v, 0)$ is an isometry from the cylinder to part of the xy-plane. For the generalised cone, Example 5.4 shows that the first fundamental form is $v^2\, du^2 + dv^2$. This is the same as the first fundamental form of $\tilde{\boldsymbol{\sigma}}\left(\frac{v}{\sqrt{2}}, u\sqrt{2}\right)$, where $\tilde{\boldsymbol{\sigma}}$ is as in Exercise 5.5. Since $\tilde{\boldsymbol{\sigma}}$ parametrises part of the plane, the generalised cone is isometric to part of the plane.

5.8 A straightforward calculation shows that the first fundamental form of $\boldsymbol{\sigma}^t$ is $\cosh^2 u(du^2 + dv^2)$; in particular, it is independent of t. Hence, $\boldsymbol{\sigma}(u, v) \mapsto \boldsymbol{\sigma}^t(u, v)$ is an isometry for all t. Taking $t = \pi/2$ gives the isometry from the catenoid to the helicoid; under this map, the parallels u = constant on the catenoid go to circular helices on the helicoid, and the meridians v = constant go to the rulings of the helicoid.

5.9 If the first fundamental forms of two surfaces are equal, they are certainly proportional, so any isometry is a conformal map. Stereographic projection is a conformal map from the unit sphere to the plane, but it is not an isometry since $\lambda \neq 1$ (see Example 5.7).

5.10 The vector $\boldsymbol{\sigma}_u$ is tangent to the rulings of the cone, so the angle θ between the curve and the rulings is given by $\cos\theta = \dot{\boldsymbol{\gamma}}.\boldsymbol{\sigma}_u / \parallel \dot{\boldsymbol{\gamma}} \parallel\parallel \boldsymbol{\sigma}_u \parallel$. ¿From $\boldsymbol{\gamma}(t) = (e^{\lambda t}\cos t, e^{\lambda t}\sin t, e^{\lambda t})$ and $\boldsymbol{\sigma}_u = (\cos t, \sin t, 1)$ at $\boldsymbol{\gamma}(t)$, we get $\cos\theta = \sqrt{2\lambda^2/(2\lambda^2 + 1)}$, which is independent of t.

5.11 The first fundamental form is $\mathrm{sech}^2 u(du^2 + dv^2)$.

5.12 The first fundamental form is $(1 + \dot{f}^2)du^2 + u^2\, dv^2$, where a dot denotes d/du. So $\boldsymbol{\sigma}$ is conformal if and only if $\dot{f} = \pm\sqrt{u^2 - 1}$, i.e. if and only if $f(u) = \pm(\frac{1}{2}u\sqrt{u^2 - 1} - \frac{1}{2}\cosh^{-1} u) + c$, where c is a constant.

5.13 The first fundamental form is $(1 + 2v\dot{\boldsymbol{\gamma}}.\boldsymbol{\delta} + v^2\dot{\boldsymbol{\delta}}.\dot{\boldsymbol{\delta}})du^2 + 2\dot{\boldsymbol{\gamma}}.\boldsymbol{\delta}\, du dv + dv^2$. So $\boldsymbol{\sigma}$ is conformal if and only if $1 + 2v\dot{\boldsymbol{\gamma}}.\dot{\boldsymbol{\delta}} + v^2\dot{\boldsymbol{\delta}}.\dot{\boldsymbol{\delta}} = 1$ and $\dot{\boldsymbol{\gamma}}.\boldsymbol{\delta} = 0$ for all u, v; the first condition gives $\dot{\boldsymbol{\delta}} = \mathbf{0}$, so $\boldsymbol{\delta}$ is constant, and the second condition then says that $\boldsymbol{\gamma}.\boldsymbol{\delta}$ is constant, say equal to d. Thus, $\boldsymbol{\sigma}$ is conformal if and only if $\boldsymbol{\delta}$ is constant and $\boldsymbol{\gamma}$ is contained in a plane $\mathbf{r}.\boldsymbol{\delta} = d$. In this case, $\boldsymbol{\sigma}$ is a generalised cylinder.

5.14 $\boldsymbol{\sigma}$ is conformal if and only if $f_u^2 + g_u^2 = f_v^2 + g_v^2$ and $f_u f_v + g_u g_v = 0$. Let $z = f_u + ig_u$, $w = f_v + ig_v$; then $\boldsymbol{\sigma}$ is conformal if and only if $z\bar{z} = w\bar{w}$ and $z\bar{w} + \bar{z}w = 0$, where the bar denotes complex conjugate; if $z = 0$, then $w = 0$ and all four equations are certainly satisfied; if $z \neq 0$, the equations give $z^2 = -w^2$, so $z = \pm iw$; these are easily seen to be equivalent to the Cauchy–Riemann equations if the sign is +, and to the 'anti-Cauchy–Riemann' equations if the sign is −.

5.15 Parametrise the paraboloid by $\boldsymbol{\sigma}(u, v) = (u, v, u^2 + v^2)$, its first fundamental form is $(1 + 4u^2)du^2 + 8uv\, dudv + (1 + 4v^2)dv^2$. Hence, the required area is $\int\int \sqrt{1 + 4(u^2 + v^2)}\, dudv$, taken over the disc $u^2 + v^2 < 1$. Let $u = r\sin\theta, v = r\cos\theta$; then the area is $2\pi\int_0^1 \sqrt{1 + 4r^2}r\, dr = \frac{\pi}{6}(5^{3/2} - 1)$.

This is less than the area 2π of the hemisphere.

5.16 Parametrize the surface by $\boldsymbol{\sigma}(u,v) = (\rho(u)\cos v, \rho(u)\sin v, \sigma(u))$, where $\boldsymbol{\gamma}(u) = (\rho(u), 0, \sigma(u))$. By Example 6.2, the first fundamental form is $du^2 + \rho(u)^2\, dv^2$, so the area is $\int\int \rho(u)\, dudv = 2\pi \int \rho(u)\, du$.
(i) Take $\rho(u) = \cos u$, $\sigma(u) = \sin u$, with $-\pi/2 \leq u \leq \pi/2$; so $2\pi \int_{-\pi/2}^{\pi/2} \cos u\, du = 4\pi$ is the area.
(ii) For the torus, the profile curve is $\boldsymbol{\gamma}(\theta) = (a + b\cos\theta, 0, b\sin\theta)$, but this is not unit-speed; a unit-speed reparametrisation is $\tilde{\boldsymbol{\gamma}}(u) = \left(a + b\cos\frac{u}{b}, 0, b\sin\frac{u}{b}\right)$ with $0 \leq u \leq 2\pi b$. So $2\pi \int_0^{2\pi b} \left(a + b\cos\frac{u}{b}\right) du = 4\pi^2 ab$ is the area.

5.17 $\boldsymbol{\sigma}$ is the tube swept out by a circle of radius a in a plane perpendicular to $\boldsymbol{\gamma}$ as its centre moves along $\boldsymbol{\gamma}$. $\boldsymbol{\sigma}_s = (1 - \kappa a\cos\theta)\mathbf{t} - \tau a\sin\theta\mathbf{n} + \tau a\cos\theta\mathbf{b}$, $\boldsymbol{\sigma}_\theta = -a\sin\theta\mathbf{n} + a\cos\theta\mathbf{b}$, giving $\boldsymbol{\sigma}_s \times \boldsymbol{\sigma}_\theta = -a(1-\kappa a\cos\theta)(\cos\theta\mathbf{n} + \sin\theta\mathbf{b})$; this is never zero since $\kappa a < 1$ implies that $1 - \kappa a\cos\theta > 0$ for all θ.
The first fundamental form is $((1 - \kappa a\cos\theta)^2 + \tau^2 a^2)\, ds^2 + 2\tau a^2\, dsd\theta + a^2\, d\theta^2$, so the area is $\int_{s_0}^{s_1}\int_0^{2\pi} a(1 - \kappa a\cos\theta) dsd\theta = 2\pi a(s_1 - s_0)$.

5.18 If $E_1 = E_2, F_1 = F_2, G_1 = G_2$, then $E_1G_1 - F_1^2 = E_2G_2 - F_2^2$, so any isometry is equiareal. The map f in Archimedes's theorem is equiareal but not an isometry (as $E_1 \neq E_2$, for example).

5.19 If $E_1 = \lambda E_2, F_1 = \lambda F_2, G_1 = \lambda G_2$, and if $E_1G_1 - F_1^2 = E_2G_2 - F_2^2$, then $\lambda^2 = 1$ so $\lambda = 1$ (since $\lambda > 0$).

5.20 By Theorem 5.5, the sum of the angles of the triangle is $\pi + A/R^2$, where A is its area and R is the radius of the earth, and so is $\geq \pi + (7\,500\,000)/(6\,500)^2 = \pi + \frac{30}{169}$ radians. Hence, at least one angle of the triangle must be at least one third of this, i.e. $\pi + \frac{10}{169}$ radians.

5.21 $3F = 2E$ because every face has three edges and every edge is an edge of exactly two faces. The sum of the angles around any vertex is 2π, so the sum of the angles of all the triangles is $2\pi V$; on the other hand, by Theorem 5.5, the sum of the angles of any triangle is π plus its area, so since there are F triangles and the sum of all their areas is 4π (the area of the sphere), the sum of all the angles is $\pi F + 4\pi$. Hence, $2V = F + 4$. Then, $V - E + F = 2 + \frac{1}{2}F - E + F = 2 + \frac{1}{2}(3F - 2E) = 2$.

5.22 Let $\boldsymbol{\sigma} : U \to \mathbf{R}^3$. Then, f is equiareal if and only if

$$\int\int_R (E_1G_1 - F_1^2)^{1/2}\, dudv = \int\int_R (E_2G_2 - F_2^2)^{1/2}\, dudv$$

for all regions $R \subseteq U$. This holds if and only if the two integrands are equal everywhere, i.e. if and only if $E_1G_1 - F_1^2 = E_2G_2 - F_2^2$.

Chapter 6

6.1 $\boldsymbol{\sigma}_u = (1, 0, 2u)$, $\boldsymbol{\sigma}_v = (0, 1, 2v)$, so $\mathbf{N} = \lambda(-2u, -2v, 1)$, where $\lambda = (1 + 4u^2 + 4v^2)^{-1/2}$; $\boldsymbol{\sigma}_{uu} = (0, 0, 2)$, $\boldsymbol{\sigma}_{uv} = \mathbf{0}$, $\boldsymbol{\sigma}_{vv} = (0, 0, 2)$, so $L = 2\lambda$, $M = 0$, $N = 2\lambda$, and the second fundamental form is $2\lambda(du^2 + dv^2)$.

6.2 $\boldsymbol{\sigma}_u.\mathbf{N}_u = -\boldsymbol{\sigma}_{uu}.\mathbf{N}$ (since $\boldsymbol{\sigma}_u.\mathbf{N} = 0$), so $\mathbf{N}_u.\boldsymbol{\sigma}_u = 0$; similarly, $\mathbf{N}_u.\boldsymbol{\sigma}_v = \mathbf{N}_v.\boldsymbol{\sigma}_u = \mathbf{N}_v.\boldsymbol{\sigma}_v = 0$; hence, $\mathbf{N}_u$ and $\mathbf{N}_v$ are perpendicular to both $\boldsymbol{\sigma}_u$ and $\boldsymbol{\sigma}_v$, and so are parallel to $\mathbf{N}$. On the other hand, $\mathbf{N}_u$ and $\mathbf{N}_v$ are perpendicular to $\mathbf{N}$ since $\mathbf{N}$ is a unit vector. Thus, $\mathbf{N}_u = \mathbf{N}_v = \mathbf{0}$, and hence $\mathbf{N}$ is constant. Then, $(\boldsymbol{\sigma}.\mathbf{N})_u = \boldsymbol{\sigma}_u.\mathbf{N} = 0$, and similarly $(\boldsymbol{\sigma}.\mathbf{N})_v = 0$, so $\boldsymbol{\sigma}.\mathbf{N}$ is constant, say equal to d, and then $\boldsymbol{\sigma}$ is part of the plane $\mathbf{r}.\mathbf{N} = d$.

6.3 From Section 4.3, $\tilde{\mathbf{N}} = \pm\mathbf{N}$, the sign being that of $\det(J)$. From $\tilde{\boldsymbol{\sigma}}_{\tilde{u}} = \boldsymbol{\sigma}_u \frac{\partial u}{\partial \tilde{u}} + \boldsymbol{\sigma}_v \frac{\partial v}{\partial \tilde{u}}$, $\tilde{\boldsymbol{\sigma}}_{\tilde{v}} = \boldsymbol{\sigma}_u \frac{\partial u}{\partial \tilde{v}} + \boldsymbol{\sigma}_v \frac{\partial v}{\partial \tilde{v}}$, we get

$$\tilde{\boldsymbol{\sigma}}_{\tilde{u}\tilde{u}} = \boldsymbol{\sigma}_u \frac{\partial^2 u}{\partial \tilde{u}^2} + \boldsymbol{\sigma}_v \frac{\partial^2 v}{\partial \tilde{u}^2} + \boldsymbol{\sigma}_{uu}\left(\frac{\partial u}{\partial \tilde{u}}\right)^2 + 2\boldsymbol{\sigma}_{uv}\frac{\partial u}{\partial \tilde{u}}\frac{\partial v}{\partial \tilde{u}} + \boldsymbol{\sigma}_{vv}\left(\frac{\partial v}{\partial \tilde{u}}\right)^2 .$$

So

$$\tilde{L} = \pm\left(L\left(\frac{\partial u}{\partial \tilde{u}}\right)^2 + 2M\frac{\partial u}{\partial \tilde{u}}\frac{\partial v}{\partial \tilde{u}} + N\left(\frac{\partial v}{\partial \tilde{u}}\right)^2\right)$$

since $\boldsymbol{\sigma}_u.\mathbf{N} = \boldsymbol{\sigma}_v.\mathbf{N} = 0$. This, together with similar formulas for $\tilde{M}$ and $\tilde{N}$, are equivalent to the matrix equation in the question.

6.4 Applying a translation to a surface patch $\boldsymbol{\sigma}$ does not change $\boldsymbol{\sigma}_u$ and $\boldsymbol{\sigma}_v$, and hence does not change $\mathbf{N}, \boldsymbol{\sigma}_{uu}, \boldsymbol{\sigma}_{uv}$ or $\boldsymbol{\sigma}_{vv}$, and hence does not change the second fundamental form. A rotation A about the origin has the following effect: $\boldsymbol{\sigma}_u \to A(\boldsymbol{\sigma}_u)$, $\boldsymbol{\sigma}_u \to A(\boldsymbol{\sigma}_v)$ and hence $\mathbf{N} \to A(\mathbf{N})$, $\boldsymbol{\sigma}_{uu} \to A(\boldsymbol{\sigma}_{uu})$, $\boldsymbol{\sigma}_{uv} \to A(\boldsymbol{\sigma}_{uv})$, $\boldsymbol{\sigma}_{vv} \to A(\boldsymbol{\sigma}_{vv})$; since $A(\mathbf{p}).A(\mathbf{q}) = \mathbf{p}.\mathbf{q}$ for any vectors $\mathbf{p}, \mathbf{q} \in \mathbf{R}^3$, the second fundamental form is again unchanged.

6.5 By Exercise 6.1, the second fundamental form of the paraboloid is $2(du^2 + dv^2)/\sqrt{1 + 4u^2 + 4v^2}$; so

$$\kappa_n = 2((-\sin t)^2 + \cos^2 t)/\sqrt{1 + 4\cos^2 t + 4\sin^2 t} = 2/\sqrt{5}.$$

6.6 $\kappa^2 = \kappa_n^2 + \kappa_g^2 = 0$, so $\kappa = 0$ and $\boldsymbol{\gamma}$ is part of a straight line.

6.7 Let $\boldsymbol{\gamma}$ be a unit-speed curve on the sphere of centre $\mathbf{a}$ and radius r. Then, $(\boldsymbol{\gamma} - \mathbf{a}).(\boldsymbol{\gamma} - \mathbf{a}) = r^2$; differentiating gives $\dot{\boldsymbol{\gamma}}.(\boldsymbol{\gamma} - \mathbf{a}) = 0$, so $\ddot{\boldsymbol{\gamma}}.(\boldsymbol{\gamma} - \mathbf{a}) = -\dot{\boldsymbol{\gamma}}.\dot{\boldsymbol{\gamma}} = -1$. At the point $\boldsymbol{\gamma}(t)$, the unit normal of the sphere is $\mathbf{N} = \pm\frac{1}{r}(\boldsymbol{\gamma}(t) - \mathbf{a})$, so $\kappa_n = \ddot{\boldsymbol{\gamma}}.\mathbf{N} = \pm\frac{1}{r}\ddot{\boldsymbol{\gamma}}.(\boldsymbol{\gamma} - \mathbf{a}) = \mp\frac{1}{r}$.

6.8 If the sphere has radius R, the parallel with latitude θ has radius $r = R\cos\theta$; if P is a point of this circle, its principal normal at P is parallel to the line through P perpendicular to the z-axis, while the unit normal to the sphere is parallel to the line through P and the centre of the sphere.

The angle ψ in Eq. (8) is therefore equal to θ or $\pi - \theta$ so $\kappa_g = \pm\frac{1}{r}\sin\theta = \pm\frac{1}{R}\tan\theta$. Note that this is zero if and only if the parallel is a great circle.

6.9 The unit normal is $\mathbf{N} = (-\dot{g}\cos v, -\dot{g}\sin v, \dot{f})$, where a dot denotes d/du. On a meridian $v =$ constant, we can use u as the parameter; since $\boldsymbol{\sigma}_u = (\dot{f}\cos v, \dot{f}\sin v, \dot{g})$ is a unit vector, u is a unit-speed parameter on the meridian and

$$\kappa_g = \boldsymbol{\sigma}_{uu}.(\mathbf{N} \times \boldsymbol{\sigma}_u) = \begin{vmatrix} \ddot{f}\cos v & \ddot{f}\sin v & \ddot{g} \\ -\dot{g}\cos v & -\dot{g}\sin v & \dot{f} \\ \dot{f}\cos v & \dot{f}\sin v & \dot{g} \end{vmatrix} = 0.$$

On a parallel $u =$ constant, we can use v as a parameter, but $\boldsymbol{\sigma}_v = (-f\sin v, f\cos v, 0)$ is not a unit vector; the arc-length s is given by $ds/dv = \parallel \boldsymbol{\sigma}_v \parallel = f(u)$, so $s = f(u)v$ (we can take the arbitrary constant to be zero). Then,

$$\kappa_g = \frac{1}{f(u)^2}\boldsymbol{\sigma}_{vv}.\left(\mathbf{N} \times \frac{1}{f(u)}\boldsymbol{\sigma}_v\right) = \frac{1}{f(u)^3}\begin{vmatrix} -f\cos v & -f\sin v & 0 \\ -\dot{g}\cos v & -\dot{g}\sin v & \dot{f} \\ -f\sin v & f\cos v & 0 \end{vmatrix} = \frac{\dot{f}}{f}.$$

6.10 $\kappa_1 = \kappa\mathbf{N}_1.\mathbf{n}$, $\kappa_2 = \kappa\mathbf{N}_2.\mathbf{n}$, so

$$\kappa_1\mathbf{N}_2 - \kappa_2\mathbf{N}_1 = \kappa((\mathbf{N}_1.\mathbf{n})\mathbf{N}_2 - (\mathbf{N}_2.\mathbf{n})\mathbf{N}_1) = \kappa(\mathbf{N}_1 \times \mathbf{N}_2) \times \mathbf{n}.$$

Taking the squared length of each side, we get

$$\kappa_1^2 + \kappa_2^2 - 2\kappa_1\kappa_2\mathbf{N}_1.\mathbf{N}_2 = \kappa^2 \parallel (\mathbf{N}_1 \times \mathbf{N}_2) \times \mathbf{n} \parallel^2 .$$

Now, $\mathbf{N}_1.\mathbf{N}_2 = \cos\alpha$; $\dot{\boldsymbol{\gamma}}$ is perpendicular to $\mathbf{N}_1$ and $\mathbf{N}_2$, so $\mathbf{N}_1 \times \mathbf{N}_2$ is parallel to $\dot{\boldsymbol{\gamma}}$, hence perpendicular to $\mathbf{n}$; hence,

$$\parallel (\mathbf{N}_1 \times \mathbf{N}_2) \times \mathbf{n} \parallel = \parallel \mathbf{N}_1 \times \mathbf{N}_2 \parallel\parallel \mathbf{n} \parallel = \sin\alpha.$$

6.11 $\mathbf{N}.\mathbf{n} = \cos\psi$, $\mathbf{N}.\mathbf{t} = 0$, so $\mathbf{N}.\mathbf{b} = \sin\psi$; hence, $\mathbf{N} = \mathbf{n}\cos\psi + \mathbf{b}\sin\psi$ and $\mathbf{B} = \mathbf{t} \times (\mathbf{n}\cos\psi + \mathbf{b}\sin\psi) = \mathbf{b}\cos\psi - \mathbf{n}\sin\psi$. Hence,

$$\begin{aligned}\dot{\mathbf{N}} &= \dot{\mathbf{n}}\cos\psi + \dot{\mathbf{b}}\sin\psi + \dot{\psi}(-\mathbf{n}\sin\psi + \mathbf{b}\cos\psi) \\ &= (-\kappa\mathbf{t} + \tau\mathbf{b})\cos\psi - \mathbf{n}\tau\sin\psi + \dot{\psi}(-\mathbf{n}\sin\psi + \mathbf{b}\cos\psi) \\ &= -\kappa\cos\psi\mathbf{t} + (\tau + \dot{\psi})(\mathbf{b}\cos\psi - \mathbf{n}\sin\psi) \\ &= -\kappa_n\mathbf{t} + \tau_g\mathbf{B}.\end{aligned}$$

The formula for $\dot{\mathbf{B}}$ is proved similarly. Since $\{\mathbf{t}, \mathbf{N}, \mathbf{B}\}$ is a right-handed orthonormal basis of $\mathbf{R}^3$, Exercise 2.22 shows that the matrix expressing $\dot{\mathbf{t}}, \dot{\mathbf{N}}, \dot{\mathbf{B}}$ in terms of $\mathbf{t}, \mathbf{N}, \mathbf{B}$ is skew-symmetric, hence the formula for $\dot{\mathbf{t}}$.

6.12 A straight line has a unit-speed parametrisation $\boldsymbol{\gamma}(t) = \mathbf{p} + \mathbf{q}t$ (with $\mathbf{q}$ a unit vector), so $\ddot{\boldsymbol{\gamma}} = \mathbf{0}$ and hence $\kappa_n = \ddot{\boldsymbol{\gamma}}.\mathbf{N} = 0$.
In general, $\kappa_n = 0 \iff \ddot{\boldsymbol{\gamma}}$ is perpendicular to $\mathbf{N} \iff \mathbf{N}$ is perpendicular to $\mathbf{n} \iff \mathbf{N}$ is parallel to $\mathbf{b}$ (since $\mathbf{N}$ is perpendicular to $\mathbf{t}$).

6.13 The second fundamental form is $(-du^2 + u^2\,dv^2)/u\sqrt{1+u^2}$, so a curve $\boldsymbol{\gamma}(t) = \boldsymbol{\sigma}(u(t), v(t))$ is asymptotic if and only if $-\dot{u}^2 + u^2\dot{v}^2 = 0$, i.e. $dv/du = \dot{v}/\dot{u} = \pm 1/u$, so $\ln u = \pm(v+c)$, where c is a constant.

6.14 By Exercise 6.12, $\mathbf{b}$ is parallel to $\mathbf{N}$, so $\mathbf{b} = \pm\mathbf{N}$; then, $\mathbf{B} = \mathbf{t} \times \mathbf{N} = \mp\mathbf{n}$. Hence, $\dot{\mathbf{B}} = \mp\dot{\mathbf{n}} = \mp(-\kappa\mathbf{t} + \tau\mathbf{b}) = \pm\kappa\mathbf{t} - \tau\mathbf{N}$; comparing with the formula for $\dot{\mathbf{B}}$ in Exercise 6.11 shows that $\tau_g = \tau$ (and $\kappa_n = \pm\kappa$).

6.15 For the helicoid $\boldsymbol{\sigma}(u,v) = (v\cos u, v\sin u, \lambda u)$, the first and second fundamental forms are $(\lambda^2 + v^2)du^2 + dv^2$ and $2\lambda du dv/\sqrt{\lambda^2+v^2}$, respectively. Hence, the principal curvatures are the roots of

$$\begin{vmatrix} -\kappa(\lambda^2+v^2) & \frac{\lambda}{\sqrt{\lambda^2+v^2}} \\ \frac{\lambda}{\sqrt{\lambda^2+v^2}} & -\kappa \end{vmatrix} = 0,$$

i.e. $\pm\lambda/(\lambda^2+v^2)$.

For the catenoid $\boldsymbol{\sigma}(u,v) = (\cosh u\cos v, \cosh u \sin v, u)$, the first and second fundamental forms are $\cosh^2 u(du^2+dv^2)$ and $-du^2+dv^2$, respectively. Hence, the principal curvatures are the roots of

$$\begin{vmatrix} -1-\kappa\cosh^2 u & 0 \\ 0 & 1-\kappa\cosh^2 u \end{vmatrix},$$

i.e. $\kappa = \pm\text{sech}^2 u$.

6.16 Let s be arc-length along $\boldsymbol{\gamma}$, and denote d/ds by a dash. Then, by Proposition 6.1,

$$\kappa_n = Lu'^2 + 2Mu'v' + Nv'^2 = \frac{L\dot{u}^2 + 2M\dot{u}\dot{v} + N\dot{v}^2}{(ds/dt)^2} = \frac{L\dot{u}^2 + 2M\dot{u}\dot{v} + N\dot{v}^2}{E\dot{u}^2 + 2F\dot{u}\dot{v} + G\dot{v}^2}.$$

6.17 By Exercises 5.4 and 6.3, we have (in an obvious notation), $\tilde{\mathcal{F}}_I = J^t\mathcal{F}_I J$, $\tilde{\mathcal{F}}_{II} = \pm J^t\mathcal{F}_{II}J$, where the sign is that of $\det(J)$. The principal curvatures of $\tilde{\boldsymbol{\sigma}}$ are the roots of $\det(\tilde{\mathcal{F}}_{II} - \tilde{\kappa}\tilde{\mathcal{F}}_I) = 0$, i.e. $\det(\pm J^t\mathcal{F}_{II}J - \tilde{\kappa}J^t\mathcal{F}_{II}J) = 0$, which (since J is invertible) are the same as the roots of $\det(\pm\mathcal{F}_{II} - \tilde{\kappa}\mathcal{F}_I) = 0$. Hence, the principal curvatures of $\tilde{\boldsymbol{\sigma}}$ are $\pm$ those of $\boldsymbol{\sigma}$.

Let $\tilde{\xi}\tilde{\boldsymbol{\sigma}}_{\tilde{u}} + \tilde{\eta}\tilde{\boldsymbol{\sigma}}_{\tilde{v}}$ be a principal vector for $\tilde{\boldsymbol{\sigma}}$ corresponding to the principal curvature $\tilde{\kappa}$. Then,

$$(\tilde{\mathcal{F}}_{II} - \tilde{\kappa}\tilde{\mathcal{F}}_I)\begin{pmatrix}\tilde{\xi}\\ \tilde{\eta}\end{pmatrix} = \begin{pmatrix}0\\0\end{pmatrix} \Longrightarrow (\mathcal{F}_{II} - \kappa\mathcal{F}_I)J\begin{pmatrix}\tilde{\xi}\\ \tilde{\eta}\end{pmatrix} = \begin{pmatrix}0\\0\end{pmatrix},$$

so if $J\begin{pmatrix}\tilde{\xi}\\ \tilde{\eta}\end{pmatrix} = \begin{pmatrix}\xi\\ \eta\end{pmatrix}$, then $\xi\boldsymbol{\sigma}_u + \eta\boldsymbol{\sigma}_v$ is a principal vector for $\boldsymbol{\sigma}$ corresponding to the principal curvature κ. But, since $\xi = \frac{\partial u}{\partial \tilde{u}}\tilde{\xi} + \frac{\partial u}{\partial \tilde{v}}\tilde{\eta}$, $\eta = \frac{\partial v}{\partial \tilde{u}}\tilde{\xi} + \frac{\partial v}{\partial \tilde{v}}\tilde{\eta}$, we have $\xi\boldsymbol{\sigma}_u + \eta\boldsymbol{\sigma}_v = \tilde{\xi}\left(\frac{\partial u}{\partial \tilde{u}}\boldsymbol{\sigma}_u + \frac{\partial v}{\partial \tilde{u}}\boldsymbol{\sigma}_v\right) + \tilde{\eta}\left(\frac{\partial u}{\partial \tilde{v}}\boldsymbol{\sigma}_u + \frac{\partial v}{\partial \tilde{v}}\boldsymbol{\sigma}_v\right) = \tilde{\xi}\tilde{\boldsymbol{\sigma}}_{\tilde{u}} + \tilde{\eta}\tilde{\boldsymbol{\sigma}}_{\tilde{v}}$, which shows that $\tilde{\xi}\tilde{\boldsymbol{\sigma}}_{\tilde{u}} + \tilde{\eta}\tilde{\boldsymbol{\sigma}}_{\tilde{v}}$ is a principal vector for $\boldsymbol{\sigma}$ corresponding to the principal curvature κ.

The second part also follows from Corollary 6.2.

6.18 $\dot{\gamma} = \dot{u}\boldsymbol{\sigma}_u + \dot{v}\boldsymbol{\sigma}_v$ is a principal vector corresponding to the principal curvature $\kappa \iff (\mathcal{F}_{II} - \kappa\mathcal{F}_I)\begin{pmatrix}\dot{u}\\ \dot{v}\end{pmatrix} = \begin{pmatrix}0\\0\end{pmatrix} \iff \mathcal{F}_I^{-1}\mathcal{F}_{II}\begin{pmatrix}\dot{u}\\ \dot{v}\end{pmatrix} = \kappa\begin{pmatrix}\dot{u}\\ \dot{v}\end{pmatrix} \iff$ $\begin{pmatrix}a & c\\ b & d\end{pmatrix}\begin{pmatrix}\dot{u}\\ \dot{v}\end{pmatrix} = -\kappa\begin{pmatrix}\dot{u}\\ \dot{v}\end{pmatrix} \iff a\dot{u} + c\dot{v} = -\kappa\dot{u}$ and $b\dot{u} + d\dot{v} = -\kappa\dot{v}$. But, $\dot{\mathbf{N}} = \dot{u}\mathbf{N}_u + \dot{v}\mathbf{N}_v = \dot{u}(a\boldsymbol{\sigma}_u + b\boldsymbol{\sigma}_v) + \dot{v}(c\boldsymbol{\sigma}_u + d\boldsymbol{\sigma}_v) = (a\dot{u} + c\dot{v})\boldsymbol{\sigma}_u + (b\dot{u} + d\dot{v})\boldsymbol{\sigma}_v$. Hence, $\dot{\gamma}$ is principal $\iff \dot{\mathbf{N}} = -\kappa(\dot{u}\boldsymbol{\sigma}_u + \dot{v}\boldsymbol{\sigma}_v) = -\kappa\dot{\gamma}$.
From Example 6.2, the first and second fundamental forms of a surface of revolution are $du^2 + f(u)^2dv^2$ and $(\dot{f}\ddot{g} - \ddot{f}\dot{g})du^2 + f\dot{g}dv^2$, respectively. Since the terms $dudv$ are absent, the vectors $\boldsymbol{\sigma}_u$ and $\boldsymbol{\sigma}_v$ are principal; but these are tangent to the meridians and parallels, respectively.

6.19 By Exercise 6.11, $\dot{\mathbf{N}} = -\kappa_n\mathbf{t} + \tau_g\mathbf{B}$, so by Exercise 6.18 $\boldsymbol{\gamma}$ is a line of curvature if and only if $\tau_g = 0$ (in which case $\lambda = \kappa_n$).

6.20 Let $\mathbf{N}_1$ and $\mathbf{N}_2$ be unit normals of the two surfaces; if $\boldsymbol{\gamma}$ is a unit-speed parametrisation of $\mathcal{C}$, then $\dot{\mathbf{N}}_1 = -\lambda_1\dot{\boldsymbol{\gamma}}$ for some scalar λ_1 by Exercise 6.18. If $\mathcal{C}$ is a line of curvature of $\mathcal{S}_2$, then $\dot{\mathbf{N}}_2 = -\lambda_2\dot{\boldsymbol{\gamma}}$ for some scalar λ_2, and then $(\mathbf{N}_1.\mathbf{N}_2)^{\cdot} = -\lambda_1\dot{\boldsymbol{\gamma}}.\mathbf{N}_2 - \lambda_2\dot{\boldsymbol{\gamma}}.\mathbf{N}_1 = 0$, so $\mathbf{N}_1.\mathbf{N}_2$ is constant along $\boldsymbol{\gamma}$, showing that the angle between $\mathcal{S}_1$ and $\mathcal{S}_2$ is constant. Conversely, if $\mathbf{N}_1.\mathbf{N}_2$ is constant, then $\mathbf{N}_1.\dot{\mathbf{N}}_2 = 0$ since $\dot{\mathbf{N}}_1.\mathbf{N}_2 = -\lambda_1\dot{\boldsymbol{\gamma}}.\mathbf{N}_2 = 0$; thus, $\dot{\mathbf{N}}_2$ is perpendicular to $\mathbf{N}_1$, and is also perpendicular to $\mathbf{N}_2$ as $\mathbf{N}_2$ is a unit vector; but $\dot{\boldsymbol{\gamma}}$ is also perpendicular to $\mathbf{N}_1$ and $\mathbf{N}_2$; hence, $\dot{\mathbf{N}}_2$ must be parallel to $\dot{\boldsymbol{\gamma}}$, so there is a scalar λ_2 (say) such that $\dot{\mathbf{N}}_2 = -\lambda_2\dot{\boldsymbol{\gamma}}$.

6.21 (i) Differentiate the three equations in (21) with respect to w, u and v, respectively; this gives

$$\boldsymbol{\sigma}_{uw}.\boldsymbol{\sigma}_v + \boldsymbol{\sigma}_u.\boldsymbol{\sigma}_{vw} = 0, \quad \boldsymbol{\sigma}_{uv}.\boldsymbol{\sigma}_w + \boldsymbol{\sigma}_v.\boldsymbol{\sigma}_{uw} = 0, \quad \boldsymbol{\sigma}_{vw}.\boldsymbol{\sigma}_u + \boldsymbol{\sigma}_w.\boldsymbol{\sigma}_{uv} = 0.$$

Subtracting the second equation from the sum of the other two gives $\boldsymbol{\sigma}_u.\boldsymbol{\sigma}_{vw} = 0$, and similarly $\boldsymbol{\sigma}_v.\boldsymbol{\sigma}_{uw} = \boldsymbol{\sigma}_w.\boldsymbol{\sigma}_{uv} = 0$.
(ii) Since $\boldsymbol{\sigma}_v.\boldsymbol{\sigma}_w = 0$, it follows that the matrix $\mathcal{F}_I$ for the $u = u_0$ surface is diagonal (and similarly for the others). Let $\mathbf{N}$ be the unit normal of the $u = u_0$ surface; $\mathbf{N}$ is parallel to $\boldsymbol{\sigma}_v \times \boldsymbol{\sigma}_w$ by definition, and hence to $\boldsymbol{\sigma}_u$ since $\boldsymbol{\sigma}_u$, $\boldsymbol{\sigma}_v$ and $\boldsymbol{\sigma}_w$ are perpendicular; by (i), $\boldsymbol{\sigma}_{vw}.\boldsymbol{\sigma}_u = 0$, hence $\boldsymbol{\sigma}_{vw}.\mathrm{N} = 0$, proving that the matrix $\mathcal{F}_{II}$ for the $u = u_0$ surface is diagonal.
(iii) By part (ii), the parameter curves of each surface $u = u_0$ are lines of curvature. But the parameter curve $v = v_0$, say, on this surface is the curve of intersection of the $u = u_0$ surface with the $v = v_0$ surface.

6.22 We have $\mathbf{N}_u = a\boldsymbol{\sigma}_u + b\boldsymbol{\sigma}_v$, $\mathbf{N}_v = c\boldsymbol{\sigma}_u + d\boldsymbol{\sigma}_v$, so

$$\begin{aligned}\mathcal{F}_{III} &= \begin{pmatrix} \mathbf{N}_u.\mathbf{N}_u & \mathbf{N}_u.\mathbf{N}_v \\ \mathbf{N}_u.\mathbf{N}_v & \mathbf{N}_v.\mathbf{N}_v \end{pmatrix} \\ &= \begin{pmatrix} Ea^2 + 2Fab + Gb^2 & Eac + F(ad+bc) + Gbd \\ Eac + F(ad+bc) + Gbd & Ec^2 + 2Fcd + Gd^2 \end{pmatrix} \\ &= \begin{pmatrix} a & b \\ c & d \end{pmatrix}\begin{pmatrix} E & F \\ F & G \end{pmatrix}\begin{pmatrix} a & c \\ b & d \end{pmatrix} = (-\mathcal{F}_I^{-1}\mathcal{F}_{II})^t\mathcal{F}_I(-\mathcal{F}_I^{-1}\mathcal{F}_{II}) \\ &= \mathcal{F}_{II}\mathcal{F}_I^{-1}\mathcal{F}_I\mathcal{F}_I^{-1}\mathcal{F}_{II} = \mathcal{F}_{II}\mathcal{F}_I^{-1}\mathcal{F}_{II}.\end{aligned}$$

6.23 By Example 6.2, the principal curvatures are $\dot{f}\ddot{g} - \ddot{f}\dot{g}$ and $\dot{g}/f$. If $\dot{g} = 0$, the surface is part of a plane $z =$ constant and no point is parabolic. Thus, $\dot{g} \neq 0$, and every point is parabolic if and only if $\dot{f}\ddot{g} - \ddot{f}\dot{g} = 0$. Multiplying through by $\dot{g}$ and using $\dot{f}^2 + \dot{g}^2 = 1$ (which implies that $\dot{f}\ddot{f} + \dot{g}\ddot{g} = 0$), we get $\ddot{f} = 0$. Hence, $f(u) = au + b$, where a and b are constants. If $a = 0$, we have a circular cylinder; if $a \neq 0$, we have a circular cone.

6.24 On the part of the ellipsoid with $z \neq 0$, we can use the parametrisation $\boldsymbol{\sigma}(x, y) = (x, y, z)$, where $z = \pm r\sqrt{1 - \frac{x^2}{p^2} - \frac{y^2}{q^2}}$. The first and second fundamental forms are $(1+z_x^2)dx^2 + 2z_xz_ydxdy + (1+z_y^2)dy^2$ and $(z_{xx}dx^2 + 2z_{xy}dxdy + z_{yy}dy^2)/\sqrt{1 + z_x^2 + z_y^2}$, respectively. By Proposition 5.3(ii), the condition for an umbilic is that $\mathcal{F}_{II} = \kappa\mathcal{F}_I$ for some scalar κ. This leads to the equations

$$z_{xx} = \lambda(1 + z_x^2), \quad z_{xy} = \lambda z_x z_y, \quad z_{yy} = \lambda(1 + z_y^2),$$

where $\lambda = \kappa\sqrt{1 + z_x^2 + z_y^2}$. If x and y are both non-zero, the middle equation gives $\lambda = 1/z$, and substituting into the first equation gives the contradiction $p^2 = r^2$. Hence, either $x = 0$ or $y = 0$. If $x = 0$, the equations have the four solutions

$$x = 0, \quad y = \pm q\sqrt{\frac{q^2 - p^2}{q^2 - r^2}}, \quad z = \pm r\sqrt{\frac{r^2 - p^2}{r^2 - q^2}}.$$

Similarly, one finds the following eight other candidates for umbilics:

$$x = \pm p\sqrt{\frac{p^2 - q^2}{p^2 - r^2}}, \quad y = 0, \quad z = \pm r\sqrt{\frac{r^2 - q^2}{r^2 - p^2}},$$

$$x = \pm p\sqrt{\frac{p^2 - r^2}{p^2 - q^2}}, \quad y = \pm q\sqrt{\frac{q^2 - r^2}{q^2 - p^2}}, \quad z = 0.$$

Of these 12 points, exactly 4 are real, depending on the relative sizes of p^2, q^2 and r^2.

Chapter 7

7.1 $\boldsymbol{\sigma}_u = (1,1,v)$, $\boldsymbol{\sigma}_v = (1,-1,u)$, $\boldsymbol{\sigma}_{uu} = \boldsymbol{\sigma}_{vv} = \mathbf{0}$, $\boldsymbol{\sigma}_{uv} = (0,0,1)$. When $u = v = 1$, we find from this that $E = 3, F = 1, G = 3$ and $L = N = 0, M = -1/\sqrt{2}$. Hence, $K = (LN - M^2)/(EG - F^2) = -1/16$, $H = (LG - 2MF + NG)/2(EG - F^2) = 1/8\sqrt{2}$.

7.2 For the helicoid $\boldsymbol{\sigma}(u,v) = (v\cos u, v\sin u, \lambda u)$, $\boldsymbol{\sigma}_u = (-v\sin u, v\cos u, \lambda)$, $\boldsymbol{\sigma}_v = (\cos u, \sin u, 0)$, $\mathbf{N} = (\lambda^2 + v^2)^{-1/2}(-\lambda\sin u, \lambda\cos u, -v)$, $\boldsymbol{\sigma}_{uu} = (-v\cos u, -v\sin u, 0)$, $\boldsymbol{\sigma}_{uv} = (-\sin u, \cos u, 0)$, $\boldsymbol{\sigma}_{vv} = \mathbf{0}$. This gives $E = \lambda^2 + v^2, F = 0, G = 1$ and $L = N = 0, M = \lambda/\sqrt{\lambda^2 + v^2}$. Hence, $K = (LN - M^2)/(EG - F^2) = -\lambda^2/(\lambda^2 + v^2)^2$.
For the catenoid $\boldsymbol{\sigma}(u,v) = (\cosh u\cos v, \cosh u\sin v, u)$, we have $\boldsymbol{\sigma}_u = (\sinh u\cos v, \sinh u\sin v, 1)$, $\boldsymbol{\sigma}_v = (-\cosh u\sin v, \cosh u\cos v, 0)$, $\mathbf{N} = \operatorname{sech} u(-\cos v, -\sin v, \sinh u)$, $\boldsymbol{\sigma}_{uu} = (\cosh u\cos v, \cosh u\sin v, 0)$, $\boldsymbol{\sigma}_{uv} = (-\sinh u\sin v, \sinh u\cos v, 0)$, $\boldsymbol{\sigma}_{vv} = (-\cosh u\cos v, -\cosh u\sin v, 0)$. This gives $E = G = \cosh^2 u, F = 0$ and $L = -1, M = 0, N = 1$. Hence, $K = (LN - M^2)/(EG - F^2) = -\operatorname{sech}^4 u$.
Alternatively, use the results of Exercise 6.15.

7.3 Parametrise the surface by $\boldsymbol{\sigma}(u,v) = (u, v, f(u,v))$. Then, $\boldsymbol{\sigma}_u = (1, 0, f_u)$, $\boldsymbol{\sigma}_v = (0, 1, f_v)$, $\mathbf{N} = (1 + f_u^2 + f_v^2)^{-1/2}(-f_u, -f_v, 1)$, $\boldsymbol{\sigma}_{uu} = (0, 0, f_{uu})$, $\boldsymbol{\sigma}_{uv} = (0, 0, f_{uv})$, $\boldsymbol{\sigma}_{vv} = (0, 0, f_{vv})$. This gives $E = 1 + f_u^2, F = f_u f_v, G = 1 + f_v^2$ and $L = (1 + f_u^2 + f_v^2)^{-1/2} f_{uu}, M = (1 + f_u^2 + f_v^2)^{-1/2} f_{uv}, N = (1 + f_u^2 + f_v^2)^{-1/2} f_{vv}$. Hence,

$$K = \frac{f_{uu}f_{vv} - f_{uv}^2}{(1 + f_u^2 + f_v^2)^2}.$$

7.4 (i) From Example 7.3, $K = 0 \iff \dot{\boldsymbol{\delta}}.\mathbf{N} = 0 \iff \dot{\boldsymbol{\delta}}.((\mathbf{t} + v\dot{\boldsymbol{\delta}}) \times \boldsymbol{\delta}) = 0 \iff \dot{\boldsymbol{\delta}}.(\mathbf{t} \times \boldsymbol{\delta}) = 0$. If $\boldsymbol{\delta} = \mathbf{n}$, $\dot{\boldsymbol{\delta}} = -\kappa\mathbf{t} + \tau\mathbf{b}$, $\mathbf{t} \times \boldsymbol{\delta} = \mathbf{b}$, so $K = 0 \iff \tau = 0 \iff \boldsymbol{\gamma}$ is planar (by Proposition 2.4). If $\boldsymbol{\delta} = \mathbf{b}$, $\dot{\boldsymbol{\delta}} = -\tau\mathbf{n}$, $\mathbf{t} \times \boldsymbol{\delta} = -\mathbf{n}$, so again $K = 0 \iff \tau = 0$.
(ii) Let $\mathbf{N}_1$ be a unit normal of $\mathcal{S}$. Then, $K = 0 \iff \dot{\mathbf{N}}_1.(\mathbf{t} \times \mathbf{N}_1) = 0$. Since $\dot{\mathbf{N}}_1$ is perpendicular to $\mathbf{N}_1$ and $\mathbf{N}_1$ is perpendicular to $\mathbf{t}$, this condition holds $\iff \dot{\mathbf{N}}_1$ is parallel to $\mathbf{t}$, i.e. $\iff \dot{\mathbf{N}}_1 = -\lambda\dot{\boldsymbol{\gamma}}$ for some scalar λ. Now use Exercise 6.18.

7.5 Using the parametrisation $\boldsymbol{\sigma}$ in Exercise 4.10, we find that $E = b^2, F = 0, G = (a + b\cos\theta)^2$ and $L = b, M = 0, N = (a + b\cos\theta)\cos\theta$. This gives $K = \cos\theta/b(a + b\cos\theta)$, $d\mathcal{A}_{\boldsymbol{\sigma}} = (EG - F^2)^{1/2} d\theta d\varphi = b(a + b\cos\theta) d\theta d\varphi$. Hence,

$$\iint K d\mathcal{A}_{\boldsymbol{\sigma}} = \int_0^{2\pi}\int_0^{2\pi} \cos\theta \, d\theta d\varphi = 0.$$

7.6 The first part follows from Exercises 5.3 and 6.4. The dilation multiplies E, F, G by a^2 and L, M, N by a, hence H by a^{-1} and K by a^{-2} (using Proposition 7.1).

7.7 Since $\boldsymbol{\sigma}$ is smooth and $\boldsymbol{\sigma}_u \times \boldsymbol{\sigma}_v$ is never zero, $\mathbf{N} = \boldsymbol{\sigma}_u \times \boldsymbol{\sigma}_v / \parallel \boldsymbol{\sigma}_u \times \boldsymbol{\sigma}_v \parallel$ is smooth. Hence, E, F, G, L, M and N are smooth. Since $EG - F^2 > 0$ (by the remark following Proposition 5.2), the formulas in Proposition 7.1(i) and (ii) show that H and K are smooth. By Proposition 7.1(iii), the principal curvatures are smooth provided $H^2 > K$, i.e. provided there are no umbilics.

7.8 At a point P of an asymptotic curve, the normal curvature is zero. By Corollary 6.2, one principal curvature $\kappa_1 \geq 0$ and the other $\kappa_2 \leq 0$. Hence $K = \kappa_1\kappa_2 \leq 0$. On a ruled surface, there is an asymptotic curve, namely a straight line, passing through every point (see Exercise 6.12).

7.9 By Exercise 6.22, $\mathcal{F}_{III} = \mathcal{F}_{II}\mathcal{F}_I^{-1}\mathcal{F}_{II}$. Multiplying on the left by $\mathcal{F}_I^{-1}$, the given equation is equivalent to

$$A^2 + 2HA + KI = 0,$$

where $A = -\mathcal{F}_I^{-1}\mathcal{F}_{II} = \begin{pmatrix} a & c \\ b & d \end{pmatrix}$. By the remarks following Definition 6.1, the principal curvatures are the eigenvalues of $-A$. Hence, $2H =$ sum of eigenvalues of $-A = -(a+d)$, $K =$ product of eigenvalues of $-A = ad - bc$. Now use the fact stated in the question.

7.10 By Eq. (9) in Chapter 6, $\dot{\boldsymbol{\gamma}}.\dot{\boldsymbol{\gamma}} = T^t\mathcal{F}_I T$; by Eq. (10) in Chapter 6, $\dot{\mathbf{N}}.\dot{\boldsymbol{\gamma}} = -\mathbf{N}.\ddot{\boldsymbol{\gamma}}$ (since $\mathbf{N}.\dot{\boldsymbol{\gamma}} = 0$) $= -\kappa_n = -T^t\mathcal{F}_{II}T$. Now, $\dot{\mathbf{N}}.\dot{\mathbf{N}} = (\dot{u}\mathbf{N}_u + \dot{v}\mathbf{N}_v).(\dot{u}\mathbf{N}_u + \dot{v}\mathbf{N}_v) = (\mathbf{N}_u.\mathbf{N}_u)\dot{u}^2 + 2(\mathbf{N}_u.\mathbf{N}_v)\dot{u}\dot{v} + (\mathbf{N}_v.\mathbf{N}_v)\dot{v}^2 = T^t\mathcal{F}_{III}T$. Hence, multiplying the equation in Exercise 7.9 on the left by T^t and on the right by T gives $\dot{\mathbf{N}}.\dot{\mathbf{N}} + 2H\dot{\mathbf{N}}.\dot{\boldsymbol{\gamma}} + K\dot{\boldsymbol{\gamma}}.\dot{\boldsymbol{\gamma}} = 0$. If $\boldsymbol{\gamma}$ is an asymptotic curve, $\kappa_n = 0$ so $\dot{\mathbf{N}}.\dot{\boldsymbol{\gamma}} = 0$. So, assuming that $\boldsymbol{\gamma}$ is unit-speed, we get $\dot{\mathbf{N}}.\dot{\mathbf{N}} = -K$. But Exercise 6.12 gives $\mathbf{N} = \pm\mathbf{b}$, so $\dot{\mathbf{N}} = \mp\tau\mathbf{n}$ and $\dot{\mathbf{N}}.\dot{\mathbf{N}} = \tau^2$.

7.11 The parametrisation is $\boldsymbol{\sigma}(u,v) = (f(u)\cos v, f(u)\sin v, g(u))$, $f(u) = e^u$, $g(u) = \sqrt{1 - e^{2u}} - \cosh^{-1}(e^{-u})$, $-\infty < u < 0$.

(i) A parallel $u =$ constant is a circle of radius $f(u) = e^u$, so has length $2\pi e^u$.

(ii) From Example 7.2, $E = 1, F = 0, G = f(u)^2$, so $d\mathcal{A}_{\boldsymbol{\sigma}} = f(u)dudv$ and the area is $\int_0^{2\pi}\int_{-\infty}^0 e^u dudv = 2\pi$.

(iii) From Example 7.2, the principal curvatures are $\kappa_1 = \dot{f}\ddot{g} - \ddot{f}\dot{g} = -\ddot{f}/\dot{g} = -(e^{-2u} - 1)^{-1/2}$, $\kappa_2 = f\dot{g}/f^2 = \dot{g}/f = (e^{-2u} - 1)^{1/2}$.

(iv) $\kappa_1 < 0$, $\kappa_2 > 0$.

7.12 (i) Setting $\tilde{u} = v, \tilde{v} = w = e^{-u}$, we have $u = -\ln\tilde{v}, v = \tilde{u}$ so, in the notation of Exercise 5.4, $J = \begin{pmatrix} 0 & -\frac{1}{\tilde{v}} \\ 1 & 0 \end{pmatrix}$. Since J is invertible, $(u,v) \mapsto$

(v, w) is a reparametrisation map. The first fundamental form in terms of v, w is given by

$$\begin{pmatrix} \tilde{E} & \tilde{F} \\ \tilde{F} & \tilde{G} \end{pmatrix} = J^t \begin{pmatrix} E & F \\ F & G \end{pmatrix} J$$

$$= \begin{pmatrix} 0 & 1 \\ -\frac{1}{\bar{v}} & 0 \end{pmatrix} \begin{pmatrix} 1 & 0 \\ 0 & f(u)^2 \end{pmatrix} \begin{pmatrix} 0 & -\frac{1}{\bar{v}} \\ 1 & 0 \end{pmatrix} = \begin{pmatrix} \frac{1}{w^2} & 0 \\ 0 & \frac{1}{w^2} \end{pmatrix},$$

so the first fundamental form is $(dv^2 + dw^2)/w^2$.

(ii) We find that the matrix

$$\tilde{J} = \begin{pmatrix} \frac{\partial v}{\partial U} & \frac{\partial v}{\partial V} \\ \frac{\partial w}{\partial U} & \frac{\partial w}{\partial V} \end{pmatrix} = \begin{pmatrix} v(w+1) & \frac{1}{2}(v^2 - (w+1)^2) \\ -\frac{1}{2}(v^2 - (w+1)^2) & v(w+1) \end{pmatrix},$$

so the first fundamental form in terms of U and V is given by

$$\tilde{J}^t \begin{pmatrix} \frac{1}{w^2} & 0 \\ 0 & \frac{1}{w^2} \end{pmatrix} \tilde{J} = \frac{(v^2 + (w+1)^2)^2}{4w^2} I = \frac{1}{(1 - U^2 - V^2)^2} I,$$

after some tedious algebra.

In (i), $u < 0$ and $-\pi < v < \pi$ corresponds to $-\pi < v < \pi$ and $w > 1$, a semi-infinite rectangle in the upper half of the vw-plane.

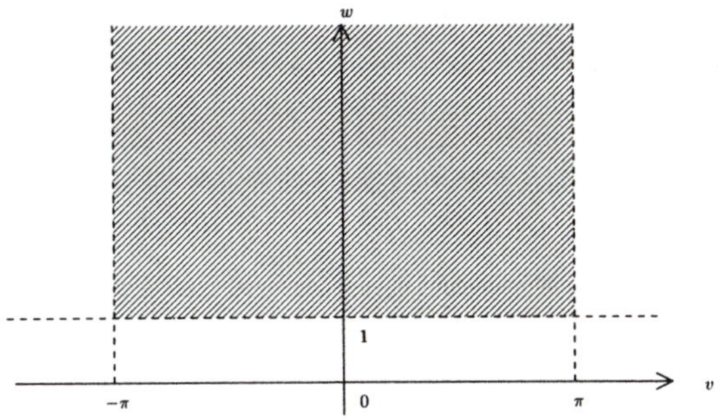

To find the corresponding region in (ii), it is convenient to introduce the complex numbers $z = v + iw$, $Z = U + iV$. Then, the equations in (ii) are equivalent to $Z = \frac{z-i}{z+i}$, $z = \frac{Z+1}{i(Z-1)}$. The line $v = \pi$ in the vw-plane corresponds to $z + \bar{z} = 2\pi$ (the bar denoting complex conjugate), i.e. $\frac{Z+1}{i(Z-1)} - \frac{\bar{Z}+1}{i(\bar{Z}-1)} = 2\pi$, which simplifies to $|Z - (1 - \frac{i}{\pi})|^2 = \frac{1}{\pi^2}$; so $v = \pi$ corresponds to the circle in the UV-plane with centre $1 - \frac{i}{\pi}$ and radius $\frac{1}{\pi}$. Similarly, $v = -\pi$ corresponds to the circle with centre $1 + \frac{i}{\pi}$ and radius $\frac{1}{\pi}$. Finally, $w = 1$ corresponds to $z - \bar{z} = 2i$, i.e. $\frac{Z+1}{i(Z-1)} + \frac{\bar{Z}+1}{i(\bar{Z}-1)} = 2i$. This simplifies to $|Z - \frac{1}{2}|^2 = \frac{1}{4}$; so $w = 1$ corresponds to the circle with centre $1/2$ and radius $1/2$ in the UV-plane. The required region in the UV-plane is that bounded by these three circles:

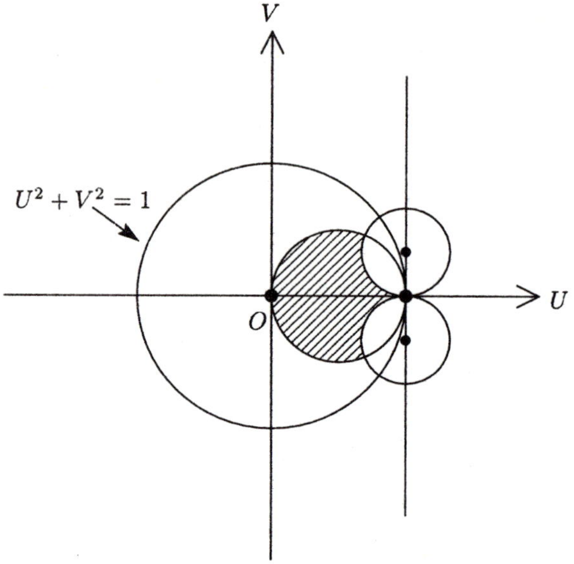

7.13 Let $\gamma(u) = (f(u), 0, g(u))$ and denote d/du by a dot; by Eq. (2), $\ddot{f} + Kf = 0$. If $K < 0$, the general solution is $f = ae^{-\sqrt{-K}u} + be^{\sqrt{-K}u}$ where a, b are constants; the condition $f(\pi/2) = f(-\pi/2) = 0$ forces $a = b = 0$, so γ coincides with the z-axis, contradicting the assumptions. If $K = 0$, $f = a + bu$ and again $a = b = 0$ is forced. So we must have $K > 0$ and $f = a\cos\sqrt{K}u + b\sin\sqrt{K}u$. This time, $f(\pi/2) = f(-\pi/2) = 0$ and a, b not both zero implies that the determinant

$$\begin{vmatrix} \cos\sqrt{K}\pi/2 & \sin\sqrt{K}\pi/2 \\ \cos\sqrt{K}\pi/2 & -\sin\sqrt{K}\pi/2 \end{vmatrix} = 0.$$

This gives $\sin\sqrt{K}\pi = 0$, so $K = n^2$ for some integer $n \neq 0$. If $n = 2k$ is even, $f = b\sin 2ku$, but then $f(0) = 0$, contradicting the assumptions. If $n = 2k+1$ is odd, $f = a\cos(2k+1)u$ and $f(\pi/2(2k+1)) = 0$, which contradicts the assumptions unless $k = 0$ or -1, i.e. unless $K = (2k+1)^2 = 1$. Thus, $f = a\cos u$, $\dot{g} = \sqrt{1 - \dot{f}^2} = \sqrt{1 - a^2\sin^2 u}$. Now, $\dot{\gamma} = (\dot{f}, 0, \dot{g})$ is perpendicular to the z-axis $\iff \dot{g} = 0$. So the assumptions give $\sqrt{1 - a^2} = 0$, i.e. $a = \pm 1$. Then, $\gamma(u) = (\pm\cos u, 0, \pm\sin u)$ (up to a translation along the z-axis) and $\mathcal{S}$ is the unit sphere.

7.14 Let $\tilde{\sigma}(\tilde{u}, \tilde{v})$ be a patch of $\mathcal{S}$ containing $P = \tilde{\sigma}(\tilde{u}_0, \tilde{v}_0)$. The gaussian curvature K of $\mathcal{S}$ is < 0 at P; since K is a smooth function of $(\tilde{u}, \tilde{v})$ (Exercise 7.7), $K(\tilde{u}, \tilde{v}) < 0$ for $(\tilde{u}, \tilde{v})$ in some open set $\tilde{U}$ containing $(\tilde{u}_0, \tilde{v}_0)$; then every point of $\tilde{\sigma}(\tilde{U})$ is hyperbolic. Let κ_1, κ_2 be the principal curvatures of $\tilde{\sigma}$, let $0 < \theta < \pi/2$ be such that $\tan\theta = \sqrt{-\kappa_1/\kappa_2}$, and let $\mathbf{e}_1$ and $\mathbf{e}_2$ be the unit tangent vectors of $\tilde{\sigma}$ making angles θ and $-\theta$, respectively, with the principal vector corresponding to κ_1 (see Corollary 6.1). Applying

Proposition 7.4 gives the result.

7.15 See the proof of Proposition 9.5 for the first part.
The first fundamental form of the given surface patch is $(1+u^2+v^2)^2(du^2+dv^2)$, so it is conformal, and $\boldsymbol{\sigma}_{uu}+\boldsymbol{\sigma}_{vv} = (-2u, 2v, 2)+(2u, -2v, -2) = \mathbf{0}$.

7.16 Parametrize the surface by $\boldsymbol{\sigma}(u,v) = (u, v, f(u,v))$. By Exercise 7.3, $E = 1+f_u^2, F = f_u f_v, G = 1+f_v^2, L = (1+f_u^2+f_v^2)^{-1/2} f_{uu}, M = (1+f_u^2+f_v^2)^{-1/2} f_{uv}, N = (1+f_u^2+f_v^2)^{-1/2} f_{vv}$. Hence,

$$H = \frac{LG - 2MF + NE}{2(EG - F^2)} = \frac{f_{uu}(1+f_v^2) - 2f_{uv}f_u f_v + f_{vv}(1+f_u^2)}{2(1+f_u^2+f_v^2)^{3/2}}.$$

Taking $f(u,v) = \ln\left(\frac{\cos v}{\cos u}\right)$ gives

$$H = \frac{\sec^2 u(1+\tan^2 v) - \sec^2 v(1+\tan^2 u)}{2(1+\tan^2 u + \tan^2 v)^{3/2}} = 0.$$

7.17 $\boldsymbol{\Sigma}_u = \boldsymbol{\sigma}_u + w\mathbf{N}_u$, $\boldsymbol{\Sigma}_v = \boldsymbol{\sigma}_v + w\mathbf{N}_v$, $\boldsymbol{\Sigma}_w = \mathbf{N}$. $\boldsymbol{\Sigma}_u.\boldsymbol{\Sigma}_w = 0$ since $\boldsymbol{\sigma}_u.\mathbf{N} = \mathbf{N}_u.\mathbf{N} = 0$, and similarly $\boldsymbol{\Sigma}_v.\boldsymbol{\Sigma}_w = 0$. Finally, $\boldsymbol{\Sigma}_u.\boldsymbol{\Sigma}_v = \boldsymbol{\sigma}_u.\boldsymbol{\sigma}_v + w(\boldsymbol{\sigma}_u.\mathbf{N}_v + \boldsymbol{\sigma}_v.\mathbf{N}_u) + w^2\mathbf{N}_u.\mathbf{N}_v = F - 2wM + w^2\mathbf{N}_u.\mathbf{N}_v = w^2\mathbf{N}_u.\mathbf{N}_v$; by Proposition 6.4, $\mathbf{N}_u = -\frac{L}{E}\boldsymbol{\sigma}_u$, $\mathbf{N}_v = -\frac{N}{G}\boldsymbol{\sigma}_v$, so $\mathbf{N}_u.\mathbf{N}_v = \frac{LN}{EG}F = 0$.
Every surface $u = u_0$ (a constant) is ruled as it is the union of the straight lines given by $v =$ constant; by Exercise 7.4(ii), this surface is flat provided the curve $\boldsymbol{\gamma}(v) = \boldsymbol{\sigma}(u_0, v)$ is a line of curvature of $\mathcal{S}$, i.e. if $\boldsymbol{\sigma}_v$ is a principal vector; but this is true since the matrices $\mathcal{F}_I$ and $\mathcal{F}_{II}$ are diagonal. Similarly for the surfaces $v =$ constant.

7.18 By Eq. (15), the area of $\boldsymbol{\sigma}(R)$ is

$$\int\int_R \| \mathbf{N}_u \times \mathbf{N}_v \| \, dudv = \int\int_R |K| \, \| \boldsymbol{\sigma}_u \times \boldsymbol{\sigma}_v \| \, dudv = \int\int_R |K| dA_{\boldsymbol{\sigma}}.$$

7.18 From the formula for K in the solution of Exercise 7.5, it follows that $\mathcal{S}^+$ and $\mathcal{S}^-$ are the annular regions on the torus given by $-\pi/2 \le u \le \pi/2$ and $\pi/2 \le u \le 3\pi/2$, respectively.

$\mathcal{S}^+$

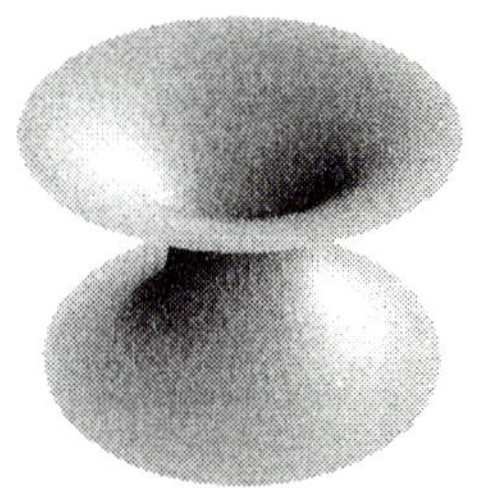

$\mathcal{S}^-$

It is clear that as a point P moves over $\mathcal{S}^+$ (resp. $\mathcal{S}^-$), the unit normal at P covers the whole of the unit sphere. Hence, $\iint_{\mathcal{S}^+} |K| d\mathcal{A} = \iint_{\mathcal{S}^-} |K| d\mathcal{A} = 4\pi$ by Exercise 7.18; since $|K| = \pm K$ on $\mathcal{S}^\pm$, this gives the result.

Chapter 8

8.1 By Exercise 4.4, there are two straight lines on the hyperboloid passing through $(1, 0, 0)$; by Proposition 8.3, they are geodesics. The circle given by $z = 0, x^2 + y^2 = 1$ and the hyperbola given by $y = 0, x^2 - z^2 = 1$ are both normal sections, hence geodesics by Proposition 8.4 (see also Proposition 8.5).

8.2 Let Π_s be the plane through $\boldsymbol{\gamma}(s)$ perpendicular to $\mathbf{t}(s)$; the parameter curve $s =$ constant is the intersection of the surface with Π_s. From the solution to Exercise 5.17, the standard unit normal of $\boldsymbol{\sigma}$ is $\mathbf{N} = -(\cos\theta\, \mathbf{n} + \sin\theta\, \mathbf{b})$. Since this is perpendicular to $\mathbf{t}$, the circles in question are normal sections.

8.3 Take the ellipsoid to be $\frac{x^2}{p^2} + \frac{y^2}{q^2} + \frac{z^2}{r^2} = 1$; the vector $(\frac{x}{p^2}, \frac{y}{q^2}, \frac{z}{r^2})$ is normal to the ellipsoid by Exercise 4.16. If $\boldsymbol{\gamma}(t) = (f(t), g(t), h(t))$ is a curve on the ellipsoid, $R = (\frac{\dot f^2}{p^2} + \frac{\dot g^2}{q^2} + \frac{\dot h^2}{r^2})^{-1/2}$, $S = (\frac{f^2}{p^4} + \frac{g^2}{q^4} + \frac{h^2}{r^4})^{-1/2}$. Now, $\boldsymbol{\gamma}$ is a geodesic $\iff$ $\ddot{\boldsymbol{\gamma}}$ is parallel to the normal $\iff$ $(\ddot f, \ddot g, \ddot h) = \lambda(\frac{f}{p^2}, \frac{g}{q^2}, \frac{h}{r^2})$ for some scalar $\lambda(t)$. From $\frac{f^2}{p^2} + \frac{g^2}{q^2} + \frac{h^2}{r^2} = 1$ we get $\frac{f\dot f}{p^2} + \frac{g\dot g}{q^2} + \frac{h\dot h}{r^2} = 0$, hence $\frac{\dot f^2}{p^2} + \frac{\dot g^2}{q^2} + \frac{\dot h^2}{r^2} + \frac{f\ddot f}{p^2} + \frac{g\ddot g}{q^2} + \frac{h\ddot h}{r^2} = 0$, i.e. $\frac{\dot f^2}{p^2} + \frac{\dot g^2}{q^2} + \frac{\dot h^2}{r^2} + \lambda\left(\frac{f^2}{p^4} + \frac{g^2}{q^4} + \frac{h^2}{r^4}\right) = 0$, which gives $\lambda = -S^2/R^2$. The curvature of $\boldsymbol{\gamma}$ is

$$\parallel \ddot{\boldsymbol{\gamma}} \parallel = (\ddot f^2 + \ddot g^2 + \ddot h^2)^{1/2} = |\lambda| \left(\frac{f^2}{p^4} + \frac{g^2}{q^4} + \frac{h^2}{r^4}\right)^{1/2} = \frac{|\lambda|}{S} = \frac{S}{R^2}.$$

Finally,

$$\begin{aligned} \frac{1}{2}\frac{d}{dt}\left(\frac{1}{R^2 S^2}\right) &= \left(\frac{f\dot f}{p^4} + \frac{g\dot g}{q^4} + \frac{h\dot h}{r^4}\right)\left(\frac{\dot f^2}{p^2} + \frac{\dot g^2}{q^2} + \frac{\dot h^2}{r^2}\right) \\ &\quad + \left(\frac{f^2}{p^4} + \frac{g^2}{q^4} + \frac{h^2}{r^4}\right)\left(\frac{\dot f \ddot f}{p^2} + \frac{\dot g \ddot g}{q^2} + \frac{\dot h \ddot h}{r^2}\right) \\ &= \frac{1}{R^2}\left(\frac{f\dot f}{p^4} + \frac{g\dot g}{q^4} + \frac{h\dot h}{r^4}\right) + \frac{\lambda}{S^2}\left(\frac{f\dot f}{p^4} + \frac{g\dot g}{q^4} + \frac{h\dot h}{r^4}\right) = 0, \end{aligned}$$

since $\lambda = -S^2/R^2$. Hence, RS is constant.

8.4 If $\boldsymbol{\gamma}$ is a geodesic, $\ddot{\boldsymbol{\gamma}} = \kappa\mathbf{n}$ is parallel to $\mathbf{N}$ (in the usual notation), so $\mathbf{n} = \pm\mathbf{N}$. In the notation of Exercise 6.11, $\mathbf{B} = \mathbf{t} \times \mathbf{N} = \pm\mathbf{b}$, so $\dot{\mathbf{B}} =$

$\kappa_g \mathbf{t} - \tau_g \mathbf{N} = \pm\dot{\mathbf{b}} = \mp\tau\mathbf{n} = -\tau\mathbf{N}$. Hence, $\tau_g = \tau$ (and $\kappa_g = 0$, which we knew already).

8.5 $\ddot{\boldsymbol{\gamma}}$ is parallel to Π since $\boldsymbol{\gamma}$ lies in Π, and $\ddot{\boldsymbol{\gamma}}$ is parallel to $\mathbf{N}$ since $\boldsymbol{\gamma}$ is a geodesic; so $\mathbf{N}$ is parallel to Π. It follows that $\dot{\mathbf{N}}$ is also parallel to Π. Since $\dot{\mathbf{N}}$ is perpendicular to $\mathbf{N}$, and $\dot{\boldsymbol{\gamma}}$ is also perpendicular to $\mathbf{N}$ and parallel to Π, $\dot{\mathbf{N}}$ is parallel to $\dot{\boldsymbol{\gamma}}$. By Exercise 6.18, $\boldsymbol{\gamma}$ is a line of curvature of $\mathcal{S}$.

8.6 If P and Q lie on the same parallel of the cylinder, there are exactly two geodesics joining them, namely the two circular arcs of the parallel of which P and Q are the endpoints. If P and Q are not on the same parallel, there are infinitely many circular helices joining P and Q (see Example 8.7).

8.7 Take the cone to be $\boldsymbol{\sigma}(u,v) = (u\cos v, u\sin v, u)$. By Exercise 5.5, $\boldsymbol{\sigma}$ is isometric to part of the xy-plane by $\boldsymbol{\sigma}(u,v) \mapsto (u\sqrt{2}\cos\frac{v}{\sqrt{2}}, u\sqrt{2}\sin\frac{v}{\sqrt{2}}, 0)$. By Corollary 8.2, the geodesics on the cone correspond to the straight lines in the xy-plane. Any such line, other than the axes $x = 0$ and $y = 0$, has equation $ax + by = 1$, where a, b are constants; this line corresponds to the curve $v \mapsto \left(\frac{\cos v}{\sqrt{2}(a\cos\frac{v}{\sqrt{2}}+b\sin\frac{v}{\sqrt{2}})}, \frac{\sin v}{\sqrt{2}(a\cos\frac{v}{\sqrt{2}}+b\sin\frac{v}{\sqrt{2}})}, \frac{1}{\sqrt{2}(a\cos\frac{v}{\sqrt{2}}+b\sin\frac{v}{\sqrt{2}})}\right)$; the x and y-axes correspond to straight lines on the cone.

8.8 From Example 5.3, $\boldsymbol{\sigma}(u,v) = \boldsymbol{\gamma}(u) + v\mathbf{a}$, where $\boldsymbol{\gamma}$ is unit-speed, $\| \mathbf{a} \| = 1$, and $\boldsymbol{\gamma}$ is contained in a plane perpendicular to $\mathbf{a}$; the map $\boldsymbol{\sigma}(u,v) \mapsto (u, v, 0)$ is an isometry from the cylinder to the xy-plane. A curve $t \mapsto \boldsymbol{\sigma}(u(t), v(t))$ is a geodesic on the cylinder $\iff t \mapsto (u(t), v(t), 0)$ is a constant-speed parametrisation of a straight line in the plane $\iff \dot{u}$ and $\dot{v}$ are constant $\iff \dot{v}$ is constant (since $\dot{u}^2 + \dot{v}^2$ is constant). Since $\mathbf{a}.\frac{d}{dt}\boldsymbol{\sigma}(u(t), v(t)) = \dot{v}$, $\dot{v}$ = constant $\iff$ the tangent vector $\frac{d}{dt}\boldsymbol{\sigma}(u,v)$ makes a fixed angle with the unit vector $\mathbf{a}$ parallel to the axis of the cylinder.

8.9 Take the cylinder to be $\boldsymbol{\sigma}(u,v) = (\cos u, \sin u, v)$. Then, $E = G = 1, F = 0$, so the geodesic equations are $\ddot{u} = \ddot{v} = 0$. Hence, $u = a + bt, v = c + dt$, where a, b, c, d are constants. If $b = 0$ this is a straight line on the cylinder; otherwise, it is a circular helix.

8.10 $E = 1, F = 0, G = 1 + u^2$, so $\boldsymbol{\gamma}$ is unit-speed $\iff \dot{u}^2 + (1 + u^2)\dot{v}^2 = 1$. The second equation in (2) gives $\frac{d}{dt}((1 + u^2)\dot{v}) = 0$, i.e. $\dot{v} = \frac{a}{1+u^2}$, where a is a constant. So $\dot{u}^2 = 1 - \frac{a^2}{(1+u^2)}$ and, along the geodesic,

$$\frac{dv}{du} = \frac{\dot{v}}{\dot{u}} = \pm\frac{a}{\sqrt{(1 - a^2 + u^2)(1 + u^2)}}.$$

If $a = 0$, then v = constant and we have a ruling. If $a = 1$, then $dv/du = \pm 1/u\sqrt{1 + u^2}$, which can be integrated to give $v = v_0 \mp \sinh^{-1}\frac{1}{u}$, where

v_0 is a constant.

8.11 We have

$$\mathbf{N}\times\boldsymbol{\sigma}_u = \frac{(\boldsymbol{\sigma}_u\times\boldsymbol{\sigma}_v)\times\boldsymbol{\sigma}_u}{\|\boldsymbol{\sigma}_u\times\boldsymbol{\sigma}_v\|} = \frac{(\boldsymbol{\sigma}_u.\boldsymbol{\sigma}_u)\boldsymbol{\sigma}_v - (\boldsymbol{\sigma}_u.\boldsymbol{\sigma}_v)\boldsymbol{\sigma}_u}{\sqrt{EG-F^2}} = \frac{E\boldsymbol{\sigma}_v - F\boldsymbol{\sigma}_u}{\sqrt{EG-F^2}},$$

and similarly for $\mathbf{N}\times\boldsymbol{\sigma}_v$. Now, $\dot{\boldsymbol{\gamma}} = \dot{u}\boldsymbol{\sigma}_u + \dot{v}\boldsymbol{\sigma}_v$, so

$$\mathbf{N}\times\dot{\boldsymbol{\gamma}} = \frac{\dot{u}(E\boldsymbol{\sigma}_v - F\boldsymbol{\sigma}_u) + \dot{v}(F\boldsymbol{\sigma}_v - G\boldsymbol{\sigma}_u)}{\sqrt{EG-F^2}},$$

$$\ddot{\boldsymbol{\gamma}} = \ddot{u}\boldsymbol{\sigma}_u + \ddot{v}\boldsymbol{\sigma}_v + \dot{u}^2\boldsymbol{\sigma}_{uu} + 2\dot{u}\dot{v}\boldsymbol{\sigma}_{uv} + \dot{v}^2\boldsymbol{\sigma}_{vv}.$$

Hence, $\kappa_g = \ddot{\boldsymbol{\gamma}}.(\mathbf{N}\times\dot{\boldsymbol{\gamma}}) = (\dot{u}\ddot{v}-\dot{v}\ddot{u})\sqrt{EG-F^2} + A\dot{u}^3 + B\dot{u}^2\dot{v} + C\dot{u}\dot{v}^2 + D\dot{v}^3$, where $A = \boldsymbol{\sigma}_{uu}.(E\boldsymbol{\sigma}_v - F\boldsymbol{\sigma}_u) = E((\boldsymbol{\sigma}_u.\boldsymbol{\sigma}_v)_u - \boldsymbol{\sigma}_u.\boldsymbol{\sigma}_{uv}) - \frac{1}{2}F(\boldsymbol{\sigma}_u.\boldsymbol{\sigma}_u)_u = E(F_u - \frac{1}{2}E_v) - \frac{1}{2}FE_u$, with similar expressions for B, C, D.

8.12 We have

$$\begin{aligned}
(E\dot{u}^2 &+ 2F\dot{u}\dot{v} + G\dot{v}^2)^{\cdot} \\
&= (E_u\dot{u} + E_v\dot{v})\,\dot{u}^2 + 2(F_u\dot{u} + F_v\dot{v})\dot{u}\dot{v} + (G_u\dot{u} + G_v\dot{v})\dot{v}^2 \\
&\qquad + 2E\dot{u}\ddot{u} + 2F(\dot{u}\ddot{v} + \ddot{u}\dot{v}) + 2G\dot{v}\ddot{v} \\
&= \dot{u}(E_u\dot{u}^2 + 2F_u\dot{u}\dot{v} + G_u\dot{v}^2) + \dot{v}(E_v\dot{u}^2 + 2F_v\dot{u}\dot{v} + G_v\dot{v}^2) \\
&\qquad + 2E\dot{u}\ddot{u} + 2F(\dot{u}\ddot{v} + \ddot{u}\dot{v}) + 2G\dot{v}\ddot{v} \\
&= 2(E_u\dot{u} + F\dot{v})^{\cdot}\,\dot{u} + 2(F\dot{u} + G\dot{v})^{\cdot}\,\dot{v} \\
&\qquad + 2(E\dot{u} + F\dot{v})\ddot{u} + 2(F\dot{u} + G\dot{v})\ddot{v} \quad \text{(by the geodesic equations)} \\
&= 2[(E\dot{u} + F\dot{v})\dot{u}]^{\cdot} + 2[(F\dot{u} + G\dot{v})\dot{v}]^{\cdot} \\
&= 2(E\dot{u}^2 + 2F\dot{u}\dot{v} + G\dot{v}^2)^{\cdot}.
\end{aligned}$$

Hence, $(E\dot{u}^2 + 2F\dot{u}\dot{v} + G\dot{v}^2)^{\cdot} = 0$ and so $\|\,\dot{\boldsymbol{\gamma}}\,\|^2 = E\dot{u}^2 + 2F\dot{u}\dot{v} + G\dot{v}^2$ is constant.

8.13 They are normal sections.

8.14 Every parallel is a geodesic $\iff$ every value of u is a stationary point of $f(u)$ (in the notation of Proposition 8.5) $\iff f = \text{constant} \iff$ the surface is a circular cylinder.

8.15 The two solutions of Eq. (13) are $v = v_0 \pm \sqrt{\frac{1}{\Omega^2} - w^2}$, so the condition for a self-intersection is that, for some $w > 1$, $2\sqrt{\frac{1}{\Omega^2} - w^2} = 2k\pi$ for some integer $k > 0$. This holds $\iff 2\sqrt{\frac{1}{\Omega^2} - 1} > 2\pi$, i.e. $\Omega < (1+\pi^2)^{-1/2}$. In this case, there are k self-intersections, where k is the largest integer such that $2k\pi < 2\sqrt{\frac{1}{\Omega^2} - 1}$.

8.16 (i) If $\boldsymbol{\gamma}(t)$ is a geodesic, so is $f(\boldsymbol{\gamma}(t))$, and if $\boldsymbol{\gamma}$ is defined for all $-\infty < t < \infty$, so is $f(\boldsymbol{\gamma}(t))$. So f takes meridians to meridians, i.e. if v is constant, so is $\tilde{v}$. Hence, $\tilde{v}$ does not depend on w.

(ii) f preserves angles and takes meridians to meridians, so must take parallels to parallels. Hence, $\tilde{w}$ does not depend on v.
(iii) The parallel $w =$ constant has length $2\pi/w$ by Exercise 7.11(i) ($w = e^{-u}$). As f preserves lengths, part (ii) implies that $2\pi/w = 2\pi/\tilde{w}$, so $w = \tilde{w}$.
(iv) We now know that $f(\boldsymbol{\sigma}(u,v)) = \boldsymbol{\sigma}(F(v),w)$ for some smooth function $F(v)$. The first fundamental form of $\boldsymbol{\sigma}(F(v),w)$ is $w^{-2}\left(\left(\frac{dF}{dv}\right)^2 dv^2 + dw^2\right)$; since f is an isometry, this is equal to $w^{-2}(dv^2+dw^2)$, hence $dF/dv = \pm 1$, so $f(v) = \pm v + \alpha$, where α is a constant. If the sign is $+$, f is rotation by α around the z-axis; if the sign is $-$, f is reflection in the plane containing the z-axis making an angle $\alpha/2$ with the xz-plane.

8.17 From the solution to Exercise 7.12, $Z = U + iV = \frac{z-i}{z+i}$, where $z = v + iw$. Since the geodesics on the pseudosphere correspond to straight lines and circles in the vw-plane which are perpendicular to the v-axis, they correspond in the UV-plane to straight lines and circles perpendicular to the image of the v-axis under the transformation $z \mapsto \frac{z-i}{z+i}$, i.e. the unit circle $U^2 + V^2 = 1$.

8.18 (i) Let the spheroid be obtained by rotating the ellipse $\frac{x^2}{a^2} + \frac{z^2}{b^2} = 1$ around the z-axis, where $a, b > 0$. Then, a is the maximum distance of a point of the spheroid from the z-axis, so the angular momentum Ω of a geodesic must be $\leq a$ (we can assume that $\Omega \geq 0$). If $\Omega = 0$, the geodesic is a meridian. If $0 < \Omega < a$, the geodesic is confined to the annular region on the spheroid contained between the circles $z = \pm b\sqrt{1 - \frac{\Omega^2}{a^2}}$, and the discussion in Example 8.9 shows that the geodesic 'bounces' between these two circles:

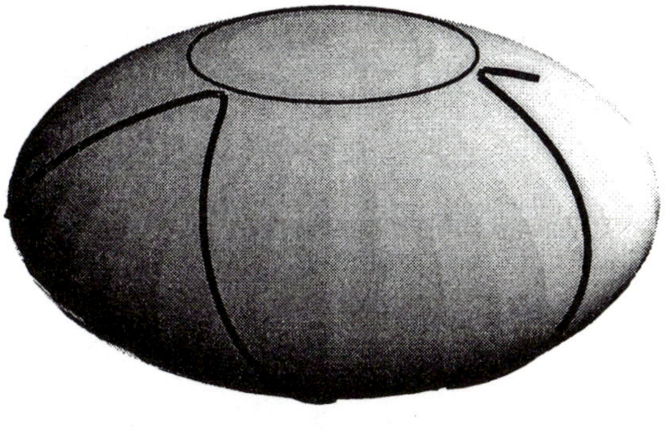

If $\Omega = a$, Eq. (10) shows that the geodesic must be the parallel $z = 0$.
(ii) Let the torus be as in Exercise 4.10. If $\Omega = 0$, the geodesic is a meridian (a circle). If $0 < \Omega < a - b$, the geodesic spirals around the torus:

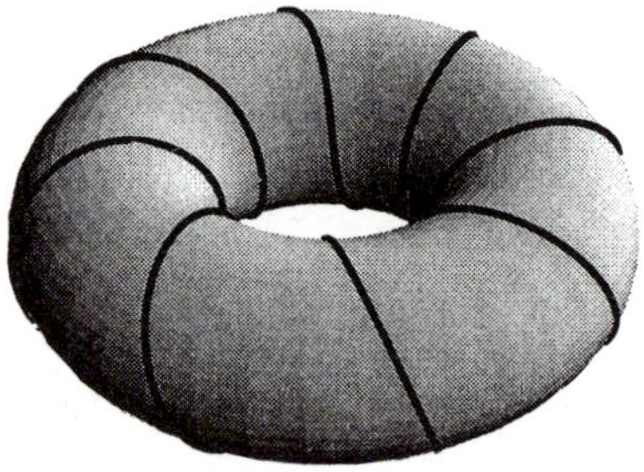

If $\Omega = a - b$, the geodesic is either the parallel of radius $a - b$ or spirals around the torus approaching this parallel asymptotically (but never crossing it):

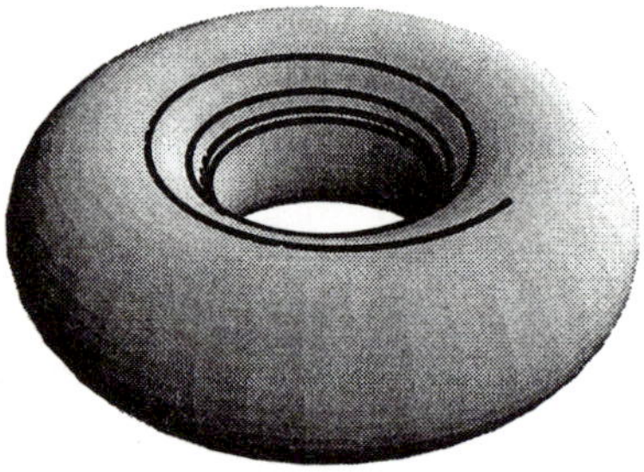

If $a - b < \Omega < a + b$, the geodesic is confined to the annular region consisting of the part of the torus a distance $\geq \Omega$ from the axis, and bounces between the two parallels which bound this region:

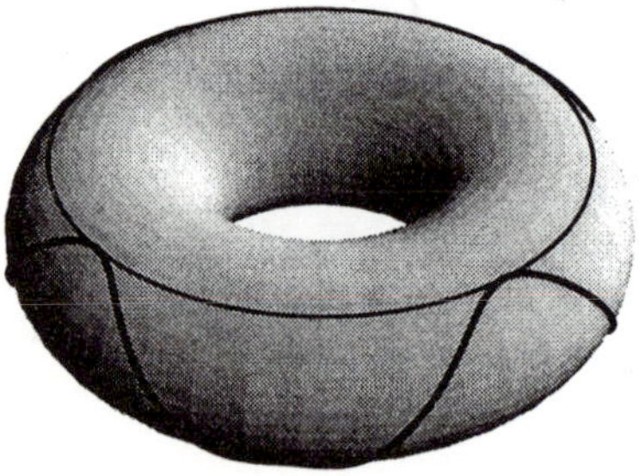

If $\Omega = a + b$, the geodesic must be the parallel of radius $a + b$.

8.19 From Exercise 5.5, the cone is isometric to the 'sector' $\mathcal{S}$ of the plane with vertex at the origin and angle $\pi\sqrt{2}$:

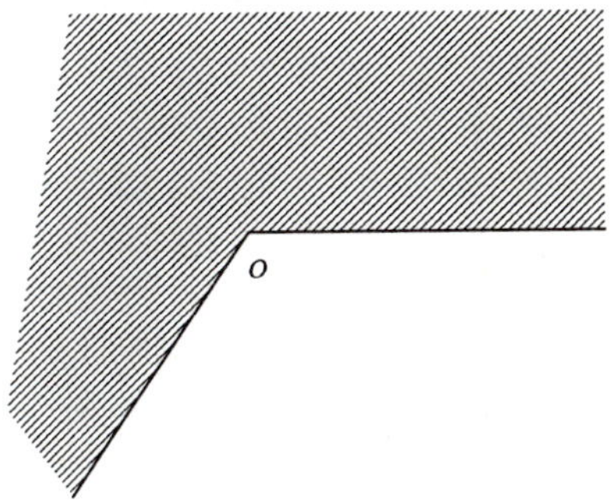

Geodesics on the cone correspond to possibly broken line segments in $\mathcal{S}$: if a line segment meets the boundary of $\mathcal{S}$ at a point A, say, it may continue from the point B on the other boundary line at the same distance as A from the origin and with the indicated angles being equal:

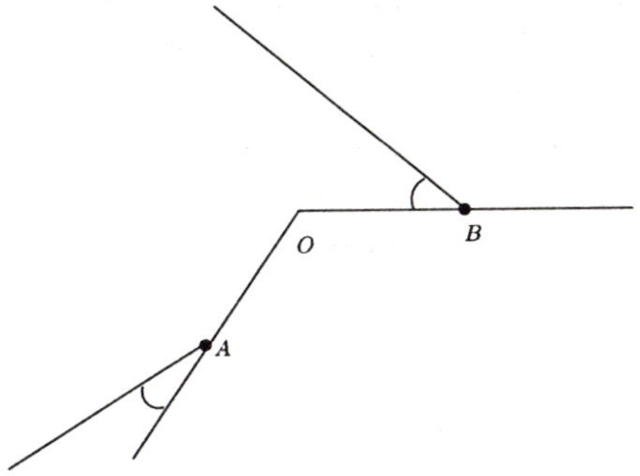

(i) TRUE: if two points P and Q can be joined by a line segment in $\mathcal{S}$ there is no problem; otherwise, P and Q can be joined by a broken line segment satisfying the conditions above:

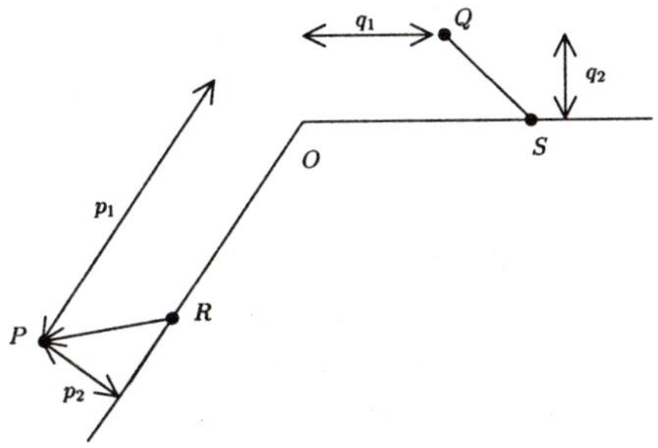

To see that this is always possible, let p_1, p_2, q_1 and q_2 be the indicated distances, and let R and S be the points on the boundary of the sector at a distance $(p_2q_1 + p_1q_2)/(p_2 + q_2)$ from the origin. Then, the broken line segment PR followed by SQ is the desired geodesic.
(ii) FALSE:

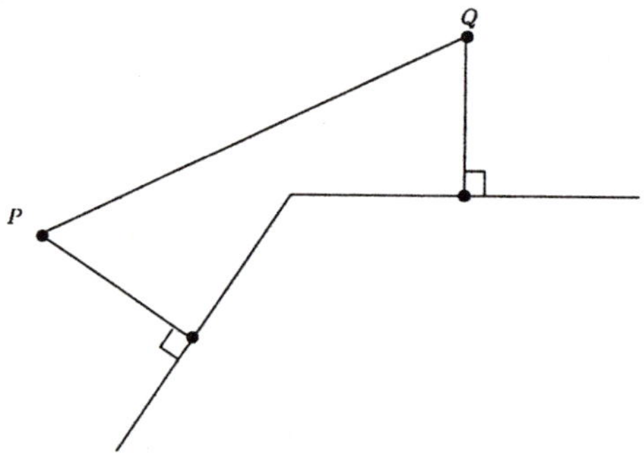

(iii) FALSE: many meet in two points, such as the two geodesics joining P and Q in the diagram in (ii).
(iv) TRUE: the meridians do not intersect (remember that the vertex of the cone has been removed), and parallel straight lines that are entirely contained in $\mathcal{S}$ do not intersect.
(v) TRUE: since (broken) line segments in $\mathcal{S}$ can clearly be continued indefinitely in both directions.
(vi) TRUE: a situation of the form

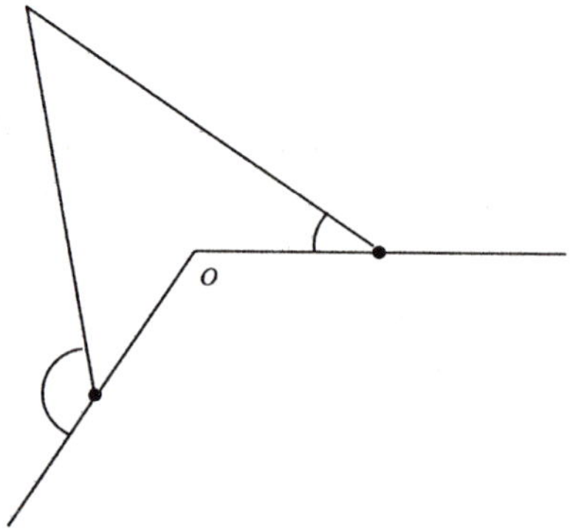

in which the indicated angles are equal is clearly impossible. But the answer to this part of the question depends on the angle of the cone: if the angle is α, instead of $\pi/4$, lines can self-intersect if $\alpha < \pi/6$, for then the corresponding sector in the plane has angle $< \pi$:

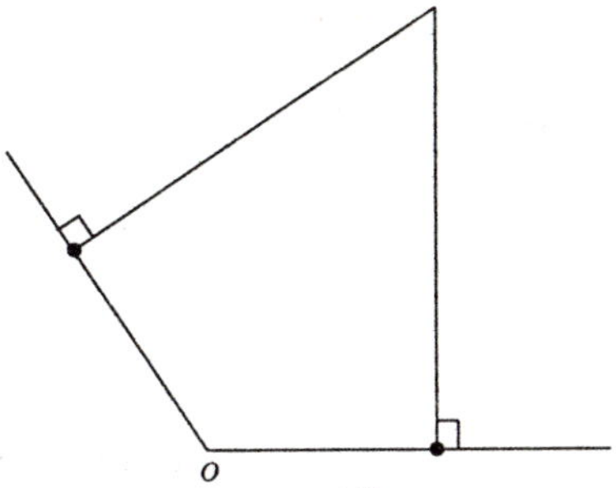

8.20 (i) This is obvious if $n \geq 0$ since $e^{-1/t^2} \to 0$ as $t \to 0$. We prove that $t^{-n}e^{-1/t^2} \to 0$ as $t \to 0$ by induction on $n \geq 0$. We know the result if $n = 0$, and if $n > 0$ we can apply L'Hopital's rule:

$$\lim_{t\to 0} \frac{t^{-n}}{e^{1/t^2}} = \lim_{t\to 0} \frac{nt^{-n-1}}{\frac{2}{t^3}e^{1/t^2}} = \lim_{t\to 0} \frac{n}{2}\frac{t^{-(n-2)}}{e^{1/t^2}},$$

which vanishes by the induction hypothesis.

(ii) We prove by induction on n that θ is n-times differentiable with

$$\frac{d^n\theta}{dt^n} = \begin{cases} \frac{P_n(t)}{t^{3n}}e^{-1/t^2} & \text{if } t \neq 0, \\ 0 & \text{if } t = 0, \end{cases}$$

where P_n is a polynomial in t. For $n = 0$, the assertion holds with $P_0 = 1$. Assuming the result for some $n \geq 0$,

$$\frac{d^{n+1}\theta}{dt^{n+1}} = \left(\frac{-3nP_n}{t^{3n+1}} + \frac{P_n'}{t^{3n}} + \frac{2P_n}{t^{3n+3}}\right)e^{-1/t^2}$$

if $t \neq 0$, so we take $P_{n+1} = (2 - 3nt^2)P_n + t^3P_n'$. If $t = 0$,

$$\frac{d^{n+1}\theta}{dt^{n+1}} = \lim_{t\to 0}\frac{P_n(t)}{t^{3n+1}}e^{-1/t^2} = P_n(0)\lim_{t\to 0}\frac{e^{-1/t^2}}{t^{3n+1}} = 0$$

by part (i).

Parts (iii) and (iv) are obvious.

8.21 Since $\boldsymbol{\gamma}_\theta$ is unit-speed, $\boldsymbol{\sigma}_r.\boldsymbol{\sigma}_r = 1$, so $\int_0^R \boldsymbol{\sigma}_r.\boldsymbol{\sigma}_r\, dr = R$. Differentiating with respect to θ gives $\int_0^R \boldsymbol{\sigma}_r.\boldsymbol{\sigma}_{r\theta}\, dr = 0$, and then integrating by parts gives

$$\boldsymbol{\sigma}_\theta.\boldsymbol{\sigma}_r\Big|_{r=0}^{r=R} - \int_0^R \boldsymbol{\sigma}_\theta.\boldsymbol{\sigma}_{rr}\, dr = 0.$$

Now $\boldsymbol{\sigma}(0,\theta) = P$ for all θ, so $\boldsymbol{\sigma}_\theta = \mathbf{0}$ when $r = 0$. So we must show that the integral in the last equation vanishes. But, $\boldsymbol{\sigma}_{rr} = \ddot{\boldsymbol{\gamma}}_\theta$, the dot denoting the derivative with respect to the parameter r of the geodesic $\boldsymbol{\gamma}_\theta$, so $\boldsymbol{\sigma}_{rr}$ is parallel to the unit normal $\mathbf{N}$ of $\boldsymbol{\sigma}$; since $\boldsymbol{\sigma}_\theta.\mathbf{N} = 0$, it follows that $\boldsymbol{\sigma}_\theta.\boldsymbol{\sigma}_{rr} = 0$.

The first fundamental form is as indicated since $\boldsymbol{\sigma}_r.\boldsymbol{\sigma}_r = 1$ and $\boldsymbol{\sigma}_r.\boldsymbol{\sigma}_\theta = 0$.

Chapter 9

9.1 This follows from Exercise 7.6.

9.2 This follows from Exercise 7.3.

9.3 $\kappa_1 + \kappa_2 = 0$ and $\kappa_1 = \kappa_2 \Longrightarrow \kappa_1 = \kappa_2 = 0$.

9.4 $\kappa_1 + \kappa_2 = 0 \Longrightarrow \kappa_2 = -\kappa_1 \Longrightarrow K = \kappa_1\kappa_2 = -\kappa_1^2 \leq 0$. $K = 0 \Longleftrightarrow \kappa_1^2 = 0 \Longleftrightarrow \kappa_1 = \kappa_2 = 0 \Longleftrightarrow$ the surface is part of a plane (by Proposition 6.5).

9.5 By Proposition 7.6, a compact minimal surface would have $K > 0$ at some point, contradicting Exercise 9.4.

9.6 By the solution of Exercise 7.2, the helicoid $\boldsymbol{\sigma}(u,v) = (v\cos u, v\sin u, \lambda u)$ has $E = \lambda^2 + v^2, F = 0, G = 1, L = 0, M = \lambda/(\lambda^2+v^2)^{1/2}, N = 0$, so

$$H = \frac{LG - 2MF + NG}{2(EG - F^2)} = 0.$$

9.7 A straightforward calculation shows that the first and second fundamental forms of $\boldsymbol{\sigma}^t$ are $\cosh^2 u(du^2 + dv^2)$ and $-\cos t\, du^2 - 2\sin t\, dudv + \cos t\, dv^2$, respectively, so

$$H = \frac{-\cos t\cosh^2 u + \cos t\cosh^2 u}{2\cosh^4 u} = 0.$$

9.8 From Example 4.10, the cylinder can be parametrised by $\boldsymbol{\sigma}(u,v) = \boldsymbol{\gamma}(u) + v\mathbf{a}$, where $\boldsymbol{\gamma}$ is unit-speed, $\parallel \mathbf{a} \parallel = 1$ and $\boldsymbol{\gamma}$ is contained in a plane Π perpendicular to $\mathbf{a}$. We have $\boldsymbol{\sigma}_u = \dot{\boldsymbol{\gamma}} = \mathbf{t}$ (a dot denoting d/du), $\boldsymbol{\sigma}_v = \mathbf{a}$, so $E = 1, F = 0, G = 1$; $\mathbf{N} = \mathbf{t} \times \mathbf{a}$, $\boldsymbol{\sigma}_{uu} = \dot{\mathbf{t}} = \kappa\mathbf{n}$, $\boldsymbol{\sigma}_{uv} = \boldsymbol{\sigma}_{vv} = \mathbf{0}$, so $L = \kappa\mathbf{n}.(\mathbf{t}\times\mathbf{a})$, $M = N = 0$. Now $\mathbf{t}\times\mathbf{a}$ is a unit vector parallel to Π and perpendicular to $\mathbf{t}$, hence parallel to $\mathbf{n}$; so $L = \pm\kappa$ and $H = \pm\kappa/2$. So $H = 0 \Longleftrightarrow \kappa = 0 \Longleftrightarrow \boldsymbol{\gamma}$ is part of a straight line $\Longleftrightarrow$ the cylinder is part of a plane.

9.9 Using Exercise 9.2, the surface is minimal $\Longleftrightarrow$

$$(1 + g'^2)\ddot{f} + (1 + \dot{f}^2)g'' = 0,$$

where a dot denotes d/dx and a dash denotes d/dy; hence the stated equation. Since the left-hand side of this equation depends only on x and the right-hand side only on y, we must have

$$\frac{\ddot{f}}{1 + \dot{f}^2} = a, \qquad \frac{g''}{1 + g'^2} = -a,$$

for some constant a. Suppose that $a \neq 0$. Let $r = \dot{f}$; then $\ddot{f} = rdr/df$ and the first equation is $rdr/df = a(1 + r^2)$, which can be integrated to give $af = \frac{1}{2}\ln(1 + r^2)$, up to adding an arbitary constant (which corresponds to translating the surface parallel to the z-axis). So $df/dx = \pm\sqrt{e^{2af} - 1}$, which integrates to give $f = -\frac{1}{a}\ln\cos a(x+b)$, where b is a constant; we can assume that $b = 0$ by translating the surface parallel to the x-axis. Similarly, $g = \frac{1}{a}\ln\cos ay$, after translating the surface parallel to the y-axis. So, up to a translation, we have

$$z = \frac{1}{a}\ln\left(\frac{\cos ay}{\cos ax}\right),$$

which is obtained from Scherk's surface by the dilation $(x,y,z) \mapsto a(x,y,z)$. If $a = 0$, then $\ddot{f} = g'' = 0$ so $f = b + cx$, $g = d + ey$, for some constants b, c, d, e, and we have the plane $z = b + d + cx + ey$.

9.10 The first fundamental form is $(\cosh v + 1)(\cosh v - \cos u)(du^2 + dv^2)$, so $\boldsymbol{\sigma}$ is conformal. By Exercise 7.15, to show that $\boldsymbol{\sigma}$ is minimal we must show that $\boldsymbol{\sigma}_{uu} + \boldsymbol{\sigma}_{vv} = \mathbf{0}$; but this is so, since

$$\boldsymbol{\sigma}_{uu} = (\sin u\cosh v, \cos u\cosh v, \sin\frac{u}{2}\sinh\frac{v}{2}),$$
$$\boldsymbol{\sigma}_{vv} = (-\sin u\cosh v, -\cos u\cosh v, -\sin\frac{u}{2}\sinh\frac{v}{2}).$$

(i) $\boldsymbol{\sigma}(0,v) = (0, 1-\cosh v, 0)$, which is the y-axis. Any straight line is a geodesic.

(ii) $\boldsymbol{\sigma}(\pi,v) = (\pi, 1+\cosh v, -4\sinh\frac{v}{2})$, which is a curve in the plane $x = \pi$ such that

$$z^2 = 16\sinh^2\frac{v}{2} = 8(\cosh v - 1) = 8(y-2),$$

i.e. a parabola. The geodesic equations are

$$\frac{d}{dt}(E\dot{u}) = \frac{1}{2}E_u(\dot{u}^2 + \dot{v}^2), \quad \frac{d}{dt}(E\dot{v}) = \frac{1}{2}E_v(\dot{u}^2 + \dot{v}^2),$$

where a dot denotes the derivative with respect to the parameter t of the geodesic and $E = (\cosh v + 1)(\cosh v - \cos u)$. When $u = \pi$, the unit-speed condition is $E\dot{v}^2 = 1$, so $\dot{v} = 1/(\cosh v + 1)$. Hence, the first geodesic equation is $0 = \frac{1}{2}E_u\dot{v}^2$, which holds because $E_u = \sin u(\cosh v + 1) = 0$ when $u = \pi$; and the second geodesic equation is

$$\frac{d}{dt}(\cosh v + 1) = (\cosh v + 1)\sinh v\,\dot{v}^2 = \sinh v\,\dot{v},$$

which obviously holds.

(iii) $\boldsymbol{\sigma}(u,0) = (u - \sin u, 1 - \cos u, 0)$, which is the cycloid of Exercise 1.7 (in the xy-plane, with $a = 1$ and with t replaced by u). The second geodesic equation is satisfied because $E_v = \sinh v(2\cosh v + 1 - \cos u) = 0$ when $v = 0$. The unit-speed condition is $2(1-\cos u)\dot{u}^2 = 1$, so $\dot{u} = 1/2\sin\frac{u}{2}$. The first geodesic equation is $\frac{d}{dt}(4\sin^2\frac{u}{2}\dot{u}) = \sin u\dot{u}^2$, i.e. $\frac{d}{dt}(2\sin\frac{u}{2}) = \cos\frac{u}{2}\dot{u}$, which obviously holds.

9.11 $\lambda = a^2 + bc = -(ad - bc)$ (since $d = -a$) $= -\det\mathcal{W} = -K$.

9.12 (i) From Example 9.1, $\mathbf{N} = (-\operatorname{sech} u \cos v, -\operatorname{sech} u \sin v, \tanh u)$. Hence, if $\mathbf{N}(u,v) = \mathbf{N}(u',v')$, then $u = u'$ since $u \mapsto \tanh u$ is injective, so $\cos v = \cos v'$ and $\sin v = \sin v'$, hence $v = v'$; thus, $\mathbf{N}$ is injective. If $\mathbf{N} = (x,y,z)$, then $x^2+y^2 = \operatorname{sech}^2 u \neq 0$, so the image of $\mathbf{N}$ does not contain the poles. Given a point (x,y,z) on the unit sphere with $x^2+y^2 \neq 0$, let $u = \pm\operatorname{sech}^{-1}\sqrt{x^2+y^2}$, the sign being that of z, and let v be such that $\cos v = -x/\sqrt{x^2+y^2}$, $\sin v = -y/\sqrt{x^2+y^2}$; then, $\mathbf{N}(u,v) = (x,y,z)$.
(ii) By the solution of Exercise 7.2, $\mathbf{N} = (\lambda^2+v^2)^{-1/2}(-\lambda\sin u, \lambda\cos u, -v)$. Since $\mathbf{N}(u,v) = \mathbf{N}(u+2k\pi, v)$ for all integers k, the infinitely many points $\boldsymbol{\sigma}(u+2k\pi, v) = \boldsymbol{\sigma}(u,v) + (0,0,2k\pi)$ of the helicoid all have the same image under the Gauss map. If $\mathbf{N} = (x,y,z)$, then $x^2+y^2 = \lambda^2/(\lambda^2+v^2) \neq 0$, so the image of $\mathbf{N}$ does not contain the poles. If (x,y,z) is on the unit sphere and $x^2+y^2 \neq 0$, let $v = -\lambda z/\sqrt{x^2+y^2}$ and let u be such that $\sin u = -x/\sqrt{x^2+y^2}$, $\cos u = -y/\sqrt{x^2+y^2}$; then $\mathbf{N}(u,v) = (x,y,z)$.

9.13 The plane can be parametrised by $\boldsymbol{\sigma}(u,v) = u\mathbf{b} + v\mathbf{c}$, where $\{\mathbf{a},\mathbf{b},\mathbf{c}\}$ is a right-handed orthonormal basis of $\mathbf{R}^3$. Then, $\boldsymbol{\varphi} = \boldsymbol{\sigma}_u - i\boldsymbol{\sigma}_v = \mathbf{b} - i\mathbf{c}$. The conjugate surface corresponds to $i\boldsymbol{\varphi} = \mathbf{c} + i\mathbf{b}$; since $\{\mathbf{a},\mathbf{c},-\mathbf{b}\}$ is also a right-handed orthonormal basis of $\mathbf{R}^3$, the plane is self-conjugate (up to a translation).

9.14 $\boldsymbol{\varphi} = (\frac{1}{2}f(1-g^2), \frac{i}{2}f(1+g^2), fg) \Longrightarrow i\boldsymbol{\varphi} = (\frac{1}{2}if(1-g^2), \frac{i}{2}if(1+g^2), ifg)$, which corresponds to the pair if and g.

9.15 By Example 9.6, $\boldsymbol{\varphi}(\zeta) = (\sinh\zeta, -i\cosh\zeta, 1)$. From the proof of Proposition 9.7, $f = \varphi_1 - i\varphi_2 = \sinh\zeta - \cosh\zeta = -e^{-\zeta}$, $g = \varphi_3/f = -e^{\zeta}$.

9.16 $\boldsymbol{\varphi} = \boldsymbol{\sigma}_u - i\boldsymbol{\sigma}_v = (1-u^2+v^2-2iuv, 2uv - i(1-v^2+u^2), 2u+2iv) = (1-\zeta^2, -i(1+\zeta^2), 2\zeta)$. So the conjugate surface is

$$\begin{aligned}\tilde{\boldsymbol{\sigma}}(u,v) &= \mathfrak{Re}\int (i(1-\zeta^2), 1+\zeta^2, 2i\zeta)\,d\zeta \\ &= \mathfrak{Re}\left(i\left(\zeta - \frac{\zeta^3}{3}\right), \zeta + \frac{\zeta^3}{3}, i\zeta^2\right) \quad \text{(up to a translation)} \\ &= \left(-v + u^2v - \frac{v^3}{3}, u + \frac{u^3}{3} - uv^2, -2uv\right).\end{aligned}$$

Let $U = (u-v)/\sqrt{2}$, $V = (u+v)/\sqrt{2}$, $\tilde{\boldsymbol{\sigma}}(U,V) = \boldsymbol{\sigma}(u,v)$; then,

$$\begin{aligned}\tilde{\boldsymbol{\sigma}}(U,V) = \Bigg(&\frac{1}{\sqrt{2}}\left(U - V + UV^2 - U^2V + \frac{1}{3}V^3 - \frac{1}{3}U^3\right), \\ &\frac{1}{\sqrt{2}}\left(U + V + UV^2 + U^2V - \frac{1}{3}V^3 - \frac{1}{3}U^3\right), U^2 - V^2\Bigg).\end{aligned}$$

Applying the $\pi/4$ rotation $(x,y,z) \mapsto \left(\frac{1}{\sqrt{2}}(x+y), \frac{1}{\sqrt{2}}(y-x), z\right)$ to $\tilde{\boldsymbol{\sigma}}(U,V)$ then gives $(U - \frac{1}{3}U^3 + UV^2, V - \frac{1}{3}V^3 + U^2V, U^2 - V^2)$, which is Enneper's surface again.

9.17 $\boldsymbol{\varphi} = \left(\frac{1}{2}(1-\zeta^{-4})(1-\zeta^2), \frac{i}{2}(1-\zeta^{-4})(1+\zeta^2), \zeta(1-\zeta^{-4})\right)$, so

$$\begin{aligned}\boldsymbol{\sigma} &= \mathfrak{Re}\left(\frac{1}{2}\left(\zeta - \frac{\zeta^3}{3} - \zeta^{-1} + \frac{\zeta^{-3}}{3}\right), \frac{i}{2}\left(\zeta + \frac{\zeta^3}{3} + \zeta^{-1} + \frac{\zeta^{-3}}{3}\right), \frac{\zeta^2}{2} + \frac{\zeta^{-2}}{2}\right)\\ &= \mathfrak{Re}\left(-\frac{1}{6}(\zeta-\zeta^{-1})^3, \frac{i}{6}(\zeta+\zeta^{-1})^3, \frac{1}{2}(\zeta+\zeta^{-1})^2\right),\end{aligned}$$

up to a translation. Put $\zeta = e^{\tilde{\zeta}}$, $\tilde{\zeta} = \tilde{u} + i\tilde{v}$. Then, $\boldsymbol{\sigma}(u,v) = \tilde{\boldsymbol{\sigma}}(\tilde{u},\tilde{v})$, where

$$\begin{aligned}\tilde{\boldsymbol{\sigma}}(\tilde{u},\tilde{v}) &= \mathfrak{Re}\left(-\frac{4}{3}\sinh^3\tilde{\zeta}, \frac{4i}{3}\cosh^3\tilde{\zeta}, 2\cosh^2\tilde{\zeta}\right)\\ &= \Bigg(4\sinh\tilde{u}\cos\tilde{v}(\cosh^2\tilde{u}\sin^2\tilde{v} - \frac{1}{3}\sinh^2\tilde{u}\cos^2\tilde{v}),\\ &\qquad 4\sinh\tilde{u}\sin\tilde{v}(\frac{1}{3}\sinh^2\tilde{u}\sin^2\tilde{v} - \cosh^2\tilde{u}\cos^2\tilde{v}),\\ &\qquad 2(\cosh^2\tilde{u}\cos^2\tilde{v} - \sinh^2\tilde{u}\sin^2\tilde{v})\Bigg).\end{aligned}$$

9.18 By Example 5.7, $\pi(x,y,z) = (u,v,0)$, where $x = 2u/(u^2+v^2+1), y = 2v/(u^2+v^2+1), z = (u^2+v^2-1)/(u^2+v^2+1)$. If $\zeta = u+iv$, then $x+iy = 2\zeta/(|\zeta|^2+1), z = (|\zeta|^2-1)/(|\zeta|^2+1)$. This gives $\zeta = (x+iy)/(1-z)$. Hence, identifying $(u,v,0)$ with $u+iv$, we have $\pi(x,y,z) = (x+iy)/(1-z)$. From Eq. (25), $\mathcal{G}(\zeta) = \frac{1}{|g|^2+1}(g+\bar{g}, -i(g-\bar{g}), |g|^2-1)$, so

$$\pi(\mathcal{G}(\zeta)) = \frac{g+\bar{g}+g-\bar{g}}{|g|^2+1-(|g|^2-1)} = g(\zeta).$$

Chapter 10

10.1 By Corollary 10.2(i),

$$K = -\frac{1}{2e^\lambda}\left(\frac{\partial}{\partial u}\left(\frac{(e^\lambda)_u}{e^\lambda}\right) + \frac{\partial}{\partial v}\left(\frac{(e^\lambda)_v}{e^\lambda}\right)\right) = -\frac{1}{2e^\lambda}(\lambda_{uu}+\lambda_{vv}).$$

10.2 By Exercise 5.4, $\begin{pmatrix}\tilde{E} & \tilde{F}\\ \tilde{F} & \tilde{G}\end{pmatrix} = J^t\begin{pmatrix}E & F\\ F & G\end{pmatrix}J$, where $J = \begin{pmatrix}\frac{\partial r}{\partial u} & \frac{\partial r}{\partial v}\\ \frac{\partial \theta}{\partial u} & \frac{\partial \theta}{\partial v}\end{pmatrix} = \begin{pmatrix}\frac{u}{r} & \frac{v}{r}\\ \frac{-v}{r^2} & \frac{u}{r^2}\end{pmatrix}$. By Exercise 8.21, $E = 1, F = 0$, and we get the stated formulas for $\tilde{E}, \tilde{F}, \tilde{G}$. From $\tilde{E} - 1 = \frac{v^2}{r^2}\left(\frac{G}{r^2}-1\right)$, $\tilde{G} - 1 = \frac{u^2}{r^2}\left(\frac{G}{r^2}-1\right)$, we get $u^2(\tilde{E}-1) = v^2(\tilde{G}-1)$. Since $\tilde{E}$ and $\tilde{G}$ are smooth functions of (u,v), they have Taylor expansions $\tilde{E} = \sum_{i+j\le 2} e_{ij}u^iv^j + o(r^2)$, $\tilde{G} = \sum_{i+j\le 2} g_{ij}u^iv^j + o(r^2)$, where $o(r^k)$ denotes terms such that $o(r^k)/r^k \to 0$ as $r \to 0$. Equating coefficients on both sides of $u^2(\tilde{E}-1) = v^2(\tilde{G}-1)$

shows that all the e's and g's are zero except $e_{02} = g_{20} = k$, say. Then, $\tilde{E} = 1 + kv^2 + o(r^2)$, which implies that $G = r^2 + kr^4 + o(r^4)$.
By Corollary 10.2(ii), $K = -\frac{1}{\sqrt{G}}\frac{\partial^2\sqrt{G}}{\partial r^2}$. From the first part, $\sqrt{G} = r + \frac{1}{2}kr^3 + o(r^3)$, hence $K = -3k + o(1)$. Taking $r = 0$ gives $K(P) = -3k$.

10.3 (i) $C_R = \int_0^{2\pi} \| \boldsymbol{\sigma}_\theta \| \, d\theta = \int_0^{2\pi} \sqrt{G}\, d\theta = \int_0^{2\pi} \left(R - \frac{1}{6}K(P)R^3 + o(R^3)\right) d\theta$ $= 2\pi\left(R - \frac{1}{6}K(P)R^3 + o(R^3)\right)$.
(ii) Since $d\mathcal{A}_{\boldsymbol{\sigma}} = \sqrt{G}\, dr d\theta$, the area $A_R = \int_0^R \int_0^{2\pi} \sqrt{G}\, dr d\theta$ is equal to $2\pi \int_0^R \left(r - \frac{1}{6}K(P)r^3 + o(r^3)\right) dr = \pi R^2 \left(1 - \frac{K(P)}{12}R^2 + o(R^2)\right)$.
Let $\mathcal{S}$ be the unit sphere, let P be the north pole, and let (θ, φ) be the usual latitude longitude parametrisation of $\mathcal{S}$. Then, the geodesic circle with centre P and radius R is the parallel $\theta = \frac{\pi}{2} - R$, which is an ordinary circle of radius $\sin R$, so that $C_R = 2\pi \sin R$; since $2\pi \sin R = 2\pi(R - \frac{1}{6}R^3 + o(R^3))$ and $K = 1$, this agrees with the formula in (i). Similarly,

$$A_R = \int_0^{2\pi} \int_{\pi/2-R}^{\pi/2} \cos\theta \, d\theta d\varphi = 2\pi(1 - \cos R) = 2\pi\left(\frac{R^2}{2} - \frac{R^4}{24} + o(R^4)\right),$$

in agreement with the formula in (ii).

10.4 (i) Let s be the arc-length of $\boldsymbol{\gamma}$, so that $ds/d\theta = \lambda$, and denote d/ds by a dot. The first of the geodesic equations ((2) in Chapter 8) applied to $\boldsymbol{\gamma}$ gives $\ddot{r} = \frac{1}{2}G_r\dot{\theta}^2$. Since $r = f(\theta)$, this gives

$$\frac{1}{\lambda}\left(\frac{1}{\lambda}f'\right)' = \frac{1}{2\lambda^2}G_r.$$

This simplifies to give the stated equation.
(ii) Since $\boldsymbol{\sigma}_r$ and $\dot{\boldsymbol{\gamma}}$ are unit vectors, $\cos\psi = \boldsymbol{\sigma}_r.\dot{\boldsymbol{\gamma}} = \frac{1}{\lambda}\boldsymbol{\sigma}_r.(f'\boldsymbol{\sigma}_r + \boldsymbol{\sigma}_\theta) = f'/\lambda$. Also, $\boldsymbol{\sigma}_r \times \dot{\boldsymbol{\gamma}} = \frac{1}{\lambda}(\boldsymbol{\sigma}_r \times \boldsymbol{\sigma}_\theta) = \frac{\sqrt{G}}{\lambda}\mathbf{N}$, so $\sin\psi = \sqrt{G}/\lambda$. Hence,

$$\left(\frac{f'}{\lambda}\right)' = -\psi' \sin\psi = -\frac{\sqrt{G}}{\lambda}\psi',$$

$$\therefore \quad \psi' = -\frac{1}{\sqrt{G}}\left(f'' - \frac{f'\lambda'}{\lambda}\right) = -\frac{1}{2\sqrt{G}}\frac{\partial G}{\partial r} = -\frac{\partial\sqrt{G}}{\partial r}.$$

(iii) Using the formula for K in Corollary 10.2(ii) and the expression for the first fundamental form of $\boldsymbol{\sigma}$ in Exercise 8.21, we get

$$\int\int_{ABC} K d\mathcal{A}_{\boldsymbol{\sigma}} = \int_0^{\angle A}\int_0^{f(\theta)} -\frac{1}{\sqrt{G}}\frac{\partial^2\sqrt{G}}{\partial r^2}\sqrt{G}\, dr d\theta$$

$$= -\int_0^{\angle A} \left.\frac{\partial\sqrt{G}}{\partial r}\right|_{r=0}^{r=f(\theta)} d\theta = \int_0^{\angle A}\left(\psi' + \left.\frac{\partial\sqrt{G}}{\partial r}\right|_{r=0}\right) d\theta.$$

By Exercise 10.2, $\sqrt{G} = r + o(r)$ so $\partial\sqrt{G}/\partial r = 1$ at $r = 0$. Hence,

$$\begin{aligned}\iint_{ABC} K dA_{\boldsymbol{\sigma}} &= \psi(\angle A) - \psi(0) + \angle A = \angle C - (\pi - \angle B) + \angle A \\ &= \angle A + \angle B + \angle C - \pi.\end{aligned}$$

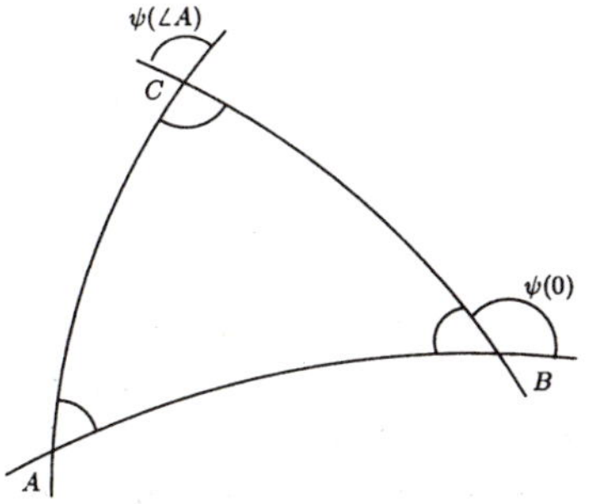

10.5 By Exercise 5.7, generalised cylinders and cones are isometric to a plane. Hence, if the sphere were isometric to a generalised cylinder or cone, the sphere would be isometric to a plane, since composites of isometries are isometries. This contradicts Proposition 10.1.

10.6 With the notation of Example 4.9 we have, on the median circle $t = 0$, $\boldsymbol{\sigma}_t = \left(-\sin\frac{\theta}{2}\cos\theta, -\sin\frac{\theta}{2}\sin\theta, \cos\frac{\theta}{2}\right)$, $\boldsymbol{\sigma}_\theta = (-\sin\theta, \cos\theta, 0)$, hence $E = 1, F = 0, G = 1$ and $\mathbf{N} = \left(-\cos\frac{\theta}{2}\cos\theta, -\sin\frac{\theta}{2}\sin\theta, -\sin\frac{\theta}{2}\right)$; $\boldsymbol{\sigma}_{tt} = \mathbf{0}$, $\boldsymbol{\sigma}_{t\theta} = \left(-\frac{1}{2}\cos\frac{\theta}{2}\cos\theta + \sin\frac{\theta}{2}\sin\theta, -\frac{1}{2}\cos\frac{\theta}{2}\sin\theta - \sin\frac{\theta}{2}\cos\theta, -\frac{1}{2}\sin\frac{\theta}{2}\right)$, from which $L = 0, M = \frac{1}{2}$. Hence, $K = (LN - M^2)/(EG - F^2) = -1/4$. Since $K \neq 0$, the Theorema Egregium implies that the Möbius band is not isometric to a plane.

10.7 The first fundamental forms of $\boldsymbol{\sigma}$ and $\tilde{\boldsymbol{\sigma}}$ are found to be $(1+\frac{1}{u^2})du^2 + u^2dv^2$ and $du^2 + (u^2+1)dv^2$, respectively. Since these are different, the map $\boldsymbol{\sigma}(u,v) \mapsto \tilde{\boldsymbol{\sigma}}(u,v)$ is not an isometry. Nevertheless, by Corollary 10.2(ii), the gaussian curvatures are equal:

$$K = \tilde{K} = -\frac{1}{2\sqrt{u^2+1}}\left(\frac{\partial}{\partial u}\left(\frac{2u}{\sqrt{1+u^2}}\right)\right) = -\frac{1}{(1+u^2)^2}.$$

If $\boldsymbol{\sigma}(u,v) \mapsto \tilde{\boldsymbol{\sigma}}(\tilde{u},\tilde{v})$ is an isometry, the Theorema Egregium tells us that $-1/(1+u^2)^2 = -1/(1+\tilde{u}^2)^2$, so $\tilde{u} = \pm u$; let $\tilde{v} = f(u,v)$. The first fundamental form of $\tilde{\boldsymbol{\sigma}}(\pm u, f(u,v))$ is $(1 + (1+u^2)f_u^2)du^2 + 2(1+u^2)f_u f_v du dv + (1+u^2)f_v^2 dv^2$; this is equal to the first fundamental form of $\boldsymbol{\sigma}(u,v) \iff 1 + (1+u^2)f_u^2 = 1 + 1/u^2$, $f_u f_v = 0$ and $(1+u^2)f_v^2 = u^2$. The middle equation gives $f_u = 0$ or $f_v = 0$, but these are both impossible by the other two equations. Hence, the isometry does not exist.

10.8 For the catenoid $\boldsymbol{\sigma}(u,v) = (\cosh u \cos v, \cosh u \sin v, u)$, the first fundamental form is $\cosh^2 u(du^2 + dv^2)$ (Exercise 5.8) and the gaussian curvature is $K = -\mathrm{sech}^4 u$ (Exercise 7.2). If $\boldsymbol{\sigma}(u,v) \mapsto \boldsymbol{\sigma}(\tilde{u},\tilde{v})$ is an isometry, then $\text{sech}^4 u = \text{sech}^4 \tilde{u}$, so $\tilde{u} = \pm u$; reflecting in the plane $z = 0$ changes u to $-u$, so assume that the sign is $+$. Let $\tilde{v} = f(u, v)$; the first fundamental form of $\boldsymbol{\sigma}(\pm u, f(u, v))$ is $(\cosh^2 u + f_u^2)du^2 + 2f_u f_v \cosh^2 u du dv + f_v^2 \cosh^2 u dv^2$; hence, $\cosh^2 u = \cosh^2 u + f_u^2$, $f_u f_v = 0$ and $f_v^2 \cosh^2 u = \cosh^2 u$, so $f_u = 0$, $f_v = \pm 1$. Thus, $f = \pm v + \alpha$, where α is a constant; if the sign is $+$, we have a rotation by α about the z-axis; if the sign is $-$ we have a reflection in the plane containing the z-axis, making an angle $\alpha/2$ with the xz-plane.

10.9 $\mathcal{W} = -\mathcal{F}_I^{-1}\mathcal{F}_{II} = -\begin{pmatrix} \cos^2 v & 0 \\ 0 & 1 \end{pmatrix}^{-1} \begin{pmatrix} -\cos^2 v & 0 \\ 0 & -1 \end{pmatrix} = I$, so $\mathbf{N}_u = \boldsymbol{\sigma}_u$, $\mathbf{N}_v = \boldsymbol{\sigma}_v$. Thus, $\mathbf{N} = \boldsymbol{\sigma} - \mathbf{a}$, where $\mathbf{a}$ is a constant vector. Hence, $\| \boldsymbol{\sigma} - \mathbf{a} \| = 1$, showing that the surface is part of the sphere of radius 1 and centre $\mathbf{a}$. The standard latitude longitude parametrisation of the unit sphere has first and second fundamental forms both given by $du^2 + \cos^2 v dv^2$, so the parametrisation $\boldsymbol{\sigma}(v, u)$ has the given first and second fundamental forms (the second fundamental form changes sign because $\boldsymbol{\sigma}_v \times \boldsymbol{\sigma}_u = -\boldsymbol{\sigma}_u \times \boldsymbol{\sigma}_v$).

10.10 $\Gamma_{22}^1 = \sin u \cos u$ and the other Christoffel symbols are zero; the second Codazzi–Mainardi equation is not satisfied.

10.11 The Christoffel symbols are $\Gamma_{11}^1 = E_u/2E$, $\Gamma_{11}^2 = -E_v/2G$, $\Gamma_{12}^1 = E_v/2E$, $\Gamma_{12}^2 = G_u/2G$, $\Gamma_{22}^1 = -G_u/2E$, $\Gamma_{22}^2 = G_v/2G$. The first Codazzi–Mainardi equation is

$$L_v = \frac{LE_v}{2E} - N\left(\frac{-E_v}{2G}\right) = \frac{1}{2}E_v\left(\frac{L}{E} + \frac{N}{G}\right),$$

and similarly for the other equation. Finally,

$$(\kappa_1)_v = \frac{E_v}{2E}\left(\frac{L}{E} + \frac{N}{G}\right) - \frac{LE_v}{E^2} = \frac{E_v}{2E}\left(\frac{N}{G} - \frac{L}{E}\right) = \frac{E_v}{2E}(\kappa_2 - \kappa_1),$$

and similarly for $(\kappa_2)_u$.

10.12 Arguing as in the proof of Theorem 10.4, we suppose that J attains its maximum value > 0 at some point P of $\mathcal{S}$ contained in a patch $\boldsymbol{\sigma}$ of $\mathcal{S}$. We can assume that the principal curvatures κ_1 and κ_2 of $\boldsymbol{\sigma}$ satisfy $\kappa_1 > \kappa_2 > 0$ everywhere. Since $H = \frac{1}{2}(\kappa_1 + \kappa_2)$, $\kappa_1 > H$ and $J = 4(\kappa_1 - H)^2$. Thus, J increases with κ_1 when $\kappa_1 > H$, so κ_1 must have a maximum at P, and then $\kappa_2 = 2H - \kappa_1$ has a minimum there. By Lemma 10.2, $K \leq 0$ at P, contradicting the assumption that $K > 0$ everywhere.

Chapter 11

11.1 If $\boldsymbol{\gamma}$ is a *simple* closed geodesic, Theorem 11.1 gives $\iint_{\text{int}(\boldsymbol{\gamma})} K dA = 2\pi$; since $K \leq 0$, this is impossible. The parallels of a cylinder are not the images under a surface patch $\boldsymbol{\sigma} : U \to \mathbf{R}^3$ of a *simple* closed curve $\boldsymbol{\pi}$ in the plane such that int($\boldsymbol{\pi}$) is contained in U. Note that the whole cylinder can actually be covered by a single patch (see Exercise 4.2) in which U is an annulus, but that the parallels correspond to circles going 'around the hole' in the annulus.

11.2 By Proposition 2.2, $\kappa_s = d\varphi/ds$, where φ is the angle through which a fixed unit vector must be rotated anti-clockwise to bring it into coincidence with the unit tangent vector of $\boldsymbol{\gamma}$. So

$$\int_0^{\ell(\boldsymbol{\gamma})} \kappa_s \, ds = \int_0^{\ell(\boldsymbol{\gamma})} \frac{d\varphi}{ds} ds = 2\pi,$$

by the Umlaufsatz.

11.3 By Corollary 11.1, the interior angles $\alpha_1, \ldots, \alpha_n$ of the polygon satisfy

$$\sum_{i=1}^{n} \alpha_i = (n-2)\pi + \iint_{\text{int}(\boldsymbol{\gamma})} K dA_{\boldsymbol{\sigma}}.$$

Since $0 < \alpha_i < 2\pi$ for all i, the left-hand side is > 0; since $K < 0$, we have $(n-2)\pi > 0$ and hence $n \geq 3$; and if $n = 3$, then $\iint_{\text{int}(\boldsymbol{\gamma})} (-K) dA_{\boldsymbol{\sigma}} < \pi$ so

$$\iint_{\text{int}(\boldsymbol{\gamma})} dA_{\boldsymbol{\sigma}} \leq \iint_{\text{int}(\boldsymbol{\gamma})} (-K) dA_{\boldsymbol{\sigma}} < \pi.$$

11.4 The parallel $u = u_1$ is the curve $\boldsymbol{\gamma}_1(v) = (f(u_1)\cos v, f(u_1)\sin v, g(u_1))$; if s is the arc-length of $\boldsymbol{\gamma}_1$, $ds/dv = f(u_1)$. Denote d/ds by a dot and d/du by a dash. Then, $\dot{\boldsymbol{\gamma}} = (-\sin v, \cos v, 0)$, $\ddot{\boldsymbol{\gamma}} = -\frac{1}{f(u_1)}(\cos v, \sin v, 0)$, and the unit normal of the surface is $\mathbf{N} = (-g'\cos v, -g'\sin v, f')$. This gives the geodesic curvature of $\boldsymbol{\gamma}$ as $\kappa_g = \ddot{\boldsymbol{\gamma}}.(\mathbf{N} \times \dot{\boldsymbol{\gamma}}) = \frac{f'(u_1)}{f(u_1)}$. Since $\ell(\boldsymbol{\gamma}_1) = 2\pi f(u_1)$, $\int_0^{\ell(\boldsymbol{\gamma}_1)} \kappa_g \, ds = 2\pi f'(u_1)$. Similarly for $\boldsymbol{\gamma}_2$. By Example 7.2, $K = -f''/f$, so

$$\iint_R K dA_{\boldsymbol{\sigma}} = \int_0^{2\pi} \int_{u_1}^{u_2} -\frac{f''}{f} f du dv = 2\pi(f'(u_1) - f'(u_2)).$$

Hence,

$$\int_0^{\ell(\boldsymbol{\gamma}_1)} \kappa_g \, ds - \int_0^{\ell(\boldsymbol{\gamma}_2)} \kappa_g \, ds = \iint_R K dA_{\boldsymbol{\sigma}}.$$

This equation is the result of applying Theorem 11.2 to the following curvilinear polygon,

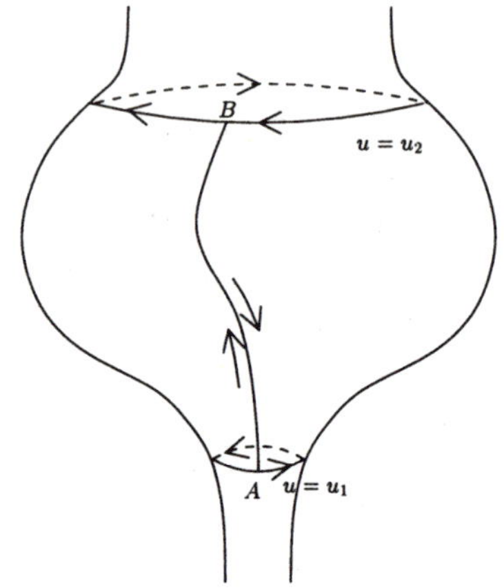

where AB is part of the meridian $v = 0$ (or $v = 2\pi$); the integrals of κ_g along AB and along BA cancel out. (Strictly speaking, this curve is not a curvilinear polygon in the sense of Definition 11.2 – condition (i) is violated – but this difficulty can be circumvented by replacing the double path AB and BA by two meridians $v = \epsilon$ and $v = 2\pi - \epsilon$ and then letting ϵ tend to zero.)

11.5 $3F = 2E$ because each face has 3 edges and each edge is an edge of 2 faces. From $\chi = V - E + F$, we get $\chi = V - E + \frac{2}{3}E$, so $E = 3(V - \chi)$. Since each edge has 2 vertices and two edges cannot intersect in more than one vertex, $E \leq \frac{1}{2}V(V-1)$; hence, $3(V-\chi) \leq \frac{1}{2}V(V-1)$, which is equivalent to $V^2 - 7V + 6\chi \geq 0$. The roots of the quadratic are $\frac{1}{2}\left(7 \pm \sqrt{49 - 7\chi}\right)$, so $V \leq \frac{1}{2}\left(7 - \sqrt{49 - 7\chi}\right)$ or $V \geq \frac{1}{2}\left(7 + \sqrt{49 - 7\chi}\right)$. Since $\chi = 2, 0, -2, \ldots$, the first condition gives $V \leq 3$, which would allow only one triangle; hence, the second condition must hold.

11.6 The n triangles have $3n$ vertices, but each vertex is counted r times as it is a vertex of r triangles, so $V = 3n/r$; similarly, $E = 3n/2$. Then, $V - E + F = 2$ gives $6/r - 4/n = 1$. This implies $6/r > 1$, so $r < 6$. Triangulations for $r = 3, 4$ and 5 can be obtained by 'inflating' a regular tetrahedron, octahedron and icosahedron, respectively.

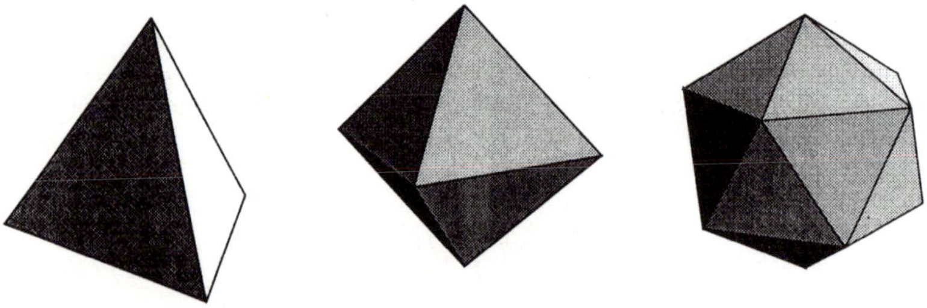

tetrahedron octahedron icosahedron

11.7 If such curves exist they would give a triangulation of the sphere with 5 vertices and $5 \times 4/2 = 10$ edges, hence $2+10-5=7$ polygons. Since each edge is an edge of two polygons and each polygon has at least 3 edges, $3F \leq 2E$; but $3 \times 7 > 2 \times 10$. If curves satisfying the same conditions exist in the plane, applying the inverse of the stereographic projection map of Example 5.7 would give curves satisfying the conditions on the sphere, which we have shown is impossible.

11.8 Such a collection of curves would give a triangulation of the sphere with $V = 6$, $E = 9$, and hence $F = 5$. The total number of edges of all the polygons in the triangulation is $2E = 18$. Since exactly 3 edges meet at each vertex, going around each polygon once counts each edge 3 times, so there should be $18/3 = 6$ polygons, not 5.

11.9 By Corollary 11.3, $\iint_{\mathcal{S}} K dA = 4\pi(1-g)$, and by Theorem 11.6, $g = 1$ since $\mathcal{S}$ is diffeomorphic to T_1. By Proposition 7.6, $K > 0$ at some point of $\mathcal{S}$.

11.10 The ellipsoid is diffeomorphic to the unit sphere by the map $(x, y, x) \mapsto (x/a, y/a, z/b)$, so the genus of the ellipsoid is zero. Hence, Corollary 11.3 gives $\iint_{\mathcal{S}} K dA = 4\pi(1-0) = 4\pi$.
Parametrising the ellipsoid by $\boldsymbol{\sigma}(\theta, \varphi) = (a\cos\theta\cos\varphi, a\cos\theta\sin\varphi, b\sin\theta)$ (cf. the latitude longitude parametrisation of the sphere), the first and second fundamental forms of $\boldsymbol{\sigma}$ are $(a^2\sin^2\theta + b^2\cos^2\theta)d\theta^2 + a^2\cos^2\theta d\varphi^2$ and $\frac{ab}{\sqrt{a^2\sin^2\theta+b^2\cos^2\theta}}(d\theta^2 + \cos^2\theta d\varphi^2)$, respectively. This gives $K = b^2/(a^2\sin^2\theta + b^2\cos^2\theta)^2$, $dA_{\boldsymbol{\sigma}} = a\cos\theta\sqrt{a^2\sin^2\theta + b^2\cos^2\theta}\, d\theta d\varphi$; hence,

$$\iint_{\mathcal{S}} K dA_{\boldsymbol{\sigma}} = \int_0^{2\pi}\int_{-\pi/2}^{\pi/2} \frac{ab^2\cos\theta\, d\theta d\varphi}{(a^2\sin^2\theta + b^2\cos^2\theta)^{3/2}}.$$

11.11 $K > 0 \Longrightarrow \iint_{\mathcal{S}} K dA > 0$. By Corollary 11.3, $g < 1$; since g is a non-negative integer, $g = 0$ so $\mathcal{S}$ is diffeomorphic to a sphere by Theorem 11.6. The converse is false: for example, a 'cigar tube' is diffeomorphic to a sphere but $K = 0$ on the cylindrical part.

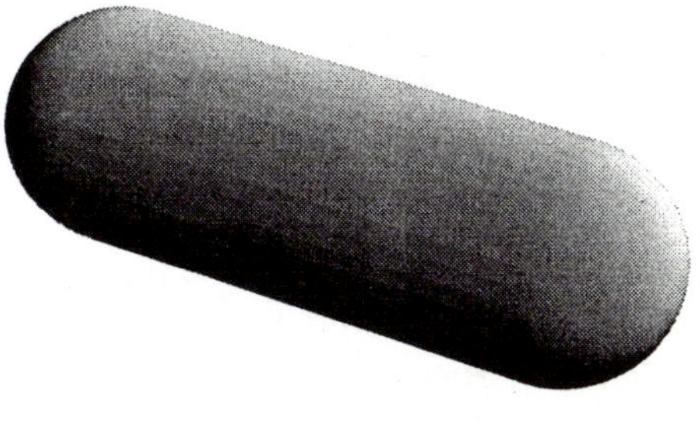

11.12 Both surfaces are closed subsets of $\mathbf{R}^3$, as they are of the form $f^{-1}(0)$, where $f : \mathbf{R}^3 \to \mathbf{R}$ is a continuous function (equal to $x^2 - y^2 + z^4 - 1$ and $x^2 + y^2 + z^4 - 1$ in the two cases). The surface in (i) is not bounded, and hence not compact, since it contains the point $(1, a^2, a)$ for all real numbers a; that in (ii) is bounded, and hence compact, since $x^2+y^2+z^4 = 1 \Longrightarrow -1 \leq x, y, z \leq 1$.
The surface in (ii) is obtained by rotating the curve $x^2 + z^4 = 1$ in the xz-plane around the z-axis:

It is clearly diffeomorphic to the sphere, so $\chi = 2$.

11.13 Take the reference tangent vector field to be $\boldsymbol{\xi} = (1, 0)$, and take the simple closed curve $\boldsymbol{\gamma}(s) = (\cos s, \sin s)$. At $\boldsymbol{\gamma}(s)$, we have $\mathbf{V} = (\alpha, \beta)$, where

$$\alpha + i\beta = \begin{cases} (\cos s + i \sin s)^k & \text{if } k > 0, \\ (\cos s - i \sin s)^{-k} & \text{if } k < 0. \end{cases}$$

By de Moivre's theorem, $\alpha = \cos ks$, $\beta = \sin ks$ in both cases. Hence, the angle ψ between $\mathbf{V}$ and $\boldsymbol{\xi}$ is equal to ks, and Definition 11.6 shows that the multiplicity is k.

11.14 If $\boldsymbol{\sigma}(u, v) = \tilde{\boldsymbol{\sigma}}(\tilde{u}, \tilde{v})$, where $(\tilde{u}, \tilde{v}) \mapsto (u, v)$ is a reparametrisation map, then $\mathbf{V} = \alpha\boldsymbol{\sigma}_u + \beta\boldsymbol{\sigma}_v = \tilde{\alpha}\tilde{\boldsymbol{\sigma}}_{\tilde{u}} + \tilde{\beta}\tilde{\boldsymbol{\sigma}}_{\tilde{v}} \Longrightarrow \tilde{\alpha} = \alpha\frac{\partial \tilde{u}}{\partial u} + \beta\frac{\partial \tilde{v}}{\partial u}$, $\tilde{\beta} = \alpha\frac{\partial \tilde{u}}{\partial v} + \beta\frac{\partial \tilde{v}}{\partial v}$. Hence, $\tilde{\alpha}$ and $\tilde{\beta}$ are smooth if α and β are smooth.
Since the components of the vectors $\boldsymbol{\sigma}_u$ and $\boldsymbol{\sigma}_v$ are smooth, if $\mathbf{V}$ is smooth so are its components. If the components of $\mathbf{V} = \alpha\boldsymbol{\sigma}_v + \beta\boldsymbol{\sigma}_v$ are smooth, then $\mathbf{V}.\boldsymbol{\sigma}_u$ and $\mathbf{V}.\boldsymbol{\sigma}_v$ are smooth functions, hence

$$\alpha = \frac{G(\mathbf{V}.\boldsymbol{\sigma}_u) - F(\mathbf{V}.\boldsymbol{\sigma}_v)}{EG - F^2}, \quad \beta = \frac{E(\mathbf{V}.\boldsymbol{\sigma}_v) - F(\mathbf{V}.\boldsymbol{\sigma}_u)}{EG - F^2}$$

are smooth functions, so $\mathbf{V}$ is smooth.

11.15 If $\tilde{\psi}$ is the angle between $\mathbf{V}$ and $\tilde{\boldsymbol{\xi}}$, we have $\tilde{\psi} - \psi = \theta$ (up to multiples of 2π); so we must show that $\int_0^{\ell(\boldsymbol{\gamma})} \dot{\theta}\, ds = 0$ (a dot denotes d/ds). This is not obvious since θ is not a well defined smooth function of s (although $d\theta/ds$ is well defined). However, $\rho = \cos\theta$ is well defined and smooth, since $\rho = \boldsymbol{\xi}.\tilde{\boldsymbol{\xi}} / \parallel \boldsymbol{\xi} \parallel\parallel \tilde{\boldsymbol{\xi}} \parallel$. Now, $\dot{\rho} = -\dot{\theta}\sin\theta$, so we must prove that

$\int_0^{\ell(\gamma)} \frac{\dot{\rho}}{\sqrt{1-\rho^2}} ds = 0$. Using Green's theorem, this integral is equal to

$$\int_{\boldsymbol{\pi}} \frac{\rho_u du + \rho_v dv}{\sqrt{1-\rho^2}} = \int_{\mathrm{int}(\boldsymbol{\pi})} \frac{\partial}{\partial u}\left(\frac{\rho_v}{\sqrt{1-\rho^2}}\right) - \frac{\partial}{\partial v}\left(\frac{\rho_u}{\sqrt{1-\rho^2}}\right),$$

where $\boldsymbol{\pi}$ is the curve in U such that $\gamma(s) = \boldsymbol{\sigma}(\boldsymbol{\pi}(s))$; and this line integral vanishes because

$$\frac{\partial}{\partial u}\left(\frac{\rho_v}{\sqrt{1-\rho^2}}\right) = \frac{\partial}{\partial v}\left(\frac{\rho_u}{\sqrt{1-\rho^2}}\right) \quad \left(= \frac{\rho_{uv}(1-\rho^2) + \rho\rho_u\rho_v}{(1-\rho^2)^{3/2}}\right).$$

11.16 Let $F : \mathcal{S} \to \mathbf{R}$ be a smooth function on a surface $\mathcal{S}$, let P be a point of $\mathcal{S}$, let $\boldsymbol{\sigma}$ and $\tilde{\boldsymbol{\sigma}}$ be patches of $\mathcal{S}$ containing P, say $\boldsymbol{\sigma}(u_0, v_0) = \tilde{\boldsymbol{\sigma}}(\tilde{u}_0, \tilde{v}_0) = P$, and let $f = F \circ \boldsymbol{\sigma}$ and $\tilde{f} = F \circ \tilde{\boldsymbol{\sigma}}$. Then, $\tilde{f}_{\tilde{u}} = f_u \frac{\partial u}{\partial \tilde{u}} + f_v \frac{\partial v}{\partial \tilde{u}}$, $\tilde{f}_{\tilde{v}} = f_u \frac{\partial u}{\partial \tilde{v}} + f_v \frac{\partial v}{\partial \tilde{v}}$, so if $f_u = f_v = 0$ at (u_0, v_0), then $\tilde{f}_{\tilde{u}} = \tilde{f}_{\tilde{v}} = 0$ at $(\tilde{u}_0, \tilde{v}_0)$. Since $f_u = f_v = 0$ at P, we have

$$\tilde{f}_{\tilde{u}\tilde{u}} = f_{uu}\left(\frac{\partial u}{\partial \tilde{u}}\right)^2 + 2f_{uv}\frac{\partial u}{\partial \tilde{u}}\frac{\partial v}{\partial \tilde{u}} + f_{vv}\left(\frac{\partial v}{\partial \tilde{u}}\right)^2,$$

with similar expressions for $\tilde{f}_{\tilde{u}\tilde{v}}$ and $\tilde{f}_{\tilde{v}\tilde{v}}$. This gives, in an obvious notation, $\tilde{\mathcal{H}} = J^t \mathcal{H} J$, where $J = \begin{pmatrix} \frac{\partial u}{\partial \tilde{u}} & \frac{\partial u}{\partial \tilde{v}} \\ \frac{\partial v}{\partial \tilde{u}} & \frac{\partial v}{\partial \tilde{v}} \end{pmatrix}$ is the jacobian matrix of the reparametrisation map $(\tilde{u}, \tilde{v}) \mapsto (u, v)$. Since J is invertible, $\tilde{\mathcal{H}}$ is invertible if $\mathcal{H}$ is invertible.
Since the matrix $\mathcal{H}$ is real and symmetric, it has eigenvectors v_1, v_2, with eigenvalues λ_1, λ_2, say, such that $v_i^t v_j = 1$ if $i = j$ and 0 if $i \neq j$. Then, if $v = \alpha_1 v_1 + \alpha_2 v_2$ is any vector, where α_1, α_2 are scalars, $v^t \mathcal{H} v = \lambda_1 \alpha_1^2 + \lambda_2 \alpha_2^2$; hence, $v^t \mathcal{H} v > 0$ (resp. < 0) for all $v \neq 0 \iff \lambda_1$ and λ_2 are both > 0 (resp. both < 0) $\iff P$ is a local minimum (resp. local maximum); and hence P is a saddle point $\iff v^t \mathcal{H} v$ can be both > 0 and < 0, depending on the choice of v. Since J is invertible, a vector $\tilde{v} \neq 0 \iff v = J\tilde{v} \neq 0$; and $\tilde{v}^t \tilde{\mathcal{H}} \tilde{v} = \tilde{v}^t J^t \mathcal{H} J \tilde{v} = v^t \mathcal{H} v$. The assertions in the last sentence of the exercise follow from this.

11.17 (i) $f_x = 2x - 2y$, $f_y = -2x + 8y$, so $f_x = f_y = 0$ at the origin. $f_{xx} = 2$, $f_{xy} = -2$, $f_{yy} = 8$, so $\mathcal{H} = \begin{pmatrix} 2 & -2 \\ -2 & 8 \end{pmatrix}$. $\mathcal{H}$ is invertible so the origin is non-degenerate; and the eigenvalues $5 \pm \sqrt{13}$ of $\mathcal{H}$ are both > 0, so it is a local minimum.
(ii) $f_x = f_y = 0$ and $\mathcal{H} = \begin{pmatrix} 2 & 4 \\ 4 & 0 \end{pmatrix}$ at the origin; $\det\mathcal{H} = -16 < 0$, so the eigenvalues of $\mathcal{H}$ are of opposite sign and the origin is a saddle point.
(iii) $f_x = f_y = 0$ and $\mathcal{H} = 0$ at the origin, which is therefore a degenerate critical point.

11.18 Using the parametrisation $\boldsymbol{\sigma}$ in Exercise 4.10 (with $a = 2, b = 1$) gives $f(\theta, \varphi) = F(\boldsymbol{\sigma}(\theta, \varphi)) = (2 + \cos\theta)\cos\varphi + 3$. Then, $f_\theta = -\sin\theta\cos\varphi$, $f_\varphi = -(2 + \cos\theta)\sin\varphi$; since $2 + \cos\theta > 0$, $f_\varphi = 0 \Longrightarrow \varphi = 0$ or π, and then $f_\theta = 0 \Longrightarrow \theta = 0$ or π; so there are four critical points, $P = (3, 0, 0)$, $Q = (1, 0, 0)$, $R = (-1, 0, 0)$ and $S = (-3, 0, 0)$. Next, $\mathcal{H} = \begin{pmatrix} -\cos\theta\cos\varphi & \sin\theta\sin\varphi \\ \sin\theta\sin\varphi & -(2 + \cos\theta)\cos\varphi \end{pmatrix} = \begin{pmatrix} -1 & 0 \\ 0 & -3 \end{pmatrix}$ at P, $= \begin{pmatrix} 1 & 0 \\ 0 & -1 \end{pmatrix}$ at Q, $= \begin{pmatrix} -1 & 0 \\ 0 & 1 \end{pmatrix}$ at R, and $= \begin{pmatrix} 1 & 0 \\ 0 & 3 \end{pmatrix}$ at S; hence, P is a local maximum, Q and R are saddle points, and S is a local minimum (all of which is geometrically obvious).

Index